Adrenoceptors
Structure, Function and Pharmacology

edited by

Robert R. Ruffolo, Jr.

SmithKline Beecham Pharmaceuticals,
King of Prussia, Pennsylvania, USA

harwood academic publishers

Australia · Austria · China · France · Germany · India · Japan · Luxembourg
Malaysia · Netherlands · Russia · Singapore · Switzerland · United Kingdom
United States

3 Boulevard Royal
L–2449 Luxembourg

These papers were first published in the journal *Pharmacology Communications*, volume 6, issues 1-3.

Cover: Molecular structure of the α_{2A}-adrenoceptor.

British Library Cataloguing in Publication Data

Adrenoceptors: Structure, Function and Pharmacology
I. Ruffolo, Robert R.
599.0188

ISBN 3–7186–5644–2

CONTENTS

PREFACE

Research in the field of the adrenoceptors has undergone explosive growth during the past decade, primarily due to the influence of the powerful techniques of molecular biology. We are now beginning to understand the molecular structures of the adrenoceptors and, how they may function within the confines of the cellular membrane. At the same time, advances in molecular biology, biochemistry and pharmacology have demonstrated clearly that multiple subtypes exist for each of the major families of the adrenoceptors and, as a result, it has become necessary to devise rational systems and schemes for subclassification of adrenoceptors. Additionally, it has been known for many years that the adrenoceptors are involved in the maintenance of a variety of homeostatic processes, and that they also play a role in a significant number of pathological states, and this has been reinforced recently by further developments in our understanding of the physiology, pharmacology and pathophysiology of the adrenoceptors.

As a result of the far-reaching developments in the field of adrenoceptors research, and the importance of the adrenoceptors in medicine, we decided that the time was appropriate to review all of the major developments in this field in an international symposium. The forum we chose for this meeting was a satellite symposium to the 1994 IUPHAR Congress, along the same lines as the three previous and highly successful symposia organized by Professor Elemer Szabadi entitled the *Pharmacology of Adrenoceptors*. The first meeting in this series was held in Paris in 1978, and the second and third meetings took place in Manchester in 1984 and 1990. The Organizing Committee for the present symposium was Professor Elemer Szabadi, Professor David B. Bylund and myself, all of us having been involved in the previous meetings in this series.

The *Pharmacology of Adrenoceptors* symposium was divided into three major themes, along the same lines proposed by Dr Bylund for the subclassification of the adrenoceptors into three major families, namely the α_1-adrenoceptors, α_2-adrenoceptors and the β-adrenoceptors. For each of these major themes, we enlisted opinion leaders in the field of adrenoceptor research to deal with the issues related to the genetic regulation of the adrenoceptors, molecular structure and functions of the receptors, receptor subclassification and nomenclature, signal transduction processes that are activated or inhibited by these receptors, receptor regulation, physiological, pharmacological and pathological responses involving these receptors and the clinical significance of the adrenoceptors. The proceedings of this symposium published in this book are organized along similar lines.

This, I believe, allows each of the major families of the adrenoceptors to be presented in a reasonably encompassing manner from the level of the gene through to that of the clinic. This integration of disciplines that are involved in adrenoceptor research was a major objective of the Organizing Committee, and it is reflected throughout these proceedings.

In a meeting such as this involving the leaders in the field of adrenoceptors, it should be no surprise that there are a significant number of highlights: far too many to summarize in this brief introduction. However, a few major highlights from this symposium should be noted. Firstly, after several years of extensive debate regarding the subclassification and nomenclature of the α_1-adrenoceptors, presentations and detailed discussions at this meeting have finally produced a consensus among the opinion leaders in the field on how the α_1-adrenoceptors should be classified and named, and we consider this to be a major accomplishment. Secondly, although very significant progress has been made in the cloning of the various α- and β-adrenoceptors, it became apparent at this meeting that additional adrenoceptors very likely exist and remain to be cloned. Furthermore, it became apparent that strict criteria need to be used in subclassifying the adrenoceptors, and many of these criteria have been outlined by several participants at the meeting, and will no doubt become standard procedures to be used for the classification of the adrenoceptors. Finally, a good deal of time was devoted to the newest member of the β-adrenoceptor family, the β_3-adrenoceptor, which is also known as the "atypical" β-adrenoceptor. It was not too long ago that significant doubts were cast over the existence of the β_3-adrenoceptor. These doubts have now been removed as a result of the recent cloning and characterization of the β_3-adrenoceptor from a variety of mammalian species, and the identification of potent and selective β_3-adrenoceptor agonists. The other highlights of the meeting, as well as those mentioned above, are presented in detail in the proceedings that follow.

I would like to thank my co-organizers, Professors Szabadi and Bylund, for their significant contributions to this meeting. Most notably, Professor Szabadi is acknowledged for conceiving of, and implementing, this series of symposia on the *Pharmacology of Adrenoceptors*, and I am grateful for his personal contribution to the present meeting. I am also indebted to Professor Bylund for the major input he provided to the scientific program and in the soliciting of funds to support the meeting. On behalf of the organizers, I would also like to thank the invited speakers, all of whom enthusiastically accepted our invitations to participate in the meeting, and the attendees who stimulated productive discussions.

Financial support for scientific symposia has begun to decline during the past several years, and with this in mind, I would like to acknowledge all of the institutions who provided support for this meeting. Major financial support for this symposium was provided through an educational grant by Dr George Poste, President of Research and Development, SmithKline Beecham Pharmaceuticals, who additionally provided the R&D facilities in King of Prussia, Pennsylvania, where the meeting was held. In addition, financial support was also generously provided by Abbott Laboratories, Allergan Inc., Syntex Discovery Research, the Upjohn Company and Zeneca Pharmaceuticals. The financial support provided by these pharmaceutical companies is especially appreciated in view of the financial difficulties that the pharmaceutical industry has experienced in the past two years. I would also like to thank Harwood Academic Publishers for working with us in the publication of the proceedings of this symposium.

Finally, I must acknowledge the dozens of SmithKline Beecham employees who devoted their time and efforts to making this symposium a success; they did so without compensation. In particular, I would like to recognize Shirley Wilson, Cheryl Gavron, Susan Tirri, Susan Gallagher and Mary Tucker for their unselfish dedication to the symposium. These individuals and many others who remain unnamed worked nights and weekends, in addition to their normal job responsibilities, to make this meeting a success. As such, it is to these SmithKline Beecham employees, and to my wife, Stephany, who spends many nights alone because of my job, that this book is dedicated.

R. R. R.

I. INTRODUCTION

THE ADRENOCEPTORS: HISTORICAL PERSPECTIVES, CURRENT STATUS AND FUTURE DIRECTIONS

ROBERT R. RUFFOLO, JR. and J. PAUL HIEBLE

Division of Pharmacological Sciences, SmithKline Beecham Pharmaceuticals, 709 Swedeland Road, King of Prussia, Pennsylvania, 19406, USA

Recently, much progress has been made in the characterization of the pharmacological properties, molecular structure and signal transduction mechanisms of α- and β-adrenoceptors. The discovery of multiple subtypes of both α_1-, α_2- and β-adrenoceptors has stimulated research into the design of new agonists and antagonists having subtype selectivity. These compounds are proving useful both as tools for characterization of the physiological roles for the adrenoceptor subtypes and as potential therapeutic agents having improved clinical profiles. For example, selective β_3-adrenoceptor agonists may offer a new approach to the treatment of diabetes and obesity, and selective α_{1C}-adrenoceptor antagonists may be one means of achieving tissue selectivity in the symptomatic treatment of benign prostatic hypertrophy. Nine pharmacologically distinct adrenoceptors have now been cloned and expressed. The availability of these recombinant adrenoceptors has allowed molecular modeling and site-directed mutagenesis to be employed to characterize the molecular interactions between ligand and receptor, and between receptor and second messengers. These techniques have been applied extensively to the β_2-adrenoceptor, and are now beginning to be directed toward the other adrenoceptor subtypes. Much remains to be learned about the molecular character of the adrenoceptors, including actual determination of its protein structure *in situ* through physical techniques. It is also possible that additional adrenoceptor subtypes will be identified, since functional studies suggest the existence of subtypes of both α_1- and α_2-adrenoceptors having pharmacological characteristics that are not entirely consistent with those of the recombinant adrenoceptors thus far identified.

KEY WORDS: α_1-adrenoceptors, β-adrenoceptors, adrenoceptor subtypes, benign prostatic hypertrophy, molecular cloning

HISTORICAL PERSPECTIVES

It has been nearly a century since the discovery that adrenal extracts, when injected into animals, would mimic many of the physiological effects of sympathetic nerve stimulation (Lewandowsky, 1898; Langley, 1901), and that ergot alkaloids would block some, but not all, of these effects (Dale, 1906). At about the same time, the concepts of chemical neurotransmission (Elliot, 1904) and the presence of "receptive substances" on effector cells (Langley, 1905) were introduced. Two decades later, the release of an excitatory transmitter from sympathetic nerves was demonstrated (Cannon and Uridil, 1921; Loewi, 1921), although this neurotransmitter was not proven to be norepinephrine, through direct chemical analysis, until 1946 (von Euler, 1946). Within another decade, several other discoveries vital to the development of adrenergic pharmacology were made: i) the responses to adrenergic stimulation were divided into α- and β-adrenoceptor classes (Ahlquist, 1948), ii) the α-adrenoceptor blocking actions of the haloalkylamines was identified

Correspondence to: Dr. Robert R. Ruffolo, Jr., SmithKline Beecham Pharmaceuticals, 709 Swedeland Road, P.O. Box 1539, King of Prussia, PA 19406-0939 USA.

(Nickerson, 1949), and iii) the first β-adrenoceptor antagonist, dichloroisoproterenol, was identified and characterized (Powell and Slater, 1958).

Our current understanding of the pharmacology of adrenoceptors has also depended on upon several theoretical concepts relating to steric interactions between the neurotransmitter, norepinephrine, and the adrenoceptors (Easson and Stedman, 1933), competition between agonists and irreversible antagonists at the adrenoceptors (Furchgott, 1954), and the formulation of "receptor theory" (Schild, 1949; Furchgott, 1964; Ariens, 1964).

The development of agents interacting with adrenoceptors into useful drugs is dependent upon selectivity between the many responses mediated *via* adrenoceptor stimulation. The discovery of selective agents has been made possible by receptor subclassification, an exercise which began with Ahlquist in 1948 and is still ongoing. The subclassification of β-adrenoceptors into the β_1- and β_2-subtypes (Lands *et al.*, 1967) has made possible the development of selective bronchodilators acting on the β_2-adrenoceptor, and the division of α-adrenoceptors into the α_1- and α_2-subtypes (Langer, 1974; Starke *et al.*, 1974) led to the development of effective antihypertensive drugs that either selectively block the α_1-adrenoceptor or selectively activate the α_2-adrenoceptor. The identification of a third β-adrenoceptor subtype (Arch *et al.*, 1984; Wilson *et al.*, 1984), and the further subclassification of α_1-adrenoceptors (Morrow and Creese, 1986; Han *et al.*, 1987) has also led to additional potential therapeutic opportunities, which will be discussed below.

As biochemical techniques for isolating and characterizing the adrenoceptors were developed, we began to learn more about the mechanisms for transduction of the changes induced by receptor occupation into the production of a physiological response. Cyclic AMP was identified as a second messenger in the metabolic response to β-adrenoceptor stimulation in the liver (Sutherland and Rall, 1958). β-Adrenoceptors were subsequently purified without disruption of their ability to bind radioligands (Caron and Lefkowitz, 1976), and guanine nucleotides were shown to selectively influence agonist binding to the β-adrenoceptors (Lefkowitz *et al.*, 1976), leading ultimately to the identification of guanine nucleotide regulatory proteins (G-proteins) that are involved with the signal transduction processes associated with the adrenoceptors (Katada and Ui, 1981).

A schematic model for many neurotransmitter receptors, including the α- and β-adrenoceptors, has been postulated, where the receptor consists of an array of seven transmembrane spanning helices, arranged around a central channel. Adrenoceptor ligands bind within the channel, approaching from the extracellular surface, while the G proteins are associated with the third cytoplasmic loop connecting helices 5 and 6 (see Reviews by Stadel, 1991 and Stadel and Nakada, 1991).

CURRENT STATUS

Despite the long history of adrenoceptor research, there is still intense theoretical and practical interest in the pharmacology of α- and β-adrenoceptors. This is based to a large extent on the identification, through molecular biological techniques, of three subtypes each of the β-, α_1- and α_2-adrenoceptors (Bylund *et al.*, 1994) and the availability of the human homologs of these 9 recombinant receptors. In some cases, such as for the β_3-adrenoceptor, the cloning of the receptor (Emorine *et al.*, 1989) in fact confirmed the presence of a novel receptor previously identified through classical pharmacological techniques (Arch *et al.*, 1984), while in the case of both α_1- and α_2-adrenoceptors, molecular biology studies have identified new subtypes which remain to be fully characterized in native tissues. In many cases, an individual adrenoceptor has been cloned from several species; these species homologs may, in some cases, have sufficiently different pharmacological characteristics to be designated as different subtypes by some investigators.

The availability of recombinant adrenoceptors has permitted molecular biology studies to characterize the interactions of receptor ligands and second messenger systems with the adrenoceptors. An extensive body of evidence from molecular modeling and site-directed mutagenesis has been used to identify potential binding sites for agonists and antagonists to the β_2-adrenoceptor. The nitrogen of the catecholamine neurotransmitters binds *via* an ionic interaction to Asp^{113} in transmembrane helix 3, while the aromatic hydroxyl groups interact *via* hydrogen bonding to Ser^{204} and Ser^{207} of transmembrane helix 6. This type of analysis has been applied to a limited degree to the other β-adrenoceptor subtypes in an attempt to determine a molecular explanations for selectivity between β_1- and β_2-adrenoceptors (Frielle, 1988: Marullo *et al.*, 1990), and is being used to optimize agonist selectivity between β_2- and β_3-adrenoceptors. α-Adrenoceptors have been studied less extensively, but evidence from site-directed mutagenesis has already provided evidence that alteration of a single amino acid (Cys^{201} to serine) is responsible for some of the different pharmacological characteristics of a human and porcine α_2-adrenoceptor (α_{2A}) from the corresponding rodent homolog (α_{2D}) (Link *et al.*, 1992). Mutagenesis studies on the α_{2A} adrenoceptor have also shown that the catecholamines may interact with this receptor in a different orientation from that previously established for the β_2-adrenoceptors (Wang *et al.*, 1991; Li *et al.*, 1994).

Many different G-proteins have been identified, and the adrenoceptors have been shown to interact with several of these in coupling to different second messenger systems. In addition to adenylate cyclase, which can be either activated (by all β-adrenoceptor subtypes) or inhibited (α_2-adrenoceptors) in response to receptor occupation, adrenoceptors have been shown to be linked, through different G-proteins, to phosphatidylinositol hydrolysis and calcium translocation (α_1-adrenoceptors). A model has been proposed in which the α_1-adrenoceptor can activate two different G proteins, resulting in release of intracellular calcium stores and/or translocation of extracellular calcium, depending on the intrinsic activity of the agonist binding to the active site (Nichols, 1991).

Site-directed mutagenesis and the preparation of chimeric receptors have established that the adrenoceptors and the G-proteins interact through the third intracellular loop of the receptors. This portion of the adrenoceptor structure represents the site of greatest diversity between the adrenoceptor classes, and between subtypes of each class. It appears that the character of the third intracellular loop of the adrenoceptor determines to which G-protein the receptor is preferentially coupled (Kobilka *et al.*, 1988). Replacing the third intracellular loop of the β_2-adrenoceptor with that of the α_{1B}-adrenoceptor resulted in the coupling of the chimeric β_2/α_{1B}-adrenoceptors to the primary second messenger system of the α_{1B}-adrenoceptor, namely phospholipase C (Cotecchia *et al.*, 1990), and replacement of only a 27 amino acid segment of the N-terminus of this loop by the α_{1B}- sequence resulted in a β_2-adrenoceptor mutant capable of activating both phosphatidylinositol hydrolysis and adenylate cyclase (Cotecchia *et al.*, 1992).

Some recent studies have shown that point mutations in the carboxyl terminus of the third intracellular loop can result in an α- or β-adrenoceptor which can activate its effector, through G-protein interaction, without the necessity of agonist binding. Thus, mutation of Ala^{293} of the α_{1B}-adrenoceptor to any of the 19 possible amino acids resulted in a receptor which stimulated basal inositol phosphate hydrolysis, by 20 to 200% (Kjelsberg *et al.*, 1992). These results can best be explained by a change in receptor conformation which allows productive interaction of the receptor and G-protein, thereby allowing activation of the effector system to occur without the necessity of an agonist-induced change in the secondary structure of the receptor. Mutation of four amino acids in the carboxy terminus of the third intracellular loop of the β_2-adrenoceptor produces a similar effect, yielding a receptor capable of activating adenylate cyclase without agonist occupation, and having enhanced agonist affinity (Samama *et al.*, 1993). Agonist independent inhibition of adenylate cyclase was induced by replacement of Thr^{373} of the α_{2A}-adrenoceptor by phenylalanine, alanine, glutamic acid, cysteine or lysine (Ren *et al.*, 1993).

Some details of agonist-induced receptor desensitization are now known. The initial desensitization process is thought to result from the phosphorylation of serine and threonine residues in the cytoplasmic tail or third intracellular loop by several protein kinases, including β-adrenergic receptor kinase (βARK), and cAMP dependent protein kinase (protein kinase A). Although identified in the β-adrenoceptor system, βARK can also phosphorylate the α_2-adrenoceptor (Liggett *et al.*, 1992). Agonist occupation of the receptor promotes phosphorylation by βARK. βARK mediated desensitization can be facilitated by the association of a specific protein, β-arrestin, with the phosphorylated receptor. (Pippig *et al.*, 1993). Binding of β-arrestin to the receptor results in inactivation, by a mechanism not yet understood.

The availability of recombinant adrenoceptor subtypes, and the ability to localize these subtypes to a particular tissue, has stimulated the design of subtype selective therapeutic agents. For example, although α_1-adrenoceptor antagonists are currently used for the treatment of both hypertension and benign prostatic hypertrophy (BPH) (Hieble and Ruffolo, 1991), blockade of vascular α_1-adrenoceptors produced by these drugs is responsible for their most commonly observed side-effects, such as dizziness, orthostatic hypotension and syncope, when they are used for therapy of BPH. The predominant adrenoceptor mRNA found in human prostate is of the α_{1C}-adrenoceptor subtype (Price *et al.*, 1993), and the ability of α_1-adrenoceptor antagonists to block norepinephrine-induced contraction in isolated human prostate correlates with their affinity for the α_{1C}-adrenoceptor (Forray *et al.*, 1994). Although the α_1-adrenoceptor subtype(s) responsible for maintenance of vascular tone by the sympathetic nervous system has not yet been established, it appears that selective α_{1C}-adrenoceptor antagonists may indeed achieve tissue selectivity for the prostate, with a reduced liability for production of orthostatic effects (Gong *et al.*, 1994).

While selectivity for a particular receptor subtype is often desirable, agents having affinity for several adrenoceptors may have a favorable therapeutic profile. For example, carvedilol, an antagonist at both β- and α_1-adrenoceptors, has been shown to be effective in the treatment of congestive heart failure (DiLenarda *et al.*, 1991; Krum *et al.*, 1993). Carvedilol is more effective than "pure" β-adrenoceptor antagonists (DiLenarda *et al.*, 1991) or selective α_1-adrenoceptor antagonists, both of which have been found to be ineffective upon prolonged therapy (Packer, 1988).

FUTURE DIRECTIONS

As noted above, characterization of the molecular interactions of ligands and second messenger systems with the adrenoceptors is only beginning, with much yet to be learned, especially with respect to the α-adrenoceptors. The molecular basis for subtype selectivity within each adrenoceptor class, and even between α- and β-adrenoceptors, is almost completely unknown. A more precise understanding of the molecular processes involved in receptor desensitization is also an important goal. While the seven-transmembrane helix model of the adrenoceptors is universally accepted, the only physical validation by crystallography to date has been done with bacteriorhodopsin, a protein having little primary sequence homology to the adrenoceptors, and which is coupled to a different second messenger system. The actual determination of the adrenoceptor structures within a biological membrane awaits further refinements in crystallographic technique.

It is likely that agents having selectivity for a particular adrenoceptor subtype will offer significant therapeutic advantages over currently available drugs, and may even open entirely new therapeutic opportunities. Agonists having high selectivity and efficacy at the human β_3-adrenoceptor are only now beginning to be identified, and may prove to be useful for the treatment of diabetes and obesity. Subtype selectivity may be able to dissociate the multiple actions produced by stimulation of central α_2-adrenoceptors, and may lead to more effective anesthetics or analgesics. The potentially useful pharmacologic actions of the α_2-adrenoceptor antagonists, such as their ability to increase cutaneous blood flow, their ability to relieve impotence, and their ability to stimulate insulin secretion, have not yet been fully exploited clinically. Again, subtype selectivity may allow the development of drugs directed toward these targets.

Finally, one cannot dismiss the possibility of additional adrenoceptor subtypes. Functional studies suggest the presence of additional subtypes beyond those which have been identified through molecular biology techniques. For example, the three α_1-adrenoceptors thus far cloned, all showing high sensitivity to prazosin, may be subdivisions of the α_{1H}-adrenoceptor subtype identified *via* functional subclassification based on prazosin sensitivity (Oshita *et al.*, 1992), whereas the α_1-adrenoceptors showing low prazosin sensitivity may not as yet be cloned. A similar situation exists for the α_2-adrenoceptors, where several functional responses to α_2-adrenoceptor stimulation show insensitivity to certain antagonists, such as SK&F 104078 and SK&F 104856, which are compounds that show nearly equivalent affinity to all of the currently known recombinant α_2-adrenoceptor subtypes (Hieble *et al.*, 1994).

CONCLUSIONS

Nearing the completion of the first century of adrenoceptor characterization, many questions have been answered, but many new questions have been raised. Several important drug classes act through activation or blockade of the adrenoceptors, and additional therapeutic opportunities may result from the recent and perhaps ongoing discovery of additional adrenoceptor subtypes. The rich historical background and growing current understanding of the adrenoceptors does not signal closure of this field, but rather presages even more excitement in the next century.

References

Ahlquist, R.P. (1948) A study of the adrenotropic receptors. *Am. J. Physiol.*, **153,** 586–600.

Arch, J.R., Ainsworth, A.T., Cawthorne, M.A., Piercy, V., Sennitt, M.V., Thod, V.E., Wilson, C. and Wilson, S. (1984) Atypical β-adrenoceptors on brown adipocytes as target for anti-obesity drugs. *Nature*, **309,** 163–165.

Ariens, E.J. (1964) *Molecular Pharmacology, Vol. 1*. Academic Press, New York.

Bylund, D.B., Eikenberg, D.C., Hieble, J.P., Langer, S.Z., Lefkowitz, R.J., Minneman, K.P., Molinoff, P.B., Ruffolo, R.R., Jr. and Trendelenburg, U. (1994) International Union of Pharmacology Nomemclature of Adrenoceptors. *Pharmacol. Rev.*, **46,** 121–136.

Cannon, W.B. and Uridil, J.E. (1921) Studies on the conditions of activity in endocrine glands. VIII. Some effects on the denervated heart of stimulating the nerves of the liver. *Am. J. Physiol.*, **58,** 353–354.

Cotecchia, S., Exum, S., Caron, M.G. and Lefkowitz, R.J. (1990) Regions of the α_1-adrenergic receptor involved in coupling to phosphatidylinositol hydrolysis and enhanced sensitivity of biological function. *Proc. Natl. Acad. Sci. USA*, **87,** 2896–2900.

Cotecchia, S., Ostrowski, J., Kjelsberg, M.A., Caron, M.G. and Lefkowitz, R.J. (1992) Discrete amino acid sequences of the α_1-adrenergic receptor determine the selectivity of coupling to phosphatidyl inositol hydrolysis. *J. Biol. Chem.*, **267,** 1633–1639.

Dale, H.H. (1906) On some physiological actions of ergot. *J. Physiol. (London)*, **34,** 163–206.

DiLenarda, A.D., Gilbert, E.M., Olsen, S.L., Mealey, P.C. and Bristow, M.R. (1991) Acute hemodynamic effects of carvedilol versus metoprolol in idiopathic dilated cardiomyopathy. *J. Am. Coll. Cardiol.*, **17,** 142A.

Easson, L.H. and Stedman, E. (1933) Studies on the relationship between chemical constitution and physiological action. V. Molecular dissymmetry and physiological activity. *Biochem. J.*, **27,** 1257–1266.

Elliot, T.R. (1905) The action of adrenalin. *J. Physiol. (London)*, **32,** 401–467.

Emorine, L.J., Marullo, S., Briend-Sutren, M.-M., Patey, G., Tate, K., Delavier-Klutchko, C. and Strosberg, A.D. (1989) Molecular characterization of the human β_3-adrenergic receptor. *Science*, **245,** 1118–1121.

Forray, C., Bard, J.A., Wetzel, J.M., Chiu, G., Shaprio, E., Tang, R., Lepor, H., Hartig, P.R., Weinshank, R.L., Branchek, T.A. and Gluchowski, C. (1994) The α_1-adrenergic receptor that mediated smooth muscle contraction in human prostate has properties of the cloned human α_{1C} subtype. *Mol. Pharmacol.*, **45,** 703–708.

Frielle, T., Daniel, K.W., Caron, M.G. and Lefkowitz, R.J. (1988) The structural basis of β-adrenergic receptor subtype selectivity: Studies with chimeric β_1/β_2-adrenergic receptors. *Proc. Natl. Acad. Sci. USA*, **84,** 920–7924.

Furchgott, R.F. (1954) Dibenamine blockade in strips of rabbit aorta and its use in differentiating receptors. *J. Pharmacol. Exp. Ther.*, **111,** 165–184.

Furchgott, R.F. (1964) Receptor mechanisms. *Ann. Rev. Pharmacol.*, **4,** 21–50.

Gong, G., Chiu, G., Gluchowski, C., Branchek, T.A., Hartig, P.R., Pettinger, W.A. and Forray, C. (1994) α_{1C}-Adrenergic antagonists and orthostatic hypotension. *FASEB J.*, **8,** A353.

Han, C., Abel, P.W. and Minneman, K.P. (1987) L: α_1-Adrenoceptor subtypes linked to different mechanisms for increasing intracellular Ca^{2+} in smooth muscle. *Nature*, **329,** 333–335.

Hieble, J.P. and Ruffolo, R.R. Jr. (1991) Therapeutic applications of agents interacting with α-adrenoceptors. In *α-Adrenoceptors: Molecular Biology, Biochemistry and Pharmacology* (ed. Ruffolo, R.R., Jr.), *Prog Basic Clin. Pharmacol.*, Basel, Karger, Vol. **8,** pp. 180–220.

Hieble, J.P., Sulpizio, A.C., Naselsky, D.P., Conway, T.M., Ellis, C., Swift, A.M., Ganguly, S. and Bergsma, D.J. (1994) Functional subclassification of α_2-adrenoceptors. *Pharmacol. Comm.*, in press.

Katada, T. and Ui, M. (1981) Islet-activating protein: A modifier of receptor-mediated regulation of rat islet adenylate cyclase. *J. Biol. Chem.*, **256,** 8310–8317.

Kjelsberg, M.A., Cotecchia, S., Ostrowski, J., Caron, M.G. and Lefkowitz, R.J. (1992) Constitutive activation of the α_{1B}-adrenergic receptor by all amino acid substitutions at a single site. *J. Biol. Chem.*, **267,** 1430–1433.

Kobilka, B.K., Kobilka, T.S., Daniel, K., Regan, J.W., Caron, M.G. and Lefkowitz, R.J. (1988) Chimeric α_2-, β_2-adrenergic receptors: Delination of domains involved in effector coupling and ligand binding specificity. *Science*, **240,** 1310–1316.

Krum, H., Schwartz, B., Sackner-Bernstein, J., Penn, J., Goldsmith, R.L., Kukin, M.L., Medina, N., Yushak, M. and Packer, M. (1993) Double-blind, placebo controlled study of the long-term efficacy of carvedilol in patients with severe heart failure treated with converting enzyme inhibitors. *J. Am. Coll. Cardiol.*, **21,** 114A.

Lands, A.M., Arnold, A., McAuliff, J.P., Luduena, F.P. and Brown, T.G. (1967) Differentiation of receptor systems activated by sympathomimetic amines. *Nature*, (London) **214,** 597–598.

Langer, S.Z. (1974) Presynaptic regulation of catecholamine release. *Brit. J. Pharmacol.*, **60,** 481–497.

Langley, J.N. (1901) The difference of behavior of central and peripheral pilomotor nerve cells. *J. Physiol. (London)*, **27,** 224–236.

Langley, J.N. (1905) On the reaction of cells and of nerve-endings to certain poisons, chiefly as regards the reaction of striated muscle to nicotine and to curare. *J. Physiol. (London)*, **33,** 374–413.

Lewandowsky, M. (1898) Uber eine Wirkung des Nebennierenextractes auf das Auge. *ZentBl. Physiol.*, **12,** 599–600.

Li, Y.-O., Hieble, J.P., Bergsma, D.J., Swift, A.M., Ganguly, S. and Ruffolo, R.R. Jr. (1994) The β-hydroxyl group of catecholamines may interact with Ser^{90} of the second transmembrane helix of the α_{2A}-adrenoceptor. *Pharmacol. Comm.*, in press.

Link, R., Daunt, D., Barsh, G., Chruscinski, A. and Kobilka, B. (1992) Cloning of two mouse genes encoding α_2-adrenergic receptor subtypes and identification of a single animo acid in the mouse α_2-C10 homolog responsible for an interspecies variation in antagonist binding. *Mol. Pharmacol.*, **42,** 16–27.

Liggett, S.B., Ostrowski, J., Chesnut, L.C., Kurose, H., Raymond, J.R., Caron, M.G. and Lefkowitz, R.J. (1992) Sites in the third intracellular loop of the α_{2A}-adrenergic receptor confer short term agonist-promoted desensitization. *J. Biol. Chem.*, **267,** 4740–4746.

Loewi, O. (1921) Uber humorale Ubertragbarkeit der Herznervenwirkung. *Pflugers Arch. Ges. Physiol.*, **214,** 678–688.

Marullo, S., Emorine, L.J., Strosberg, A.D. and Delvier-Klutchko, C. (1990) Selective binding of ligand to β_1, β_2 or chimeric β_1/β_2 adrenergic receptors involves multiple subsites. *EMBO Journal*, **9,** 1471–1476.

Morrow, A.L. and Creese, I. (1986) Characterization of α_1-adrenergic receptor subtypes in rat brain: a reevaluation of ^{3}H-WB 4101 and ^{3}H-prazosin binding. *Mol. Pharmacol.*, **29,** 321–330.

Nichols, A.J. (1991) α-Adrenoceptor signal transduction mechanisms. In: Ruffolo, R.R., Jr, ed.: α-Adrenoceptors: Molecular Biology, Biochemistry and Pharmacology. *Prog. Basic Clin. Pharmacol.*, Basel, Karger, Vol. **8,** pp. 44–74

Nickerson, M. (1949) The pharmacology of adrenergic blockade. *Pharmacol. Rev.*, **1,** 27–101.

Oshita, M., Takita, M., Kigoshi, S. and Muramatsu, L. (1992) α_1-Adrenoceptor subtypes in rabbit thoracic aorta. *Japan J. Pharmacol.*, **58** (Suppl II), 276P.

Packer, M. (1988) Vasodilator and inotropic drugs for the treatment of chronic heart failure: Distinguishing hype from hope. *J. Am. Coll. Cardiol.*, **12,** 1299–1317.

Pippig, S., Andexinger, S., Daniel, K., Puzicha, M., Caron, M.G., Lefkowitz, R.J. and Lohse, M.J. (1993) Overexpression of β-arrrestin and β-adrenergic receptor kinase augment densensitization of β_2-adrenergic receptors. *J. Biol. Chem.*, **268,** 3201–3208.

Powell, C.E. and Slater, L.H. (1958) Blocking of inhibitory adrenergic receptors by a dichloro analogue of isoproterenol. *J. Pharmacol. Exp. Ther.*, **122,** 480–488.

Price, D.T., Schwinn, D.A., Lomasney, J.W., Allen, L.F., Caron, M.G. and Lefkowitz, R.J. (1993) Indentification, quantification and localization of mRNA for three distinct α_1-adrenergic receptor subtypes in human prostate. *J. Urol.*, **150,** 546–551.

Ren, Q., Kurose, H., Lefkowitz, R.J. and Cotecchia, S. (1993) Constitutively active mutants of the α_2-adrenergic receptor. *J. Biol. Chem.*, **268,** 16483–16487.

Samama, P., Cotecchia, S., Costa, T. and Lefkowitz, R.J. (1993) A mutation-induced activated state of the β_2-adrenergic receptor. *J. Biol. Chem.*, **268,** 4625–4636.

Schild, H.O. (1949) pAx and competitive drug antagonism. *Brit. J. Pharmacol.*, **4,** 277–280.

Stadel, J.M. (1991) Molecular biology of α-adrenoceptors. In: Ruffolo, R.R., Jr, ed.: α-Adrenoceptors: Molecular Biology, Biochemistry and Pharmacology. *Prog. Basic Clin. Pharmacol.*, Basel, Karger, Vol. **8,** pp. 24–43.

Stadel, J.M. and Nakada, M.T. (1991) Molecular biology of β-adrenoceptors. In: Ruffolo, R.R., Jr, ed.: β-Adrenoceptors: Molecular Biology, Biochemistry and Pharmacology. *Prog. Basic Clin. Pharmacol.*, Basel, Karger, Vol. **7,** pp. 26–66.

Starke, K., Montel, H., Gayk, W. and Merker, R. (1974) Comparison of the effects of clonidine on pre- and postsynaptic adrenoceptors in the rabbit pulmonary artery. *Naunyn-Schmiedeberg's Arch. Pharmacol.*, **285,** 133–150.

Sutherland, E.W. and Rall, T.W. (1958) Fractionation and characterization of a cyclic adenosine ribonucleotide formed by tissue particles. *J. Biol. Chem.*, **232,** 1077–1091.

von Euler, U.S. (1946) A specific sympathomimetic ergone in adrenergic nerve fibres (sympathin) and its relations to adrenaline and nor-adrenaline. *Acta Physiol. Scand.*, **12,** 73–97.

Wang, C.-D., Buck, M.A. and Fraser, C.M. (1991) Site-directed mutagenesis of α_{2A}-adrenergic receptors: Identification of amino acids involved in ligand binding and receptor activation by agonists. *Mol. Pharmacol.*, **40,** 168–179.

Wilson, C., Wilson, S., Piercy, V., Sennitt, M.V. and Arch, J.R.S. (1984) The rat lipolytic β-adrenoceptor: studies using novel β-adrenoceptor agonists. *Europ. J. Pharmacol.*, **100,** 309–319.

II. α_1-ADRENOCEPTORS

PHARMACOLOGICAL EVIDENCE FOR THE EQUIVALENCE OF THE α_{1A}-ADRENOCEPTOR OF RAT AND THE CLONED BOVINE α_{1C}-ADRENOCEPTOR: A REVISED α_1-ADRENOCEPTOR CLASSIFICATION SCHEME

DAVID E. CLARKE[1], ANTHONY P. D. W. FORD, TIMOTHY J. WILLIAMS, DOUGLAS BONHAUS, RACHEL VIMONT and DAVID R. BLUE, JR.

The proposal is advanced that the pharmacologically-defined α_{1A}-AR in rat is equivalent to the cloned bovine α_{1C}-AR and suggests that the latter appellation be discontinued.

KEY WORDS: α_{1A}-adrenoceptor, α_{1C}-adrenoceptor, adrenoceptor classification, rat, bovine

INTRODUCTION

The α_{1A}-adrenoceptor (α_{1A}-AR) of rat, as defined by pharmacological studies, has lacked a molecular counterpart. However, we have suggested recently that this α_{1A}-AR represents the rat homologue of the cloned bovine α_{1C}-AR (Ford *et al.*, 1994). Quantitative pharmacological evidence is now presented in support of this contention.

METHODS

Isolated renal and mesenteric arterial vascular beds of rat were perfused at a constant rate (6 ml min^{-1}) and changes in perfusion pressure measured. The perfusion medium was composed of Krebs' solution containing: cocaine (30 μM), corticosterone (30 μM), propranolol (1 μM), and indomethacin (10 μM). Non-cumulative concentration-effect curves to agonists were obtained by perfusing the kidneys and mesentery with each concentration of agonist until a steady-state vasoconstrictor response was obtained. Affinity estimates (pA_2 values) for antagonists were made after a 1 hr equilibration period. In some experiments with mesentery, the vasoconstrictor response to noradrenergic nerve stimulation was also studied.

Radioligand binding assays were performed using membrane preparations of submaxillary gland of rat and rat-1 fibroblasts expressing cloned α_1-ARs (hamster α_{1B}-; bovine α_{1C}-; rat α_{1D}-; purchased from Duke University, North Carolina, U.S.A). Competition assays were performed using [^{3}H]-prazosin (0.1nM) and phentolamine (10 μM) for estimation of specific binding.

[1]Correspondence to: David E. Clarke, SYNTEX RESEARCH, Department of Pharmacology, 3401 Hillview Avenue, Palo Alto, California 94303.

RESULTS

Studies in kidney and mesentery revealed the absence of α_2-AR-mediated vasoconstriction and only a 2-fold or less dextral shift in the concentration-effect curve to norepinephrine (NE) after treatment with chloroethylclonidine (CEC, 100 μM for 20 min followed by a 40 min perfusion with CEC-free solution).

Isolated Perfused Kidney of Rat

Prazosin, phentolamine, and WB 4101 acted as reversible competitive antagonists (Table 1) in that the NE concentration-effect curves were shifted dextrally in a parallel manner with no change in maximal response. However, 5-methyl-urapidil (5-MU) disclosed evidence for two sites (Fig. 1). Induced biphasicity in the concentration-effect curves to NE was removed by nitrendipine (1 μM), diminished by pretreatment with CEC (100 μM for 20 min), and was essentially absent with methoxamine as agonist. Therefore, subsequent characterization of the high affinity site for 5-MU (using 5-MU, oxymetazoline, HV 723, and spiperone) was made in the presence of nitrendipine (1 μM). S(+)-Niguldipine and to a lesser extent, R(-)-niguldipine, acted as insurmountable antagonists but putative affinity estimates could be obtained from initial dextral shifts (Table 1). The composite pharmacological profile of antagonist affinity estimates in kidney fits closely (a) those abstracted from the literature (native α_{1A}-ARs), (b) ligand binding data from submaxillary gland of rat (a well characterized, homogeneous α_{1A}-AR population, see Michel *et al.*, 1989), and (c) data from the cloned α_{1C}-AR (Table 1). Correlation parameters between affinity estimates are given in Table 2.

Isolated Perfused Mesentery of Rat

Pharmacological studies in this tissue were less extensive than in the kidney of rat. pA_2 estimates of 9.3, 9.6, 9.0, 7.7 were obtained for prazosin, WB 4101, 5-MU, and spiperone respectively. These same antagonists inhibited the vasoconstrictor response to periarterial, noradrenergic nerve stimulation yielding pIC_{40} estimates of: 8.7, 9.1, 8.5, and 7.3 respectively. S(+)-Niguldipine was a potent, insurmountable antagonist against both NE and nerve stimulation (approximate affinity estimate versus NE of 11.0). Prazosin (10nM) protected against the inhibitory action of S(+)-niguldipine (i.e. restored curve maximum) suggesting that inhibition with S(+)-niguldipine was α_1-AR-mediated.

DISCUSSION

The present findings show clearly that the pharmacologically defined α_{1A}-AR in kidney of rat exhibits an antagonist profile almost identical to that of the cloned bovine α_{1C}-AR.

That vasoconstriction to NE is mediated primarily via an α_{1A}-AR in kidney of rat is consistent with previous observations (Blue and Clarke, 1990; Eltze *et al.*, 1991) and with its high correlative association with the α_{1A}-AR in submaxillary gland of rat (Michel *et al.*, 1989; this study), as well as with affinity estimates for α_{1A}-ARs abstracted from the literature (Tables 1 and 2). Likewise, the latter two data sets (submaxillary gland and literature) correlate well with affinity estimates made at the bovine α_{1C}-AR (Table 2). A recent pub-

TABLE 1
Summary of affinity estimates for α_1-adrenoceptor ligands in isolated perfused kidney of rat, submaxillary gland of rat, α_{1A}-adrenoceptors in tissues of rat taken from the literature and cloned hamster α_{1B}-, bovine α_{1C}- and rat α_{1D}-adrenoceptors.

Ligand	*Kidney*[a]	*Submaxillary Gland* (α_{1A})[b]	*Literature* (α_{1A})[bc]	α_{1B}	*Cloned*[b] α_{1C}	α_{1D}	*References*[d]
Prazosin	9.5	10.1	10.1	10.1	9.9	9.9	1–18, 20–27
WB 4101	10.3	10.1	9.9	8.6	10.0	9.6	1, 2, 5, 7, 9, 12–16, 18, 20–23, 26, 27
Phentolamine	8.6	8.8	8.7	7.9	8.9	8.1	1–3, 5, 7, 9, 11, 15, 16, 22, 24, 26, 27
S(+)-Niguldipine	10.5	9.7	10.3	7.0	9.8	7.1	5, 6, 11, 12, 14, 17, 19, 27
R(–)-Niguldipine	9.1	ND	8.9	ND	ND	ND	5, 11
5-Methyl-urapidil	9.2	9.2	9.0	7.5	9.4	8.0	4, 5, 7, 10–12, 14, 18–20, 22, 24–27
Oxymetazoline	7.9	8.5	8.3	6.4	7.8	6.5	7, 9, 14, 24
HV 723	9.3	ND	8.9	8.4	9.6	9.4	2, 15, 16, 20
Spiperone	8.1	8.1	8.2	8.9	8.5	8.5	9, 17, 25, 26

ND = Not determined. [a]pA_2 values versus norepinephrine in isolated perfused kidney of rat. [b]Binding data (pK_i) using [^{3}H]-prazosin as ligand. [c]pK_i values from high affinity binding sites in rat brain, heart, lung, kidney, vas deferens and submaxillary gland. [d]Literature source for α_{1A}-adrenoceptor affinity estimates: 1, Morrow *et al.*, (1985) *Eur. J. Pharmacol.*, 109, 285; 2, Morrow and Creese (1986) *Mol. Pharmacol.*, 29, 321; 3, Honda *et al.* (1987) *Naunyn-Schmiedeberg's Arch. Pharmacol.*, 336, 295; 4, Gross *et al.*, (1988) *Eur. J. Pharmacol.*, 151, 333; 5, Boer *et al.*, (1989) *Eur. J. Pharmacol.*, 172, 131; 6, Hanft and Gross (1989) *Br. J. Pharmacol.*, 98, 652P; 7, Hanft and Gross (1989) *Br. J. Pharmacol.*, 97, 691; 8, Hanft *et al.*, (1989) *J. Pharm. Pharmacol.*, 41, 714; 9, Michel *et al.*, (1989) *Br. J. Pharmacol.*, 98, 883; 10, Hanft and Gross (1990) *Br. J. Clin. Pharmacol.*, 30, 125S; 11, Michel *et al.*, (1990) *Naunyn-Schmiedeberg's Arch. Pharmacol.*, 341, 385; 12, Eltze *et al.*, (1991) *Eur. J. Pharmacol.*, 202, 33; 13, Feng *et al.*, (1991) *J. Pharmacol. Exp. Ther.*, 258, 263; 14, Klign *et al.*, (1991) *Mol. Pharmacol.*, 39, 407; 15, Oshita *et al.*, (1991) *Br. J. Pharmacol.*, 104, 961; 16, Tsuchihashi *et al.*, (1991) *Japan. J. Pharmacol.*, 56, 523; 17, Eltze and Boer (1992) *Eur. J. Pharmacol.*, 224, 125; 18, Hiramatsu *et al.*, (1992) *Br. J. Pharmacol.*, 105, 6; 19, Jackson *et al.*, (1992) *J. Cardiovasc. Pharmacol.*, 19, 857; 20, Muramatsu (1992) In *α-Adrenoceptors.*, Tokyo: Excerpta Medica, Ltd., 193; 21, Ohmura *et al.*, (1992) *Br. J. Pharmacol.*., 107, 697; 22, Yazawa *et al.*, (1992) *J. Pharmacol. Exp. Ther.*, 263, 201; 23, Ivorra *et al.*, (1993) *J. Cardiovasc. Pharmacol.*, 21, 931; 24, Michel *et al.*, (1993) *Naunyn-Schmiedeberg's Arch. Pharmacol.*, 348, 385; 25, Schwietert *et al.*, (1993) *Br. J. Pharmacol.*, 108, 139P; 26, Sleight *et al.*, (1993) *Eur. J. Pharmacol.*, 238, 407; 27, Salles and Badia (1994) *Eur. J. Pharmacol.*, 266, 301.

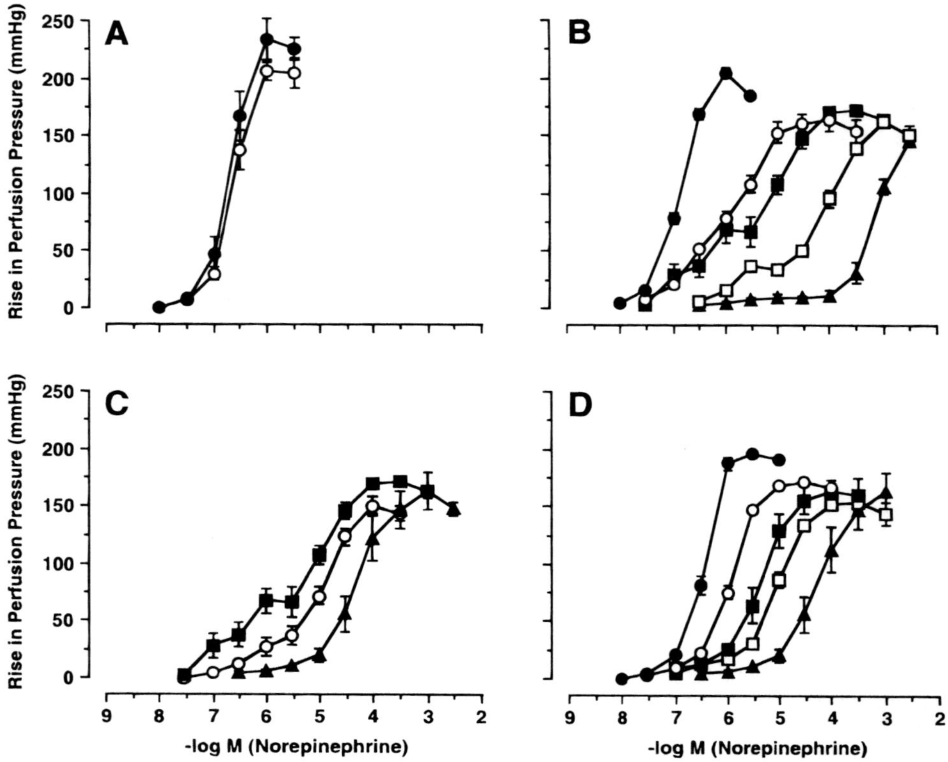

FIGURE 1 The effect of 5-methyl-urapidil (5-MU) on the vasoconstrictor response to infused norepinephrine (NE) in the isolated perfused kidney of rat. A, Time Control; NE at zero time (closed circles); NE 60 min after completion of the first curve (open circles). B, NE Control (closed circles); NE after treatment with 5-MU, 0.03 μM (open circles), 0.1 μM (closed squares), 1.0 μM (open squares), and 3.0 μM (closed triangles). C, NE after treatment with 5-MU 0.1 μM (closed squares); NE after treatment with 5-MU 0.1 μM in kidneys pretreated with chloroethylclonidine (100 μM for 20 min and wash for 40 min, open circles); NE after treatment with 5-MU 0.1 μM in the presence of nitrendipine 1.0 μM (closed triangles). D, NE control in the presence of nitrendipine 1.0 μM (closed circles); after treatment with 5-MU 0.003 μM (open circles), 0.01 μM (closed squares), 0.03 μM (open squares), and 0.1 μM (closed triangles), all in the presence of nitrendipine 1.0 μM. Shown are mean values ± SEM obtained from at least 4 kidneys.

lication by Forray *et al.* (1994) supports the notion that the functionally-defined α_{1A}-AR of rat is equivalent to the α_{1C}-AR of man. Furthermore, accumulating molecular biological evidence supports the existence of the α_{1C}-AR in rat (Alonso-Lamazares *et al.*, 1993; Laz *et al.*, 1993; Rokosh *et al.*, 1994) despite an earlier negative report (for discussion see Ford *et al.*, 1994). In view of these data it is now unnecessary to postulate the existence of a fourth α_1-AR to explain the pharmacology of the native α_{1A}-AR in rat (Schwinn and Lomasney, 1992). The reasons for this postulate have been addressed by Ford *et al.*, (1994), but it should be noted that the original ligand binding profile of the so-called α_{1A}-AR clone (Lomasney *et al.*, 1991), now designated the α_{1D}-AR (Perez *et al.*, 1991), never aligned well with published pharmacological data for the native α_{1A}-AR in rat (Perez *et al.*, 1991).

TABLE 2

Correlations of affinity estimates for α_1-adrenoceptor ligands in isolated perfused kidney of rat, submaxillary gland of rat, cloned hamster α_{1B}-, bovine α_{1C}- and rat α_{1D}-adrenoceptors, and α_{1A}-adrenoceptors of rat taken from the literature.

	R-Squared (Slope)					
	Kidney (α_{1A})	*Submaxillary Gland (α_{1A})*	*Literature (α_{1A})*	*Cloned α_{1B}*	*Cloned α_{1C}*	*Cloned α_{1D}*
Kidney (α_{1A})	—	0.79	0.89	0.02*	0.85	0.12*
	—	(1.15)	(1.09)	(0.12)*	(1.14)	(0.28)*
Submaxillary Gland(α_{1A})	0.79	—	0.89	0.13*	0.84	0.29*
	(1.15)	—	(0.85)	(0.22)*	(0.95)	(0.35)*
Literature (α_{1A})	0.89	0.89	—	0.08*	0.81	0.16*
	(1.09)	(0.85)	—	(0.20)*	(0.84)	(0.29)*
Cloned α_{1C}	0.85	0.82	0.82	0.22*	—	0.39*
	(1.15)	(0.94)	(0.85)	(0.31)*	—	(0.42)*

*Significantly different from 1 ($p<0.05$). Other R-squared values and slopes of regression lines are not significantly different from 1 and slopes do not differ significantly from the line of identity (95% confidence limits).

As in kidney, the vasoconstrictor response to NE in the perfused mesenteric arterial bed of rat also appears to fit the α_{1A}-AR, α_{1C}-AR classification. The pA_2 estimates for antagonists versus exogenously applied NE correlate well with their pIC_{40} values for inhibition of sympathetic nerve-induced vasoconstriction. This result suggests that α_{1A}-ARs in the mesenteric arteries of rat are innervated by postganglionic sympathetic noradrenergic fibers, a conclusion drawn previously for the renovascular bed (Blue *et al.*, 1992).

The biphasicity in the concentration-effect curves to NE induced by 5-MU suggests heterogeneity of α_1-ARs involved in renovascular vasoconstriction. However, it is unclear whether the low affinity site for 5-MU is an α_{1B}- or α_{1D}-AR. The inability of methoxamine to act at the site favors the α_{1B}-AR (see Oshita *et al.*, 1993) whereas only partial sensitivity to inhibition by a high concentration of CEC (100uM for 20min) suggests an α_{1D}-AR. An α_{1D}-AR is also suggested by the fact that WB 4101 failed to show biphasicity (see text and Table 1).

The equivalency of the pharmacologically-defined α_{1A}-AR of rat (in kidney, mesentery, submaxillary gland, and from literature data) with the bovine α_{1C}-AR (Table 2) allows a more parsimonious classification scheme for α_1-ARs. For historical reasons, we suggest that the α_{1A}-AR appellation is retained and that the α_{1C}-AR notation is dropped (see Ford *et al.*, 1994). How α_{1L}-ARs fit this scheme is currently under examination but preliminary molecular biological data suggest that they may be related to the α_{1A}-AR as classified here.

References

Alonso-Llamazares, A., Zamanillo, D., Fernandez, A., and Calvo, P. (1993) Differential expression of the α_{1C} adrenergic receptor subtype in rat tissues. *NeuroReport*, **4,** 1266–1268.

Blue, D.R. and Clarke, D.E. (1990) Characterisation and modulation of vascular α-adrenoceptor responses in rat kidney. *Br. J. Pharmacol.*, **99,** 204P.

Blue, D.R., Jr., Vimont, R.L. and Clarke, D.E. (1992) Evidence for a noradrenergic innervation to α_1-adrenoceptors in rat kidney. *Br. J. Pharmacol.*, **107,** 414–417.

Eltze, M., Boer, R., Sanders, K.H. and Kolassa, N. (1991) Vasodilatation elicited by 5-HT_{1A} receptor agonists in constant-pressure-perfused rat kidney is mediated by blockade of α_{1A}-adrenoceptors. *Eur. J. Pharmacol.*, **202,** 33–44.

Ford, A.P.D.W., Williams, T.J., Blue, D.R. and Clarke, D.E. (1994) α_1-adrenoceptor Classification: Sharpening Occam's Razor. *Trends. Pharmacol.Sci.*, **15,** 167–170.

Forray, C., Bard, J.A., Wetzel, J.M., Chiu, G., Shapiro, E., Tang, R., Lepor, H., Hartig, P.R., Weinshank, R.L., Branchek, T.A. and Gluchowski, C. (1994) The α_1-Adrenergic Receptor that Mediates Smooth Muscle Contraciton in Human Prostate Has the Pharmacological Properties of the Cloned Human α_{1C}-Subtype. *Mol. Pharmacol.*, **45,** 703–708.

Laz, T.M., Forray, C., Smith, K.E., Vaysse, P.J.J., Hartig, P.R., Gluchowski, C., Branchek, T.A. and Wienshank, R.L. (1993) Cloned rat homolog of the bovine α_{1C}-adrenergic receptor exhibits an α_{1A}-like receptor pharmacology. *Soc. Neruosci. Abstracts*, **19,** 1788.

Michel, A.D., Loury, D.N. and Whiting, R.L. (1989) Identification of a single α_1-adrenoceptor corresponding to the α_{1A}-subtype in rat submaxillary gland. *Br. J. Pharmacol.*, **98,** 883–889.

Oshita, M., Kigoshi, S. and Muramatsu, I. (1993) Pharmacological characterization of two distinct α_1-adrenoceptor subtypes in rabbit thoracic aorta. *Br. J. Pharmacol.*, **108,** 1071–1076.

Perez, D.M., Piascik, M.T. and Graham, R.M. (1991) Solution-Phase Library Screening for the Identification of Rare Clones: Isolation of an α_{1D}-Adrenergic Receptor cDNA. *Mol. Pharmacol.*, **40,** 876–883.

Rokosh, D.G., Bailey, B.A., Stewart, A.F.R., Karns, L.R., Long, C.S. and Simpson, P.C. (1994) Distribution of α1C-Adrenergic Receptor mRNA in Adult Rat Tissues by RNase Potection Assay and Comparison with α1B and α1D. *Biochem. Biophys. Res. Comm.*, **200,** 1177–1184.

Schwinn, D.A., Page, S.O., Middleton, J.P., Lorenz, W., Liggett, S.B., Yamamoto, K., Lapetina, E.G., Caron, M.G., Lefkowitz, R.J. and Cotecchia, S. (1991) The α_{1C}-Adrenergic Receptor: Characterization of Signal Transduction Pathways and Mammalian Tissue Heterogeneity. *Mol. Pharmacol.*, **40,** 619–626.

CHARACTERISATION OF α_1-ADRENERGIC RECEPTOR SUBTYPES

ROBERT M. GRAHAM†‡¶, DIANNE M. PEREZ*, MICHAEL T. PIASCIK§, R. PETER RIEK† and JOHN HWA‡*

*†The Victor Chang Cardiac Research Institute, St Vincent's Hospital, Sydney, New South Wales, 2010 Australia; *Department of Molecular Cardiology, Research Institute, Cleveland Clinic Foundation, Cleveland, Ohio 44195; §Department of Pharmacology, University of Kentucky, Lexington, Kentucky, 40536, and ‡Department of Physiology and Biophysics, Case Western Reserve University, Cleveland, Ohio 44106*

α_1-Adrenergic receptors are members of the G-protein-coupled receptor superfamily. Functional and biochemical data have provided evidence for significant heterogeneity within the α_1-receptor family, which is supported by the molecular cloning of three distinct subtypes. The classification of these cloned subtypes has been confounded by a lack of subtype selective compounds and by their low levels of expression. However, as discussed here, it is clear now that the cloned α_{1c}- and α_{1b}-receptors correspond to the pharmacologically defined, tissue expressed α_{1A}- and α_{1B}-subtypes, respectively. The α_{1d} ($\alpha_{1a/d}$)-receptor is a unique subtype that mediates constriction of the rat aorta, but does not correspond to the "classical" α_{1A}-receptor. Thus, this receptor should now be designated, the α_{1D}-subtype.

KEY WORDS: α_{1A}-adrenoceptors, α_{1B}-adrenoceptors, α_{1D}-adrenoceptors, adrenoceptor cloning, receptor classification

I. INTRODUCTION

α_1-Adrenergic receptors are intrinsic membrane glycoproteins that mediate a variety of important sympathetic nervous system responses, particularly those involved in cardiovascular homeostasis, such as arteriolar smooth muscle constriction (Graham, 1990; Hwa *et al.*). Like other adrenergic receptors, α_1-receptors are activated by the catecholamines, norepinephrine and epinephrine, and are members of the G-protein-coupled (Strader *et al.*, 1994; Iismaa *et al.*, 1994) or heptatopic (Riek *et al.*) receptor superfamily, which are likely responsible for signal transduction by 80% of all neurotransmitters and hormones (Birnbaumer *et al.*, 1990). Stimulation of α_1-adrenergic receptors is coupled to a variety of intracellular responses, including activation of phospholipases C, A_2 and D, as well as activation of calcium channels, Na^+-H^+ and Na^+-Ca^{2+} exchange, and inhibition of potassium channels (Hwa *et al.*). Additionally, α_1-receptor activation may lead to transcriptional activation of early- (c-fos, c-jun, c-myc) and late (Egr-1, MLC-2, ANF)-response genes (Hwa *et al.*). Activation or inhibition of these receptor-coupled effectors involves coupling via

¶ Correspondence to: The Victor Chang Cardiac Research Institute, Level 4, Suite 1, 376 Victoria Street, Darlinghurst, Sydney, NSW 2010 Australia.

Over the past 10–15 years it has become evident, initially from functional, radioligand and biochemical studies, that α_1-adrenergic receptors are a heterogeneous group of distinct but related proteins. This conclusion has been confirmed with the molecular cloning of the cDNAs (Cotecchia *et al.*, 1988; Schwinn *et al.*, 1990; Lomasney *et al.*, 1991; Perez *et al.*, 1991; Hirasawa *et al.*, 1993; Bruno *et al.*, 1991; Voigt *et al.*, 1990; Perez *et al.*, Weinberg *et al.*, 1994) and genes (Perez *et al.*, Weinberg *et al.*, 1994; Ramarao *et al.*, 1992) for distinct α_1-receptor subtypes, although until recently discrepancies between the properties of the cloned, expressed receptors, and those characterised pharmacologically and biochemically, has lead to confusion in the classification of α_1-receptor subtypes and their coupled effector-responses.

As detailed here, much of this confusion has now been clarified for the three cloned α_1-adrenergic receptors. However, recent pharmacological data suggest that additional α_1-adrenergic subtypes may also exist, termed α_L-receptors, that can be distinguished by their low affinity for the α_1-receptor-selective antagonist, prazosin (Porter *et al.*, 1992; Ohuma *et al.*, 1993; Oshita *et al.*, 1993).

II. HISTORICAL PERSPECTIVES

Evidence for α-receptor-mediated responses dates back to 1906 when Dale (Dale, 1906) demonstrated that ergot alkaloids inhibit the pressor effects of adrenaline. However, it was not until 1948 that the concept of distinct receptors, α and β, was first put forward by Ahlquist (Ahlquist, 1948), based on his analysis of responses to a variety of sympathomimetic amines. Characterisation of α-receptors at the biochemical level was initially reported by Williams and Lefkowitz (Williams *et al.*, 1976), and Snyder and colleagues (Greenberg *et al.*, 1976) in 1976 using radiolabelled analogues of ergot alkaloids, and of benzodioxans, respectively. These and subsequent radioligand studies greatly aided in the characterisation of the binding properties of α-receptors.

Although distinct β-adrenergic receptor subtypes were reported in 1967, based on functional responses in different tissues to a variety of sympathomimetic amines (Lands *et al.*, 1967; Furchgott, 1967), a subdivision of α-adrenergic receptors into α_1- and α_2-adrenergic subtypes was not proposed until 1974 (Langer, 1974). Initially, this classification was restricted to distinguishing postjunctional α_1-receptors, mediating vasoconstriction, from prejunctional α_2-receptors that inhibit stimulus-induced norepinephrine release from postganglionic sympathetic nerve terminals. Subsequently, Berthelson and Pettinger (Berthelsen and Pettinger, 1977) proposed a more widely applicable, functional classification of α-receptors, based on the relative potencies of adrenergic agonists and antagonists to stimulate or inhibit α-receptor responses.

Over the next decade studies from a number of laboratories, particularly those of McGrath (McGrath, 1982) and Ruffolo (Ruffolo, 1985), provided evidence that α_1-adrenergic receptors are a heterogeneous group of receptor-proteins, and that postjunctional responses, mediated by α_1-receptors, could not be explained adequately on the basis of a single receptor subtype. This concept was further advanced by the radioligand binding studies of Morrow and Cresse (Morrow and Creese, 1986) who suggested a subdivision of α_1-receptors into α_{1A} and α_{1B} subtypes. This was based on the finding of a differential affinity of these subtypes for a variety of agents, with the agonist, oxymetazoline, and the antagonists, WB 4101 and phentolamine, being recognised with higher affinity at α_{1A}- than at α_{1B}-receptors. This subdivision was supported by the findings by Minneman and coworkers (Han *et al.*, 1987; Han *et al.*, 1987) that α_1-receptors with the characteristics of an α_{1B}-subtype could be completely inactivated by the alkylating agent, chlorethylclonidine (CEC), whereas α_{1A}-receptors, although competitively antagonised by CEC, were resistant to irreversible inactivation by this compound. Also, Piascik and coworkers (Piascik *et al.*, 1989; Kusiak *et al.*, 1989) demonstrated that α_{1A}-receptors were more sensitive to irreversible inactivation by SZL-49 [1-(4-amino-6,7-dimethoxy-2-quinazolnyl)-4-(2-bicyclo [2,2,2] octa-2,5-dienylcarbonyl)-piperazine] than α_{1B}-receptors, and Terman *et al.*, (Terman *et al.*, 1990) demonstrated that α_{1B}- but not α_{1A}-receptors could be photolabelled with the prazosin analogue, 2-[4-(4-azido-3-[^{125}I]iodobenzoyl)piperazin-1-yl]-4-amino-6,7-dimethoxyquinazoline (azidoprazosin). Differences in the biochemical mechanisms of receptor-mediated signalling, and in the requirements for extracellular calcium in signal transduction by α_1-receptor subtypes, were also reported during this time by Garcia-Sainz and coworkers (Garcia-Sainz *et al.*, 1985), and by Tsujimoto and colleagues (Tsujimoto *et al.*, 1989). More recently, heterogeneity of α_1-receptor subtypes has been confirmed by the cloning of the cDNA and genes for three distinct receptor proteins (Cotecchia *et al.*, 1988; Schwinn *et al.*, 1990; Lomasney *et al.*, 1991; Perez *et al.*, 1991; Hirasawa *et al.*, 1993; Bruno *et al.*, 1991; Voigt *et al.*, 1990; Perez *et al.*, Weinberg *et al.*, 1994).

III. CHARACTERISTICS OF THE CLONED α_1-ADRENERGIC RECEPTORS

The properties of the three cloned α_1-adrenergic receptors are summarised in Table 1. Classification of these receptors and their relationships to the pharmacologically characterised subtypes has been problematic for several of reasons. First, few drugs are currently available that are markedly subtype selective ie. are recognised by one subtype with at least a 100-fold higher affinity than by the other two subtypes. This has limited the ability of functional and radioligand binding studies to discriminate between subtypes. Moreover, this problem has been compounded by the fact that most tissues and even cell lines express more than one subtype. For example, no tissue or cell line has been identified that expresses only the α_{1A}-receptor (Esbenshade *et al.*, 1993). Second, cloning of α_1-receptors has revealed the existence of previous uncharacterised subtypes e.g. the α_{1D}-adrenergic receptor (Perez *et al.*, 1991). Third, each subtype couples similarly to multiple effector pathways via distinct G-proteins (Schwinn *et al.*, 1991; Perez *et al.*, 1993). Fourth, the mRNA for adrenergic receptors are frequently expressed at very low levels, making their detection very difficult. Indeed, there are examples of α_1-receptor cloning from a cell line in which the mRNA for this receptor could not be detected by Northern blot analysis (Perez *et al.*, 1991). Finally, there are marked species differences in their tissue distribution. For example, the α_{1B}-receptor is the predominant subtype expressed in rat (Rokosh *et al.*, 1994), but not in rabbit (Schwinn *et al.*, 1991) liver (Table 1).

The first subtype to be cloned was isolated from the DDT1-MF2 syrian hamster smooth muscle cell line (Cotecchia *et al.*, 1988). The pharmacology of the expressed, encoded receptor, determined using a variety of agonists and antagonists, was exactly as predicted by

TABLE 1

Characteristics of the Cloned α_1-Adrenergic Receptors

Characteristics	*Receptor Subtype*		
Subtype Classification§			
Proposed	$\alpha_{1A}(\alpha_{1A/c})$	α_{1B}	α_{1D}
Current	α_{1c}	α_{1b}	$\alpha_{1a};\alpha_{1d};\alpha_{1a/d}$
Mr*	-	80	_65
Amino acids	466	515	560
Gylcosylation sites (N-terminus)	3	4	2
Phosphorylation sites	PKA	PKA	-
Genomic organisation			
Introns	1	1	1
Exons	2	2	2
Chromosonal localisation†	8	5	20 p13
Pharmacological selectivity			
Subtype-selective agents	5MU; (+)niguldipine; oxymetazoline; A-61603; SNAP-5150	spiperone; respiridone	(+)norepinephrine; BMY 7378
Non-selective agents‡		prazosin; WB 4101; phentolamine; benoxathian; abanoquil	
Susceptibility to covalent modification			
CEC	+	+++	++
SZL-49	+++	+	-
Azidoprazosin	n.d.	+++	-
Prototypic tissue(s)	rat kidney & submaxillary gland; rabbit liver; human prostate	rat spleen, liver & heart.	rat aorta & cerebral cortex
Receptor-coupled signalling‡		calcium mobilisation; PLC; PLA_2; PLD	
G-Protein coupling	$G_{q/11/14}$	$G_{q/11/14/16}$; G_h	$G_{q/11}$

§Lower case subscripts are used to designate the cloned subtypes, and upper case subscripts to indicate that a pharmacologically characterised correlate has been demonstrated for the respective cloned receptor.
*The apparent molecular weights (Mr) on SDS-PAGE are shown, that for the α_{1D}-receptor is based on the work of Kost *et al.*.[52]
†Refers to the human genome.
‡These characteristics are the same for all subtypes.

all previous pharmacological studies of α_{1B}-receptors expressed endogenously in tissues, and its tissue distribution matched that of the pharmacologically defined α_{1B}-receptor. Also, it can be completely inactivated by CEC, is relatively resistant to inactivation by SZL-49, and can be photolabelled with azidoprazosin (Perez *et al.*, 1991). The pharmacologically properties of the hamster α_{1b}-receptor are also shared by the receptor encoded by a cDNA for the human homologue of this receptor, which was cloned subsequently (Weinberg *et al.*, 1994; Ramarao *et al.*, 1992). Thus, it is clear that the cloned α_{1b}-receptor is identical to the pharmacologically-characterised α_{1B}-receptor.

The next subtype to be cloned, was isolated from a bovine brain cDNA library (Schwinn *et al.*, 1990). Its pharmacology resembled that of the "classical" α_{1A}-receptor with a 10-fold higher affinity for WB4101, phentolamine and oxymetazoline, than the α_{1B}-adrenergic receptor. However, its sensitivity to CEC (65% inactivated by 100μm CEC for 20 min at 37°C), and the apparent lack of expression of an mRNA for this receptor, in any rat or bovine tissue, and in most human tissues except a discrete area in the brain (Schwinn *et al.*, 1990), seemed to indicate that this receptor represented a new subtype. It was, therefore, designated as the α_{1c}-adrenergic receptor. However, with the isolation of a partial cDNA for the human homolog of the bovine α_{1c}-receptor from the SK-N-MC cell line (Perez *et al.*), which predominantly expresses α_1-receptors with an α_{1A}-subtype pharmacology (Esbenshade *et al.*, 1993), together with the finding that the pharmacology of the expressed bovine α_{1c}-receptor is similar to that of the "classical" α_{1A}-subtype (Schwinn *et al.*, 1990), led us to speculate that the cloned α_{1c}-receptor is identical to the "classical" α_{1A}-receptor subtype (Perez *et al.*, 1994). This postulate was confirmed with the isolation and characterisation of a full-length cDNA for the rat homologue of the bovine α_{1c}-receptor (Perez *et al.*). Thus, we have been able to show by three different techniques, RNase protection assays, RT-PCR Northerns and *in situ* hybridization histochemistry, that its mRNA is expressed in typical α_{1A}-abundant tissues, such as submaxillary gland, hippocampus, and vas deferens. Additionally, the ligand binding profile of the receptor encoded by this cDNA is similar to that for endogenously expressed tissue α_{1A}-receptors, and compared to the other cloned α_1-receptors, it is both the least sensitive to alkylation by CEC and the most sensitive to inactivation by SZL-49 (Perez *et al.*). Taken together these data clearly indicate that this cloned rat α_1-receptor is the "classical" α_{1A}-subtype; a conclusion that is supported by the findings of three other laboratories (Rokesh *et al.*, 1994; Ford *et al.*, 1994; Price *et al.*, 1994). Accordingly, we suggest that this receptor should now be reclassified as either the α_{1A} or the $\alpha_{1A/c}$ subtype (Table 1). Interestingly, the $\alpha_{1A/c}$-receptor has recently been shown to be abundantly expressed in various human tissues including the heart, prostate, liver, cerebellum and cerebral cortex (Price *et al.*, 1994).

A third subtype, cloned from rat cerebral cortex was initially designated as the α_{1A}-receptor, because of its higher affinity for WB4101, its partial resistance to inactivation by CEC, and the expression of its mRNA in vas deferens, hippocampus and cerebral cortex (Lomasney *et al.*, 1991). However, further analysis of an almost identical cDNA isolated independently from rat hippocampus (Perez *et al.*, 1991), indicated that the receptor encoded by this cDNA was a previously uncharacterised subtype, since its pharmacology was quite distinct from that of the "classical" α_{1A}-receptor, and under appropriate conditions it was actually very sensitive to CEC (72–74% inactivation with 100μMCEC). Based on these findings it has been designated as the α_{1d}-receptor (Perez *et al.*, 1991). Subsequently, the designation $\alpha_{1a/d}$, was also proposed (Schwinn and Lomasney, 1992). However, since the rat cDNA discussed above corresponds to the "classical" α_{1A}-receptor, it is evident that the α_{1d}-receptor is a unique subtype, and should now be referred to as the α_{1D}-subtype (Table 1). A final issue with regards to the α_{1D}-receptor, is which prototypical tissue expresses the functional correlate of this subtype? This issue has also now been resolved, since, in an analysis of 21 compounds, it has been shown recently that the pK_B values, determined using the rat aorta, correlate very closely with the pKi values for these compounds, determined from radioligand-binding studies with the expressed α_{1D}-receptor, but not with the pKi values determined with the other two cloned subtypes (Saussy). Additionally, expression of mRNA for the α_{1D}-adrenergic receptor has been demonstrated by *in situ* hybridization histochemistry in rat aorta, as well as renal and mesenteric resistance arteries (Piascik *et al.*, 1994). The α_{1D}-receptor is also the major functional subtype mediating vasoconstriction in the rat iliac artery, although other subtypes appear to predominate in rat caudal, renal and mesenteric resistance arteries.[1] With the availability of α_{1D}-subtype-selective compounds, such as BMY 7378, it is likely that other functional tissue correlates for this cloned subtype will soon be defined.

As shown in Table 1, all three cloned α_1-receptor subtypes have been shown to mobilise intracellular calcium, and to activate phospholipases C, and A_2 (Schwinn *et al.*, 1991; Perez *et al.*, 1993). Additionally, there is preliminary evidence that these receptors couple to activation of phospholipase D (Lomasney *et al.*, 1993). For phospholipase C activation, receptor coupling may involve a differential interaction with various members of the G_q G-protein family (Wu *et al.*, 1992). For the α_{1B}-receptor a novel G-protein, G_h, has also recently been shown to support receptor-stimulated polyphosphoinositide turnover (Nakaoka *et al.*, 1994). However, it is presently unclear if all three α_1-receptors couple via both G_q and G_h, or if there is a differential selectivity of the various receptor subtypes for these two G-protein families.

ACKNOWLEDGMENTS

We thank Ms Elaine Martin for expert secretarial assistance. Studies from the authors laboratories were supported in part by grants from the National Institutes of Health (HL 19583 and HL 38120), grants-in-aid from the American Heart Association (AHA) and AHA Northeast Ohio Affiliate, and an Eccles Award from the National Health and Medical Research Council, Australia.

References

Ahlquist, R.P. (1948) A study of the adrenotropic receptors. *Am. J. Physiol.*, **153,** 586–600.

Berthelsen, S. and Pettinger, W.A. (1977) A functional basis for classification of α-adrenergic receptors. *Life Sci.*, **21,** 595–606.

Birnbaumer, L., Abramowitz, J. and Brown, A.M. (1990) Receptor-effector coupling by G-proteins. *Biochem. Biophys. Acta.*, **1031,** 163–224.

Bruno, J.F., Whittaker, J., Song, J. and Berelowitz, M. (1991) Molecular cloning and sequencing of a cDNA encoding a human α_{1A}-adrenergic receptor. *Biochem. Biophys. Res. Commun.*, **179,** 1485–1490.

[1]M. T. Piasick, unpublished data.

Cotecchia, S., Schwinn, D.A., Randall, R.R., Lefkowitz, R.J., Caron, M.G. and Kobilka, B.K. (1988) Molecular cloning and expression of the cDNA for the hamster α_1-adrenergic receptor. *Proc. Natl. Acad. Sci. USA*, **85**, 7159–7163.

Dale, H.H. (1906) On some biological actions of ergot. *J. Physiol.*, **34**, 163–206.

Esbenshade, T.A., Han, C., Murphy, T.J. and Minneman, K.P. (1993) Comparison of α_1-adrenergic receptor subtypes and signal transduction in BK-N-MC and NB 41A3 neuronal cell lines. *Mol. Pharmacol.*, **44**, 76–86.

Ford, A.P.D.W., Williams, T.J., Blue, D.R. and Clarke, D.E. (1994) α_1-Adrenoceptor classification: sharpening Occam's razor. *Trends Pharmacol. Sci.*, **15**, 167–170, 1994.

Furchgott, R.F. (1967) The pharmacological differentiation of adrenergic receptors. *Ann. NY Acad. Sci.*, **139**, 553–570.

Garcia-Sainz, J.A. and Hernandez-Sotomayer, S.M. (1985) Adrenergic regulation of gluconeogenesis: possible involvement of two mechanisms of signal transduction in α_1-adrenergic action. *Proc. Natl. Acad. Sci. (USA)*, **82**, 6727–6730.

Graham, R.M. (1990) Adrenergic receptors: structure and function. *Cleve. Clin. J. Med.*, **57**, 481–491.

Greenberg, D.A., U'Prichard, D.C. and Snyder, S.H. (1976) Alpha-noradrenergic receptor binding in mammalian brain: differential labelling of agonist and antagonist states. *Life Sci.*, **19**, 69–76.

Han, C.P., Abel, W. and Minnemann, K.P. (1987) α_1-Adrenoceptor subtypes linked to different mechanisms for increasing intracellular Ca^{2+} in smooth muscle. *Nature (Lond)*, **329**, 333–335.

Han, C.P., Abel, W. and Minnemann K.P. (1987) Heterogeneity of α_1-adrenoceptors revealed by chlorethylclonidine. *Mol. Pharmacol.*, **32**, 506–510.

Hirasawa, A., Horie, K., Tanaka, T., Takagaki, K., Murai, M., Yano, J. and Tsujimoto, G. (1993) Cloning, functional expression and tissue distribution of human cDNA for the α_{1C}-adrenergic receptor. *Biochem. Biophys. Res. Commun.*, **195**, 902–909.

Hwa, J., De Young, M.D., Perez, D.M. and Graham, R.M. Autonomic control of myocardium: alpha-adrenoceptor mechanisms. In: *The Autonomic Nervous System*, G. Burnstock, ed. *Volume VIII: The Nervous Control of the Heart*, Volume eds. J. Shepherd, S.F. Vatner. Harvard Academic Press. (in press).

Iismaa, T., Biden, T.J. and Shine, J. G-protein-coupled receptors. (1994) *New York, CRC Press.*

Kost, T., Kadwell, S., Watson, M., Clay, W., Goestz, A., King, H., Wisely, B., Lewis, K., Patel, I., Su, J.L., Graham, R.M., Tseng-Crank, J., Saussy, D. and Buckholz, R. (1994) Control of α_1-adrenergic receptor expression by NH_2-terminal extracellular domain sequences. Pharmacology of Adrenoceptors. *Satellite Symposium of the XIIth UIPHAR Congress. King of Prussia*, PA July 21–23.

Kusiak, J.W., Pitha, J. and Piascik, M.T. (1989) Interaction of a chemically reactive prazosin analog with α-adrenoceptors of rat tissues. *J. Pharmacol. Exp. Therr.*, **249**, 70–77.

Lands, A.M., Arnold, A., McAuliff, J.P., Luduena, F.P. and Brown, T.G. (1967) Differentiation of receptor systems activated by sympathomimetic amines. *Nature*, **214**, 597–598.

Langer, S.Z. (1974) Presynaptic regulation of catecholamine release. *Biochem. Pharmacol.*, **23**, 1788–1800.

Lomasney, J.W., Allen, L.F., Capel, D. and Lefkowitz, R.J. (1993) Distinct transmembrane signalling pathways coupled to different cloned subtypes of the α_1-adrenergic receptor. **41**, 264A.

Lomasney, J.W., Cotecchia, S., Lorenz, W., Leung, W.Y., Schwinn, D.A., Yang-Feng, T.L., Braunstein, M., Lefkowitz, R.J. and Caron, M.G. (1991) Molecular cloning and expression of the cDNA for the α_{1A}-adrenergic receptor. *J. Biol. Chem.*, **266**, 6365–6369.

McGrath, J.C. (1982) Evidence for more than one type of post-junctional α-adrenoceptor. *Biochem. Pharmacol.*, **31**, 467–484.

Morrow, A.L. and Creese, I. (1986) Characterization of α_1-adrenoceptor subtypes in rat brain: a reevaluation of [^{3}H]WB4101 and [^{3}H]prazosin binding. *Mol. Pharmacol.*, **29**, 321–330.

Nakaoka, H., Perez, D.M., Baek, K.J., Das, T., Husain, A., Misono, K., Im, M.-J. and Graham, R.M. (1994) G_h: A GTP-binding protein with transglutaminase activity and receptor signalling function. *Science*, **264**, 1593–1596.

Ohuma, T., Sakamoto, S., Hayashi, H., Kigoshi, S. and Muramatsu, I. (1993) Identification of α_1-adrenoceptor subtypes in the dog prostate. *Urol. Res.*, **21**, 211–215.

Oshita, M., Kigoshi, S. and Muramatsu, I. (1993) Pharmacological characterization of two distinct α_1-adrenergic subtypes in the rabbit thoracic aorta. *Br. J. Pharmacol.*, **108**, 1071–1076.

Perez, D.M., Chen, J-L., Piasick, M.T., Malik, N. and Graham, R.M. (1994) Is the α_{1c}-adrenergic receptor the α_{1A}-subtype. *FASEB J*, **8**, A353.

Perez, D.M., De Young, M.B. and Graham, R.M. (1993) Coupling of expressed α_{1B} and α_{1D}-adrenergic receptors to multiple signalling pathways is both G protein and cell type specific. *Mol. Pharmacol.*, **44**, 784–795.

Perez, D.M., Piasick, M.T. and Graham, R.M. (1991) Solution-phase library screening for the identification of rare clones: isolation of an α_{1D}-adrenergic receptor cDNA. *Mol. Pharmacol.*, **40**, 876_883.

Perez, D.M., Piascik, M.T., Malik, N., Gaivin, R. and Graham, R.M. Cloning, expression and tissue distribution of the rat homolog of the bovine α_{1c}-adrenergic receptor provide evidence for its classification as the α_{1A}-subtype. *Mol. Pharmacol.* (in press).

Piascik, M.T., Butler, B.T., Kusiak, J.W., Pitha, J. and Holtmann, J.R. (1989) Effect of an alkylating analog of prazosin on α_1-adrenoceptor subtypes and arterial blood pressure. *J. Pharmacol. Exp. Ther.*, **251,** 878–883.

Piascik, M.T., Smith, M.S., Soltis, E.E. and Perez, D.M. (1994) Identification of the mRNA for the novel $alpha_{1D}$ and two other $alpha_1$-adrenoceptors in vascular smooth muscle. *Mol. Pharmacol.*, **46,** 30–40.

Porter, J.E., Dowd, F.J. and Abel, P.W. (1992) Atypical $alpha_1$-adrenergic receptors on the rat parotid gland acinar cells. *J. Pharmacol. Exp. Ther.*, **263,** 1062–1067.

Price, D.T., Chari, R.S., Berkowitz, D.E., Meyers, W. and Schwinn, D.A. (1994) Expression of α_1-adrenergic receptor subtype mRNA in rat tissues and human SK-N-MC neuronal cells: implications of α_1-adrenergic receptor subtype classification. Pharmacology of Adrenoceptors. *Satellite Symposium of the XII IUPHAR Congress King of Prussia*, PA July 21–23.

Price, D.T., Lefkowitz, R.J., Caron, M.G., Berkowitz, D. and Schwinn, D.A. (1994) Localization of mRNA for three distinct α_1-adrenergic receptor subtypes in human tissues: Implications for human α-adrenergic physiology. *Mol. Pharmacol.*, **45,** 171–175.

Ramarao, C.S., Kincade, J.M., Perez, D.M., Garvin, R.J., Riek, R.P. and Graham, R.M. (1992) Genomic organization and expression of the human α_{1B}-adrenergic receptor. *J. Biol. Chem.*, **267,** 21936–21945.

Riek, R.P., Handschumacher, M.K., Sung S.-S., Tan, M., Glynias, M.J., Schluchter, M.D., Novotney, J. and Graham, R.M. Evolutionary conservation of both the hydrophilic and hydrophobic nature of transmembrane residues. *J. Theoretical. Biol.*, (in press).

Rokosh, D.G., Bailey, B.A., Stewart, A.F.R., Karns, L.R., Long, C.S. and Simpson, P.C. (1994) Distribution of α_{1C}-adrenergic receptor mRNA in adult rat tissue by RNase protection assay and comparison with α_{1B} and α_{1D}. *Biophys. Res. Commun.*, **200,** 1177–1184.

Ruffolo R.R. (1985) Relative agonist potency as a means of differentiating alpha-adrenoceptors and alpha-adrenergic mechanisms. *Clin. Sci.*, **10,** 9s–14s.

Saussy, D. Personal communication.

Schwinn, D.A. and Lomasney, J.W. (1992) Pharmacologic characterisation of cloned α_1-adrenoceptor subtypes: selective antagonists suggest the existence of a fourth subtype. *Eur. J. Pharmacol.*, **227,** 433–436.

Schwinn, D.A., Lomasney, J.N., Loreng, W., Szklut, P.J., Fremeau, R.T., Yang-Feng, T.L., Caron, M.G., Lefkowitz, R.J. and Cotecchia, S. (1990) Molecular cloning and expression of the cDNA for a novel α_1-adrenergic receptor subtype. *J. Biol. Chem.*, **265,** 8183–8189.

Schwinn, D.A., Page, S.O., Middleton, J.P., Lorenz, W., Liggett, S.B., Yamamoto, K., Lapetina, E.G., Caron, M.G., Lefkowitz, R.J. and Cotecchia, S. (1991) The α_{1C}-adrenergic receptor: characterisation of signal transduction pathways and mammalian tissue heterogeneity. *Mol. Pharmacol.*, **40,** 619–626.

Strader, C.D., Fong, T.M., Tota, M.R. and Underwood, D. (1994) Structure and function of G-protein-coupled receptors. *Annu. Rev. Biochem.*, **63,** 101–132.

Terman, B.I., Riek, R.P., Grodski, A., Hess, H.J. and Graham, R.M. (1990) Identification and characterization of α_1-adrenergic receptor subtypes. *Mol. Pharmacol.*, **37,** 526–534.

Tsujimoto, G., Tsujimoto, A., Suzuki, E. and Hashimoto, H. (1989) Glycogen phosphorylase activation by two different α_1-adrenergic receptor subtypes: methoxamine selectively stimulates a putative α_1-adrenergic receptor subtype (α_{1A}) that couples with Ca^{2+} influx. *Mol. Pharmacol.*, **36,** 166–176.

Voigt, M.M., Kispert, J. and Chin, H.M. (1990) Sequence of a rat brain cDNA encoding an α_{1B}-andrenergic receptor. *Nucl. Acids. Res.*, **18,** 1053.

Weinberg, D.H., Trivedi, P., Tan, C.P., Mitra, S., Perkins-Barrow, A., Barkowski, D., Strader, C.D. and Bayne, M. (1994) Cloning, expression and characterization of human α-adrenergic receptors α_{1A}, α_{1B} and α_{1C}. *Biochem. Biophys. Res. Commun.*, **201,** 1296–1304.

Williams, L.T. and Lefkowitz, R.J. (1976) Alpha-adrenergic receptor identification by [^{3}H]dihydroergoisytine binding. *Science*, **192,** 791–793.

Wu, D., Katz, A., Lee, C.H. and Simon, M.I. (1992) Activation of phospholipase C by α_1-adrenergic receptors is mediated by the α subunits of Gq family. *J. Biol. Chem.*, **267,** 36, 25798–25802.

FUNCTIONAL SUBCLASSIFICATION OF VASCULAR α_1-ADRENOCEPTORS

IKUNOBU MURAMATSU, TSUYOSHI OHMURA, SHIGEKI HASHIMOTO and MASAFUMI OSHITA

Department of Pharmacology, Fukui Medical School, Matsuoka, Fukui 910-11, Japan

KEY WORDS: α_{1H}-adrenoceptors, α_{1L}-adrenoceptors, α_{1N}-adrenoceptors, vascular adrenoceptors, functional classification of adrenoceptors

INTRODUCTION

α_1-Adrenoceptors are not homogeneous, and different affinities for agonists and antagonists have been reported in the functional and binding studies. In 1985, Drew first suggested that the variety of pA_2 values for prazosin and yohimbine in many tissues may reflect a possible heterogeneity of α_1-adrenoceptors. Next year, Flavahan and Vanhoutte (1986) developed the Drew's idea and differentiated α_1-adrenoceptors into two groups; prazosin-high (α_{1H}) and -low (α_{1L}). Thereafter, Muramatsu *et al.*, (1990) extended this functional subclassification by the addition of a third subtype, α_{1N}, which was recently categorized as a member of α_{1L} group because of its low affinity for prazosin. In addition to this functional subclassification, another subclassification has also been proposed. This second subclassification is based on the data obtained from molecular biological and binding studies, in which classical α_{1A} (detected by binding study) and four cloned subtypes (α_{1A}, α_{1B}, α_{1C} and α_{1D}) are included (Minneman, 1988; Lomasney *et al.*, 1991; Bylund *et al.*, 1994). Since these five subtypes have high affinity for prazosin, the second subclassification is essentially different from the functional subclassification mentioned above. However, it was recently postulated that the classical α_{1A} and cloned α_1-adrenoceptors are subtypes of the α_{1H} group in the functional subclassification, as mentioned below (Table 1) (Muramatsu *et al.*, 1991). In this section, the evidence on the functional subclassification is described.

FUNCTIONAL STUDY

Difference in antagonist affinity gives a useful clue for pharmacological subclassification of α_1-adrenoceptors. However, the affinity constant may apparently vary under various experimental conditions. In order to get the precise affinity of α-adrenoceptor antagonists, we should eliminate or correct the factors which would affect functional response as possible.

Correspondence to: Ikunobu Muramatsu, Department of Pharmacology, Fukui Medical School, Matsuoka, Fukui 910-11, Japan.

TABLE 1

α_1-Adrenoceptor subclassification

Antagonist	*classical* α_{1A}	*cloned* α_{1A}	α_{1H} *group* *cloned* α_{1B}	*cloned* α_{1C}	*cloned* α_{1D}	α_{1L} *group* α_{1L}	α_{1N}
Prazosin	H (9~10)	H (9~10)	H (9~10)	H (9~10)	H (9~10)	L (< 9)	L (< 9)
WB4101	H (9~10)	H (9)	L (7~8)	H (9~10)	H (9)	L (8)	L (8)
5-Methylurapidil	H (9~10)	L (< 7)	L (< 7)	H (9)	L (8)	L (8)	L (8)

H: High affinity, L: Low affinity, ( ): pK_B α_{1N}: HV723-high affinity

In the functional study with isolated blood vessels, the factors are endothelium cells, coexisting β- and α_2-adrenoceptors, neuronal and extraneuronal uptake of noradrenaline, treatment period of antagonist, concentration of antagonist, reproducibility of concentration-response curve and others. Figure 1A shows the actual results in which the concentration-response curves for noradrenaline were five times repeatedly obtained from the same strip of endothelium-denuded rat thoracic aorta. Propranolol, desipramine and deoxycorticosterone were added in the bathing solution to block β-adrenoceptors, neuronal and extraneuronal uptake of noradrenaline. In this preparation and other vessels such as rabbit thoracic aorta, the response produced by first trial of noradrenaline was apparently different from the responses produced by the subsequent trials (Figure 1B). In general, good reproducibility is seen in the responses from the third to fifth trials; therefore we have used the third response as control and the fourth and fifth responses as tested responses. Postjunctional α_2-adrenoceptor may be stimulated by noradrenaline in some tissues such as dog saphenous vein. In this case, α_2-antagonist must be treated or α_1-selective agonist must be used instead of noradrenaline. There are several reports in which yohimbine was added to the bathing solution. However, in the rat thoracic aorta where noradrenaline acts on α_1-adrenoceptor only, yohimbine at 100 nM significantly inhibited the concentration-response curve for noradrenaline (Figure 1C). This result indicates that yohimbine can bind to the α_1-adrenoceptor at relatively low concentrations, recommending the use of more selective α_2-antagonist in the tissues having postjunctional α_2-adrenoceptors. The affinity for α_1-antagonist may be underestimated in the presence of 100 nM yohimbine. Furthermore, antagonist must be treated for at least 30 min before recording the test response. In this way, many precautions should be taken in order to obtain the precise affinity and to eliminate artificial variation in the functional study.

Affinity constants (pK_B) for prazosin in various tissues are roughly subdivided into two groups (Table 2). The high pK_B value is close to 10, while the low value is less than 9. In some tissues such as rabbit thoracic aorta, both affinities for prazosin are detected, suggesting the existence of prazosin-high (α_{1H}) and -low (α_{1L})-adrenoceptors. In such cases, the slope of Schild plot for prazosin significantly deviates from the unity (Muramatsu, 1991; Oshita *et al.*, 1993).

In the dog carotid artery, a single and high pK_B value (9.7) for prazosin was estimated. However, the pK_B values for WB4101 and 5-methylurapidil were relatively low (8.2 and 7.2, respectively). Pretreatment with chloroethylclonidine (CEC, which produces relatively selective inactivation of the cloned α_{1B} and α_{1C} subtypes) remarkably attenuated the contractile response to noradrenaline. These results suggest that the α_1-adrenoceptor of dog carotid

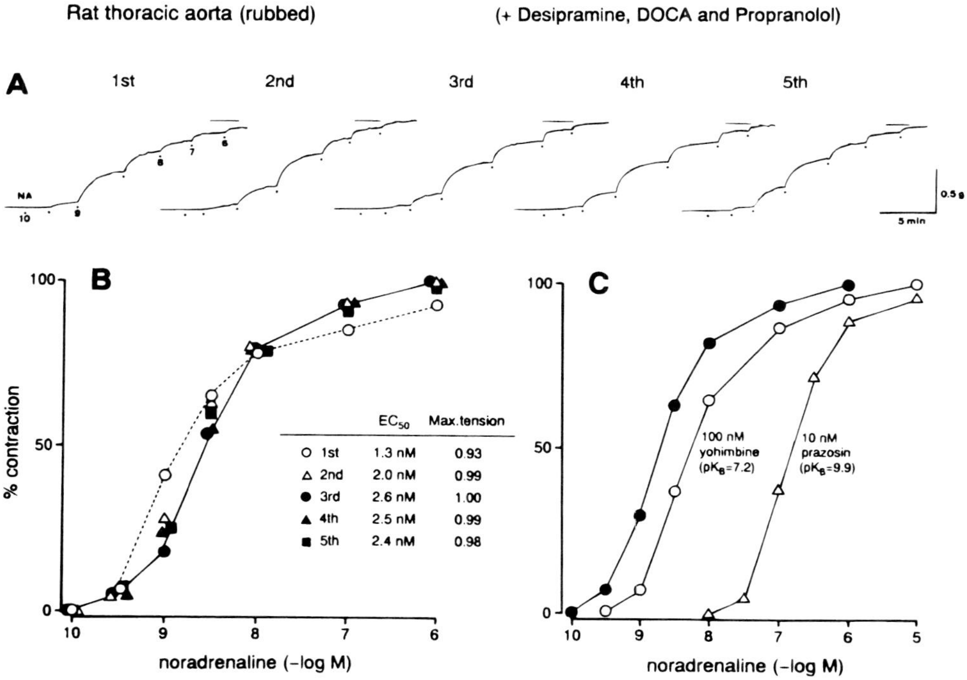

FIGURE 1 Reproducibility of contractile responses to noradrenaline in endothelium-denuded rat thoracic aorta. A, actual recordings. The concentration-response curves for noradrenaline (NA, 0.1 nM – 1 μM) were five times repeatedly recorded from a same strip. B, concentration-response curves for noradrenaline recorded in A. Maximal contraction induced by noradrenaline at third trial was taken as 100%. C, effects of yohimbine (100 nM) and prazosin (10 nM). The antagonists were treated for 30 min before and during the test response. Desipramine (0.1 μM), deoxycorticosterone (DOCA, 5 μM) and propranolol (3 μM) were present throughout the experiments.

artery belongs in the α_{1H} group of the functional subclassification and further corresponds to the cloned α_{1B} subtype. The α_1-adrenoceptor with prazosin-high affinity (α_{1H}) of rabbit thoracic aorta is similar subtype to that of dog carotid artery (Oshita *et al.*, 1993). On the other hand, the features of α_{1H}-adrenoceptor ($pK_B = 9.7$) of rat thoracic aorta are a little different from those of dog carotid artery and rabbit thoracic aorta. The affinities (pK_B) for WB4101 and 5-methylurapidil are respectively 9.2 and 7.9 in the rat thoracic aorta, and the α_{1H}-adrenoceptor is activated by CEC (Muramatsu *et al.*, 1991). These results indicate that the pharmacological features of α_{1H}-adrenoceptors in various tissues are not homogeneous, suggesting that the α_{1H}-adrenoceptors may be subdivided into several subtypes according to different affinities of several antagonists, as shown Table 1. The pharmacological features of α_{1H}-adrenoceptor of rat thoracic aorta resemble to those of cloned α_{1A} or α_{1D} subtype.

Functional α_1-adrenoceptors with low affinity for prazosin (α_{1L}-adrenoceptors) are also detected in many tissues (Table 2). The pK_B values for prazosin are approximately 8.5, which is 10 time less in affinity than α_{1H}-adrenoceptors. The α_{1L}-adrenoceptors are also less sensitive to WB4101 and 5-methylurapidil (pK_B: approximately 8) and are relatively resistant to CEC. The α_{1L}-adrenoceptors in blood vessels of dog visceral region show higher

TABLE 2
Functional α_1-adrenoceptor subtypes involved in noradrenaline-responses

Animal	Tissue	Subtype	pK_B (prazosin)	
rat	thoracic aorta	α_{1H} (α_{1A} or α_{1D})	9.7	
	heart	α_{1L}		8.4
	vas deferens	α_{1L}		8.3
rabbit	thoracic aorta	α_{1H} (α_{1B}) + α_{1L}	9.7	8.2
	carotid artery	α_{1H} (α_{1B}) + α_{1L}	9.9	8.3
	prostate	α_{1L}		8.0
dog	carotid artery	α_{1H} (α_{1B})	9.7	
	vertebral artery	α_{1H} (α_{1B})	9.3	
	mesenteric artery	α_{1N}		8.1
	mesenteric vein	α_{1N}		8.0
	portal vein	α_{1N}		8.1
	renal artery	α_{1N}		8.1
	femoral artery	α_{1L}		8.3
	femoral vein	α_{1L}		8.3
	prostate	α_{1L}		7.9
human	cerebral artery	α_{1H} + α_{1L}	10.0	8.5
	coronary artery	α_{1H} (α_{1B})	10.0	
	coronary vein	α_{1L}		8.2
	mesenteric artery	α_{1H}	9.5	
	mesenteric vein	α_{1H}	9.7	
	prostate	α_{1L}		8.3

affinity for a new α_1-antagonist, HV723 (Kohno *et al.*, 1994), as compared with those in other regions tested (Table 2). Thus, it seems that α_{1L}-adrenoceptors are also not be homogeneous. At present, we have tentatively subdivided the α_{1L}-adrenoceptors into two subtypes with the use of HV723, α_{1L} (HV723-low) and α_{1N} (HV723-high) (Table 1).

BINDING STUDY

α_1-Adrenoceptors are recognized as specific binding sites of ^{3}H-prazosin or ^{125}I-HEAT. In most binding studies, the concentrations of ^{3}H-prazosin used were up to 1 nM and then a single high affinity constant for prazosin (pK_D close to 10) has been estimated. However, it is interesting to note that affinity constant (K_B) of the α_{1L}-adrenoceptors found in the functional study is more than 1 nM. Thus, it is more probable that prazosin-low affinity sites (α_{1L}-adrenoceptors) were overlooked when low concentrations (< 1nM) of the ^{3}H-ligand were used. In fact, specific ^{3}H-prazosin binding to α_{1H}-adrenoceptors of dog carotid artery enoughly saturated at concentrations less than 1 nM of ^{3}H-prazosin, whereas higher concentrations of ^{3}H-prazosin were necessary to complete the saturation in the rabbit thoracic aorta (Figure 2). Computer-aided analysis of these data clearly resulted in a high pK_D value (10.1) in the dog carotid artery and two distinct affinities (9.9, 8.6) in the rabbit thoracic aorta (Oshita *et al.*, 1993). Displacement study reveals that the binding of low concentrations (~ 200 pM) of ^{3}H-prazosin to α_1-adrenoceptors of various tissues was inhibited by prazosin with a high affinity (pK_I: approximately 10) but by other antagonists (such as WB4101 and 5-methylurapidil) with different affinities, suggesting heterogeneity of prazosin-high affinity sites (α_{1H}) among tissues. The binding results also suggest that the pharmacological features of α_1-adrenoceptors are resemble to those of cloned α_{1B} subtype in

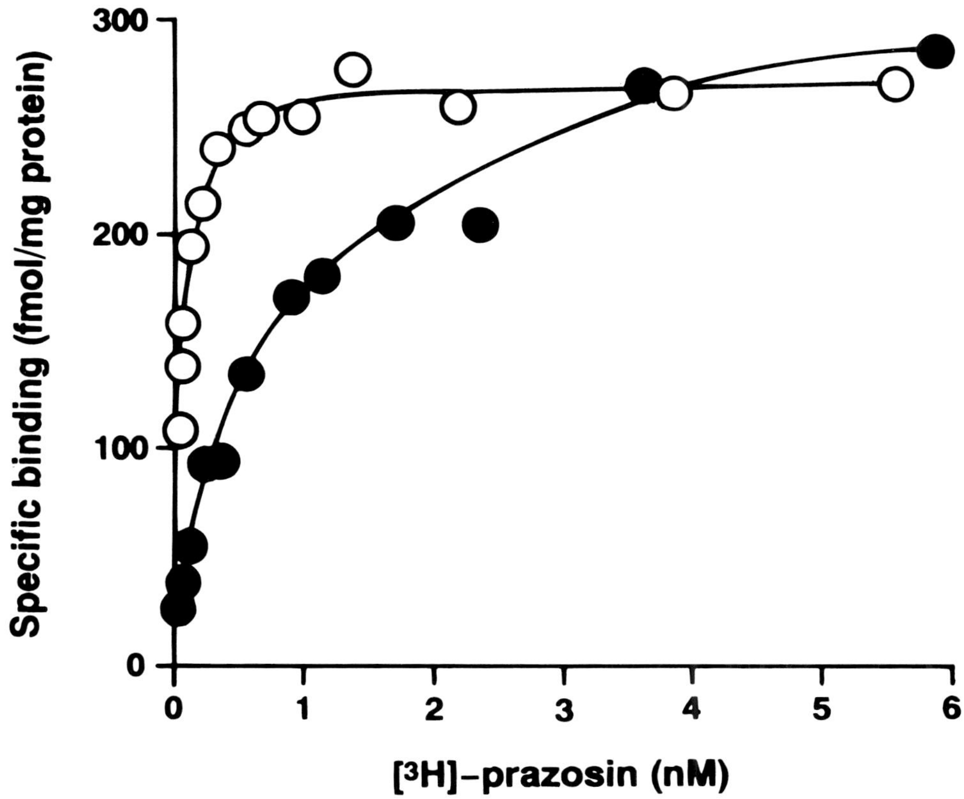

FIGURE 2 Comparison between the saturation curves for ^{3}H-prazosin binding to rabbit thoracic aorta and dog carotid artery membranes.: ●: rabbit thoracic aorta membranes, ○: dog carotid artery membranes

the dog carotid artery and to those of cloned α_{1A} or α_{1D} subtype in the rat thoracic aorta, respectively (Oshita and Muramatsu, unpublished observations).

In contrast to functional study, more subtypes may be detected in the binding study. For example, in the human prostate only single α_1-adrenoceptor subtype (presumably α_{1L}) with low affinity for prazosin was identified from the functional study, whereas two distinct affinity sites for prazosin (presumably cloned α_{1C} and α_{1L}) were detected from the binding study (Muramatsu *et al.*, 1994). These results indicate that, even though more than single α_1-subtype are expressed in a given tissue, all the subtypes are not necessarily involved in the contractile response to noradrenaline.

CONCLUSION

Prazosin-high and -low affinity sites (α_{1H} and α_{1L}) are detected in the functional and binding studies. The pharmacological features of each site are not homogeneous among the tissues, suggesting the presence of subtypes in each α_{1H} or α_{1L}-adrenoceptors. Cloned α_{1A},

α_{1B}, α_{1C} and α_{1D} subtypes with a high affinity for prazosin may be categolized into the α_{1H} group. However, there is no molecular biological evidence on the α_{1L} group. Development of each subtype-selective drugs will make tissue-specific control of sympathetic responses to a possibility and will warrant their therapeutic utility.

ACKNOWLEDGMENTS

We thank N. Aoki for secretarial assistance. This work was supported in part by a grant from the Ministry of Education, Science and Culture, Japan.

References

Bylund, D.B., Eikenberg, D.C., Hieble, J.P., Langer, S.Z., Lefkowitz, R.J., Minneman, K.P., Molinoff, P.B., Ruffolo, R.R. Jr. and Trendelenberg, U. (1994) International Union of Pharmacology Nomenclature of Adrenoceptors. *Pharmacological Reviews*, **46,** 121–136.

Drew, G.M. (1985) What do antagonists tell us about α_1-adrenoceptors? *Clinical Science*, **68,** 15s–19s.

Flavahan, N.A. and Vanhoutte, P.M. (1986) α_1-Adrenoceptor subclassification in vascular smooth muscle. Trends in Pharmacological Sciences, **7,** 347–349.

Kohno, Y., Saito, H., Takita, M., Kigoshi, S. and Muramatsu, I. (1994) Heterogeneity of α_1-adrenoceptor subtypes involved in adrenergic contractions of dog blood vessels. *British Journal of Pharmacology*, in press.

Lomasney, J.W., Cotecchia, S., Lefkowitz, R.J. and Caron, M.G. (1991) Molecular biology of α-adrenergic receptors: implications for receptor classification and for structure-function relationships. *Biochimica et Biophysica Acta*, **1095,** 127–139.

Minneman, K.P. (1988) α_1-Adrenergic receptor subtypes, inositol-phosphate, and sources of cell Ca^{++}. *Pharmacological Review*, **40,** 87–119.

Muramatsu, I. (1991) Relation between adrenergic neurogenic contraction and α_1-adrenoceptor subtypes in dog mesenteric and carotid arteries and rabbit carotid arteries. *British Journal of Pharmacology*, **102,** 210–214.

Muramatsu, I., Kigoshi, S. and Ohmura, T. (1991) Subtypes of α_1-adrenoceptors involved in noradrenaline-induced contractions of rat thoracic aorta and dog carotid artery. *Japanese Journal of Pharmacology*, **57,** 535–544.

Muramatsu, I., Ohmura, T., Kigoshi, S., Hashimoto, S. and Oshita, M. (1990) Pharmacological subclassification of α_1-adrenoceptors in vascular smooth muscle. *British Journal of Pharmacology*, **99,** 197–201.

Muramatsu, I., Oshita, M., Ohmura, T., Kigoshi, S., Akino, H., Gobara, M. and Okada, K. (1994) Pharmacological characterization of α_1-adrenoceptor subtypes in the human prostate: functional and binding studies. *British Journal of Urology*, **74,** 572–578.

Oshita, M., Kigoshi, S. and Muramatsu, I. (1993) Pharmacological characterization of two distinct α_1-adrenoceptor subtypes in rabbit thoracic aorta. *British Journal of Pharmacology*, **108,** 1071–1076.

THE ATYPICAL α_1-ADRENOCEPTOR

PETER W. ABEL, WANYUN ZENG, JAMES E. PORTER, MARGARET A. SCOFIELD, FENGMING LIU, ISABEL GONZALEZ-CABRERA, FRANK J. DOWD and WILLIAM B. JEFFRIES

Creighton University School of Medicine, Department of Pharmacology, 2500 California Plaza, Omaha, NE 68178

It is now known that the α_1-adrenoceptors comprise a heterogenous population of receptor subtypes. Three of the α_1-adrenoceptors have now been cloned including the $\alpha_{1a/d}$-adrenoceptor, the α_{1b}-adrenoceptor and the α_{1c}-adrenoceptor. Radioligand binding assays and functional studies in native tissues support the existence of α_1-adrenoceptor subtypes; however, complete agreement between the pharmacological properties of the cloned receptors and receptors from native tissues has not yet been achieved. We have reported that the rat parotid gland contains the α_{1A}-adrenoceptor and an α_1-adrenoceptor that has not been identified previously. Using radioligand binding we will describe the pharmacological characteristics of this atypical α_1-adrenoceptor subtype. We will also provide evidence using functional measurements of smooth muscle contraction that this receptor subtype is also present in other tissues from other species. Finally, the similarity between the atypical α_1-adrenoceptor and α_1-adrenoceptors with low affinity for prazosin will be discussed.

KEY WORDS: α_1-adrenoceptors, rat parotid gland, guinea-pig ileum, guinea-pig nasal mucosa, polymerase chain reaction

INTRODUCTION

It is now clear that the α_1-adrenoceptors do not represent a homogenous population of receptors. Radioligand binding measurements in native tissues have identified two distinct α_1-adrenoceptor subtypes called α_{1A}- and α_{1B}-adrenoceptors (Morrow and Creese, 1986; Han *et al.*, 1987). The α_{1A}-adrenoceptor has a high affinity for the competitive antagonists phentolamine, WB-4101, 5-methyl-urapidil and (+)-niguldipine while the α_{1B}-adrenoceptor has a lower affinity for these antagonists. These receptor subtypes can also be discriminated by the alkylating drug chloroethylclonidine (CEC). The α_{1A}-adrenoceptor is resistant to inactivation by CEC while the α_{1B}-adrenoceptor is highly sensitive to the alkylating effects of this drug (Han *et al.*, 1987b).

Molecular cloning has also identified subtypes of the α_1-adrenoceptor. The $\alpha_{1a/d}$-adrenoceptor was cloned independently by Lomasney *et al.* (1991) who called this receptor the α_{1a}-adrenoceptor and by Perez *et al.* (1991) who called it the α_{1d}-adrenoceptor. This receptor subtype does not appear to represent either the α_{1A}- or α_{1B}-adrenoceptor identified in native tissues. The α_{1b}-adrenoceptor was first cloned from hamster (Cotecchia *et al.*,

Correspondence to: Dr. Peter W. Abel, Creighton University School of Medicine, Department of Pharmacology, 2500 California Plaza, Omaha, NE 68178.

1988) and has the same pharmacological characteristics as the α_{1B}-adrenoceptor identified in native tissues. The α_{1c}-adrenoceptor was originally cloned from bovine brain (Schwinn *et al.*, 1990). Although controversial, it now appears that the cloned α_{1c}-adrenoceptor may represent the α_{1A}-adrenoceptor that has been identified previously in native tissues (Forray *et al.*, 1994; Ford *et al.*, 1994).

While the classification scheme for α_1-adrenoceptors described above is generally accepted by most investigators, other classification schemes have also been proposed. Based on isolated tissue studies in various blood vessels, Flavahan and Vanhoutte (1986) suggested that α_1-adrenoceptors can be subdivided into two subtypes based on their affinity for the competitive antagonist prazosin. The α_{1H}-adrenoceptor has a high affinity for prazosin (< 1 nM) while the α_{1L}-adrenoceptor has a low affinity (> 1 nM) for prazosin. Using the competitive antagonist HV723 and other drugs, Muramatsu *et al.* (1990) have extended this classification scheme adding an additional low affinity prazosin receptor that has a high affinity (< 1 nM) for HV723 which they call the α_{1N}-adrenoceptor. While some similarities exist between this classification scheme and the generally accepted α_1-adrenoceptor classification described above, the relationship between these subclassifications is not completely clear.

We have recently identified a novel α_1-adrenoceptor in the rat parotid gland which we have called the atypical α_1-adrenoceptor (Porter *et al.*, 1992). We will describe the pharmacological characteristics of this α_1-adrenoceptor and discuss the similarity between the atypical α_1-adrenoceptor and α_1-adrenoceptors with low affinity for prazosin.

METHODS

For radioligand binding, rat parotid glands were removed from male Sprague Dawley rats and acinar clumps were isolated from visible fat, connective tissue, blood vessels and parotid ducts. A crude membrane fraction was prepared and α_1-adrenoceptors were labeled with ^{125}IBE2254 (^{125}IBE). Competition binding curves for α_1-adrenoceptor antagonists were generated and fit to a one or two site binding model using nonlinear regression analysis. K_i values were calculated and the fits were compared using an F test. For a detailed description of the procedures used, see Porter *et al.*, 1992.

For measurement of α_1-adrenoceptor mediated contraction of guinea-pig ileum or guinea-pig nasal mucosa, the tissues were removed from Hartley guinea pigs. Using standard *in vitro* techniques, segments of guinea-pig ileum or strips of nasal mucosa were placed in glass muscle chambers maintained at 37°C containing Krebs' solution. Concentration-response curves for norepinephrine or phenylephrine were generated and pA_2 or K_B values for competitive antagonists were obtained using standard techniques. The procedures used were a modification of those described in Minneman *et al.*, 1982.

To identify and quantify mRNA for the $\alpha_{1a/d}$- α_{1b}- and α_{1c}-adrenoceptors, a quantitative reverse transcriptase (RT) polymerase chain reaction (PCR) procedure similar to that of Pitak *et al.* (1993) was developed. Sense and antisense oligonucleotide primers were designed that specifically amplified the third intracellular loop for either $\alpha_{1a/d}$- α_{1b}- or α_{1c}-adrenoceptor genes. Competing RNA templates that shared identical primer binding sequences with each of the respective adrenoceptor gene transcripts were synthesized by *in vitro* transcription. These templates were used as internal standards for the RT-PCR reactions and generated PCR products either slightly larger or smaller than the products from the native sequences. Cellular RNA extracted from intact rat parotid glands and known amounts of competing RNA templates were cotranscribed and coamplified in the presence of α^{32}P-dCTP. The PCR products were separated on a 4% agarose/ethidium bromide gel and the appropriately sized bands derived from cellular RNA and competing RNA templates were excised and quantified by scintillation spectrometry.

RESULTS

α_1-adrenoceptor subtype mRNA in rat parotid gland

To quantify the steady state quantity of $\alpha_{1a/d}$- α_{1b}- and α_{1c}-adrenoceptor mRNA in rat parotid gland a competitive RT-PCR method was used. In preliminary studies we found that the competing RNA templates and the corresponding cellular RNA are reverse transcribed and amplified with similar efficiency. These results suggest that our RT-PCR procedure can accurately measure the amount of α_1-adrenoceptor mRNA in rat parotid gland. Figure 1 shows the molecules of $\alpha_{1a/d}$- α_{1b}- and α_{1c}-adrenoceptor mRNA identified in rat parotid gland. The α_{1c}-adrenoceptor mRNA was present in greatest abundance (328,000 ± 25,000 molecules/μg total RNA) representing 80% of the total α_1-adrenoceptor message. $\alpha_{1a/d}$-Adrenoceptor message was much lower (81,000 ± 4000 molecules/μg) while α_{1b}-adrenoceptor mRNA (3000 ± 1000 molecules/μg) was approximately 1% of total α_1-adrenoceptor message.

Characteristics of atypical α_1-adrenoceptors in rat parotid gland

The α_1-adrenoceptors in membranes of rat parotid gland were labeled with ^{125}IBE. ^{125}IBE labeled a homogenous population of high affinity binding sites with a K_D of 0.24 ± 0.03 nM and a B_{max} of 41 ± 4 fmol of ^{125}IBE bound per mg of membrane protein. Competition curves for prazosin inhibition of ^{125}IBE binding fit best to a single binding site with a K_i of 1.0 ± 0.4 nM (Table 1). The α_{1A} subtype selective antagonists WB-4101 and 5-methylurapidil also inhibited ^{125}IBE binding. Competition curves for both drugs were best fit to a two-site binding model with approximately 40% high affinity and 60% low affinity binding sites (Table 1). Treatment of rat parotid membranes with the α_{1B}-selective alkylating drug CEC (10 μM for 10 minutes or 30 μM for 20 minutes) had no significant effect on the K_D or B_{max} for ^{125}IBE binding. K_i values for WB-4101 and 5-methylurapidil were not affected by pretreatment of parotid gland membranes with CEC (Table 1).

The atypical α_1-adrenoceptor in the guinea-pig

The α_1-adrenoceptor mediated contractile response of the isolated guinea-pig ileum was used to examine a functional response caused by stimulation of the atypical α_1-adrenoceptor. In Figure 2, control concentration-response curves for phenylephrine were obtained in guinea-pig ileum before and after treatment with 100 μM of the α_1- adrenoceptor alkylating drug CEC for 30 minutes. CEC had no significant effect on the concentration-response curve for phenylephrine. We also obtained K_B values for the α_1-adrenoceptor subtype selective antagonists WB-4101 and 5-methylurapidil to inhibit phenylephrine-induced con-

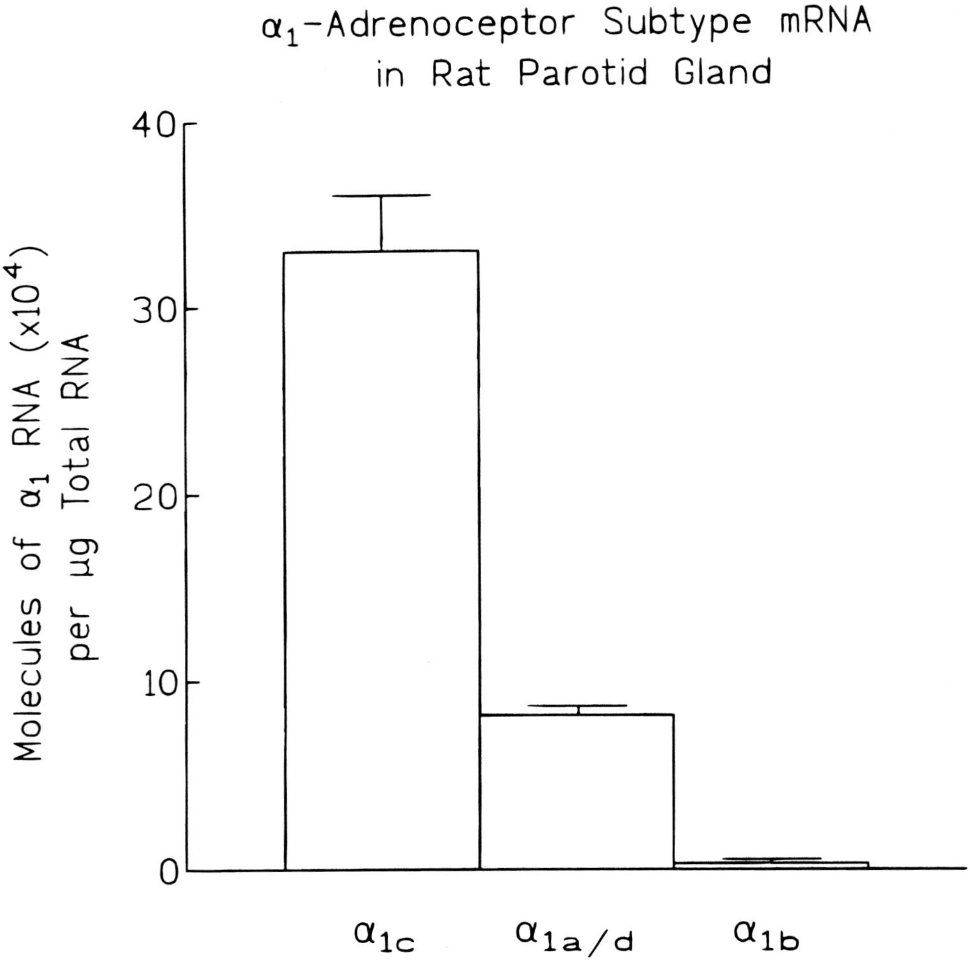

FIGURE 1 Quantitative competitive RT-PCR was used to quantify mRNA for $\alpha_{1a/d}$- α_{1b}- and α_{1c}-adrenoceptors in the rat parotid gland. α_{1c}-adrenoceptor mRNA was present in the highest percentage (80%) followed by $\alpha_{1a/d}$ (19%) with little (< 1%) α_{1b}-adrenoceptor mRNA. Each bar represents the mean ± S.E.M. of 3 experiments, using parotid glands from different animals.

TABLE 1
K_i values (nM) and percent binding sites (%) for antagonists at α_1-adrenoceptors in rat parotid gland.

	K_{IH}	K_{IL}	*%High*	*%Low*
Prazosin	1.0 ± 0.4	—	—	—
WB-4101	0.45 ± 0.2	18 ± 3	37 ± 7	63 ± 7
+ CEC[a]	0.47 ± 0.2	15 ± 4	40 ± 8	60 ± 9
5-methylurapidil	0.29 ± 0.06	75 ± 18	33 ± 8	67 ± 9
+ CEC[a]	0.41 ± 0.2	79 ± 22	50 ± 8	50 ± 8

[a]10 µM CEC for 10 minutes.
Each value represents the mean ± S.E.M. from 4–7 separate radioligand binding experiments.

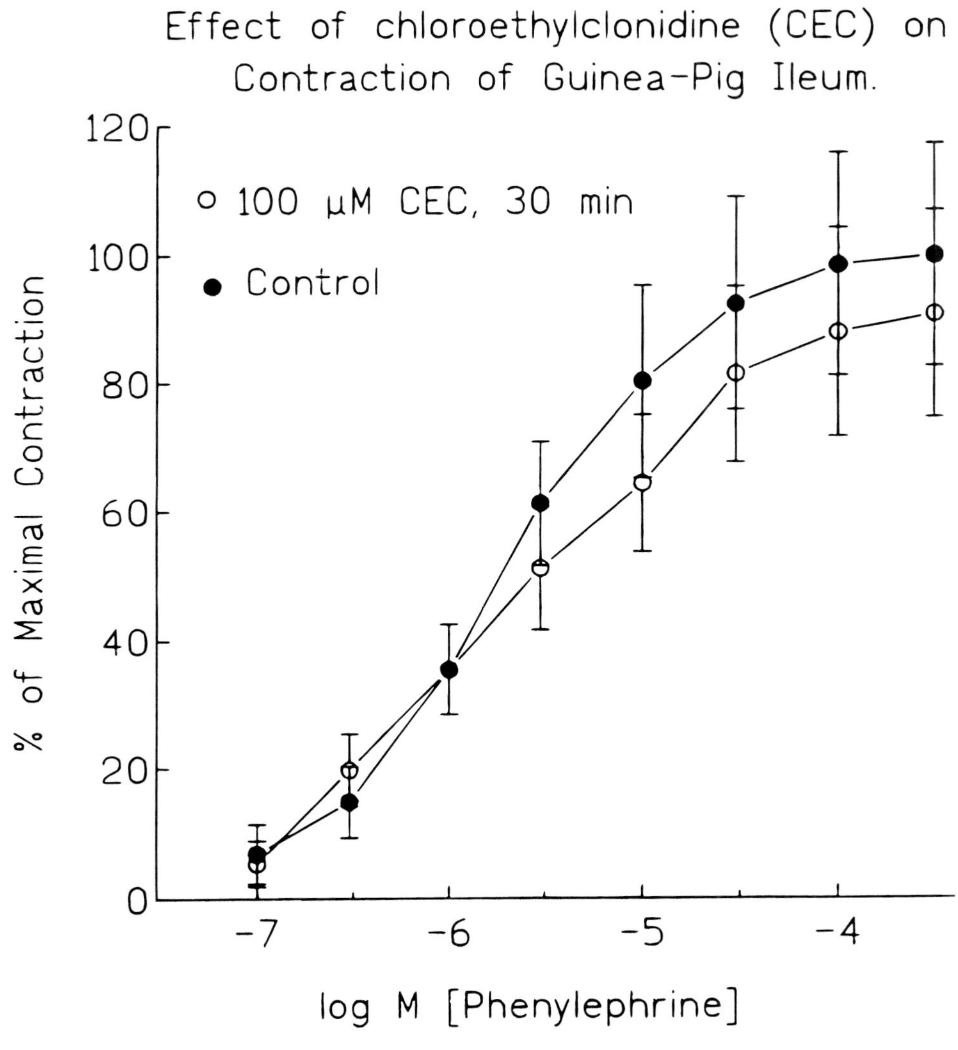

FIGURE 2 Mean concentration-response curves for phenylephrine before (Control) and after treatment with the alkylating drug chloroethylclonidine (CEC). Tissues were treated with 100 μM CEC for 30 minutes followed by washout of the drug for an additional 30–45 minutes. Treatment with CEC had no significant effect on the concentration-response curve for phenylephrine. Each point is the mean ± S.E.M. of 5–7 experiments, using tissue from different animals.

traction of guinea-pig ileum. The K_B values for these drugs listed in Table 2, show that both WB-4101 and 5-methylurapidil had a low affinity to inhibit phenylephrine-induced contraction. We also obtained K_B values for the non-subtype selective α_1-adrenoceptor antagonist prazosin, which also had a low affinity (>1 nM) for the α_1-adrenoceptors causing contraction of this tissue (Table 2).

TABLE 2

Affinities (nM) of α_1-adrenoceptor antagonists in rat parotid gland, guinea-pig ileum and guinea-pig nasal mucosa.

Tissue	*WB-4101*	*5-Methylurapidil*	*Prazosin*
[a]Rat Parotid Gland (Atypical α_1)	18 ± 3	75 ± 18	1.0 ± .4
[b]Guinea-Pig Ileum	13 ± 3	72 ± 12	4.1 ± .9
[b]Guinea-Pig Nasal Mucosa	5 ± .9	97 ± 11	2.3 ± .1

[a]Mean K_i values ± S.E.M. from radioligand binding experiments (n = 4–7).
[b]Mean K_B or pA_2 values ± S.E.M. calculated from functional experiments for antagonist inhibition of contraction (n = 4–6).

The α_1-adrenoceptor mediated contractile response of the isolated guinea-pig nasal mucosa was studied to determine whether or not activation of the atypical α_1-adrenoceptor caused contraction of this tissue. Phenylephrine-induced contraction was unaffected by treatment of nasal mucosa with 10 μM CEC for 30 minutes. K_B values for WB-4101 and 5-methylurapidil inhibition of phenylephrine-induced contraction (Table 2) were very similar to the low affinity K_B values for these drugs obtained in the guinea-pig ileum. Like the guinea-pig ileum, prazosin also had a low affinity to inhibit phenylephrine-induced contraction of this tissue.

DISCUSSION

With radioligand binding assays using ^{125}IBE, we have previously identified two α_1-adrenoceptor subtypes in the rat parotid gland (Porter *et al.*, 1992). One of these subtypes is insensitive to inactivation by the α_1-adrenoceptor alkylating drug CEC, and has a high affinity for the competitive antagonists WB-4101 and 5-methylurapidil. This pharmacological profile is the same as that reported for the α_{1A}-adrenoceptor that has been identified in native tissue from the rat and other species. Although this receptor has been suggested to represent an α_1-adrenoceptor whose gene has not been identified (Schwinn and Lomasney, 1992), recent evidence suggests that the gene encoding the pharmacologically defined α_{1A}-adrenoceptor has, in fact, been cloned.

Forray *et al.* (1994) have expressed the three cloned human α_1-adrenoceptors (α_{1a}, α_{1b}, α_{1c}) in LM(tk$^-$) or CHO cells and examined the pharmacological characteristics of those receptors. All three receptor subtypes had a high affinity (< 1 nM) for the non-subtype selective antagonist prazosin. Compared to the $\alpha_{1a/d}$- and α_{1b}-adrenoceptors, the α_{1c}-adrenoceptor was resistant to inactivation by the alkylating drug CEC. In competition binding experiments using ^{3}[H]prazosin, the α_{1c}-adrenoceptor had the highest affinity for the subtype selective drugs WB-4101 (0.4 nM) and 5-methylurapidil (2 nM). K_i values for these drugs, as well as others, were highly similar to their K_i values at α_{1A}-adrenoceptors in native tissues (Han *et al.*, 1987a). These results led Forray *et al.*, (1994) to suggest that the α_{1c}-adrenoceptor gene encodes the pharmacologically defined α_{1A}-adrenoceptor.

Schwinn *et al.*, (1990) using northern blot analysis were unable to identify mRNA for the α_{1c}-adrenoceptor in rat tissue using probes based on the bovine α_{1c}-adrenoceptor sequence. Using different hybridization conditions, Faure *et al.*, (1994) have reported the hybridization of both bovine and rat α_{1c}-adrenoceptor probes to a 3.7 kb mRNA species in a variety of tissues known to contain the α_{1A}-adrenoceptor. In a recent study a partial α_{1c}-adrenoceptor cDNA was obtained from rat cardiac muscle cells and using the more sensitive method of RNase protection, the rat α_{1c}-adrenoceptor transcript was localized in rat tissues (Rokosh *et al.*, 1994). Thus α_{1c}-adrenoceptor mRNA was found in the rat parotid gland and in other rat tissues that had previously been reported to contain the pharmacologically defined α_{1A}-adrenoceptor.

In our studies we developed oligonucleotide primer pairs based on the partial cDNA sequence for the rat α_{1c}-adrenoceptor reported by Alonso-Llamazares *et al.*, (1993). Using these α_{1c}-adrenoceptor primers and primer pairs that we have previously shown to specifically anneal to and amplify $\alpha_{1a/d}$- and α_{1b}-adrenoceptor mRNA (Feng *et al.*, 1993), we used competitive RT-PCR to quantify mRNA for the $\alpha_{1a/d}$- α_{1b}- and α_{1c}-adrenoceptors in rat parotid gland. We found that 80% of the total α_1-adrenoceptor mRNA was message for the α_{1c}-adrenoceptor subtype. Taken together, the results of Forray *et al.*, (1994), Rokosh *et al.*, (1994) and our RT-PCR and radioligand binding studies suggest that in the rat parotid gland an α_{1c}-adrenoceptor gene encodes a receptor with the pharmacological characteristics of the α_{1A}-adrenoceptor identified in native tissue.

In the rat parotid gland we also identified a second α_1-adrenoceptor that has pharmacological characteristics which differ from those of the cloned ($\alpha_{1a/d}$, α_{1b}, α_{1c}) or pharmacologically defined (α_{1A}, α_{1B}) α_1-adrenoceptor subtypes. We have previously referred to this receptor as the atypical α_1-adrenoceptor subtype. This receptor subtype was resistant to inactivation by CEC and had a relatively low affinity for the α_1-adrenoceptor subtype selective antagonists WB-4101 and 5-methylurapidil. K_i values for these antagonists in rat parotid gland are very similar to those reported for the cloned α_{1b}- and native tissue α_{1B}-adrenoceptor. However, both $\alpha_{1B,b}$-adrenoceptors are highly sensitive to inactivation by the alkylating drug CEC, while the α_1-adrenoceptors in the rat parotid gland were resistant to inactivation by this drug. In addition, our quantitative RT-PCR studies found that < 1% of the total α_1-adrenoceptor mRNA was for the α_{1b}-adrenoceptor subtype. While the atypical α_1-adrenoceptor has some features in common with α_{1B}-adrenoceptors, the insensitivity to CEC and the lack of a substantial amount of α_{1b}-adrenoceptor message, suggests that the atypical α_1-adrenoceptor is not the $\alpha_{1B,b}$-adrenoceptor subtype.

Using quantitative RT-PCR we found that 19% of the total α_1-adrenoceptor mRNA was message for the $\alpha_{1a/d}$-adrenoceptor. However, the pharmacological characteristics of the atypical α_1-adrenoceptor are not entirely consistent with the $\alpha_{1a/d}$-adrenoceptor subtype. Perez *et al.*, (1991) have reported that treatment of COS cell membranes transfected with rat $\alpha_{1a/d}$-adrenoceptor cDNA, with 100 μM CEC for 10 minutes at 37°C, inactivated 72% of the $\alpha_{1a/d}$-adrenoceptors present. A similar degree of $\alpha_{1a/d}$-adrenoceptor inactivation (74%) was reported by Forray *et al.*, (1994) using a similar CEC treatment protocol of LM(tk$^-$) cells transfected with the human $\alpha_{1a/d}$-adrenoceptor. In a previous study (Feng *et al.*, 1993) we found that rat renal medullary thick ascending limb cells, which express α_{1d}-adrenoceptors, were also extensively inactivated (79%) by CEC. In the rat parotid gland we used a similar CEC treatment and found that the atypical α_1-adrenoceptor was completely insensitive to inactivation by CEC.

In addition to our finding with CEC, the affinities of the subtype selective antagonists WB-4101, 5-methylurapidil and (+) niguldipine do not support the idea that the atypical

α_1-adrenoceptor is the $\alpha_{1a/d}$-adrenoceptor subtype. K_i values for WB-4101 at the $\alpha_{1a/d}$-adrenoceptor are approximately 1 nM while the atypical α_1-adrenoceptor in the rat parotid gland gave a mean K_i value of 18 nM. A similar pattern is found for 5-methylurapidil with average K_i values of approximately 15 nM at the $\alpha_{1a/d}$-adrenoceptor while we found a mean K_i value of 75 nM for this drug at the atypical α_1-adrenoceptor. In preliminary studies we have also found that the α_1-adrenoceptor selective antagonist (+) niguldipine has a high affinity (approximately 1 nM) for the atypical α_1-adrenoceptor. In contrast to our findings with this drug in the rat parotid gland, (+) niguldipine has a low affinity (50–200 nM) for the cloned $\alpha_{1a/d}$-adrenoceptor (Perez *et al.*, 1991; Forray *et al.*, 1994). Taken together, these results suggest that the atypical α_1-adrenoceptor is not the $\alpha_{1a/d}$-adrenoceptor subtype.

Although RT-PCR studies have identified message for the $\alpha_{1a/d}$-adrenoceptor in the intact rat parotid gland, radioligand binding in membranes from acinar clumps has not identified the $\alpha_{1a/d}$-adrenoceptor subtype. The reason for this discrepancy is unknown. It may be that although there is message for $\alpha_{1a/d}$-adrenoceptors in the parotid gland that this message is not translated into receptor protein. A second explanation may be related to methodological differences between the way in which tissues were prepared for use in the two studies. For radioligand binding assays fat, connective tissue, blood vessels and parotid ducts were carefully dissected from the parotid gland leaving primarily acinar clumps to be made into membranes for radioligand binding assays. In contrast, for RT-PCR studies, the intact gland was rapidly removed and frozen for subsequent isolation of cellular RNA. The $\alpha_{1a/d}$-adrenoceptors could be present on fat cells, connective tissue, blood vessels or parotid duct cells which were present for RNA isolation but were less concentrated in membranes prepared for radioligand binding. A definitive resolution of this discrepancy will require additional experiments.

The pharmacological characteristics of the atypical α_1-adrenoceptor have been previously defined by us using radioligand binding assays in membranes of rat parotid gland. In this report we have also identified a functional response that is activated by the atypical α_1-adrenoceptor. Although α_1-adrenoceptor stimulation can cause contraction of the guinea-pig ileum, the α_1-adrenoceptor subtype contracting this tissue is not known. In our studies α_1-adrenoceptor mediated contraction caused by phenylephrine was not modified by treatment of this tissue with the α_1-adrenoceptor alkylating drug CEC. In addition, the α_1-adrenoceptor subtype selective drugs WB-4101 and 5-methylurapidil had a low affinity to antagonize contraction of this tissue. Comparison of K_i values for these drugs in the rat parotid gland with K_B values for inhibition of contraction of guinea-pig ileum show that the receptors in these two tissues have the same pharmacological characteristics.

The same pattern was seen when α_1-adrenoceptor mediated contraction of guinea-pig nasal mucosa by phenylephrine was examined. Phenylephrine-induced contraction of nasal mucosa is thought to be caused by activation of α_1-adrenoceptors that are present on the dense network of blood vessels located in this tissue. α_1-Adrenoceptor mediated contraction was not affected by treatment with CEC. In addition both WB-4101 and 5-methylurapidil antagonized the α_1-adrenoceptor mediated contraction with low affinity. K_B values for these drugs for inhibition of α_1-adrenoceptor mediated contraction of guinea-pig nasal mucosa were very similar to both K_B values for inhibition of contraction in the guinea pig ileum and to K_i values from radioligand binding studies in the rat parotid gland. Therefore, three tissues, the rat parotid gland, the guinea-pig ileum and the guinea-pig nasal mucosa all appear to contain the atypical α_1-adrenoceptor. This receptor type can be identified by both radioligand binding (rat parotid gland) and by functional measurements of smooth muscle contraction (guinea-pig ileum and nasal mucosa). It is found in at least two different species, the rat and the guinea-pig and it is present in both smooth muscle (guinea-pig ileum and nasal mucosa) and in exocrine glands (rat parotid gland).

It is interesting to note that the cloned ($\alpha_{1a/d}$, α_{1b}, α_{1c}) or pharmacologically defined (α_{1A}, α_{1B}) α_1-adrenoceptor subtypes all have a relatively high affinity (< 1 nM) for the non-subtype selective antagonist prazosin. However, the atypical α_1-adrenoceptor has a somewhat lower affinity for prazosin (>1 nM) compared to the previously identified α_1-adrenoceptor subtypes. α_1-Adrenoceptors with a relatively low affinity for prazosin have been previously identified by measurement of the affinity of prazosin to antagonize α_1-adrenoceptor mediated contraction of different blood vessels (Flavahan and Vanhoutte, 1986). These authors suggested that prazosin affinity can be used to subdivide α_1-adrenoceptors on vascular smooth muscle into receptor subtypes with high (α_{1H}; < 1 nM) and low (α_{1L}; > 1 nM) affinity for prazosin. This classification scheme for vascular smooth muscle has been extended to include other drugs and an additional low affinity prazosin receptor with a high affinity (< 1 nM) for the competitive antagonist HV 723 (Muramatsu *et al.*, 1990).

The atypical α_1-adrenoceptor described by us has a number of similarities to the α_{1L}-adrenoceptor described by other investigators. The α_{1L}-adrenoceptor is found in vascular smooth muscle (the thoracic aorta) of the guinea-pig and is insensitive to inactivation by CEC. In addition to having a low affinity for prazosin (> 1 nM) the α_{1L}-adrenoceptor also has a relatively low affinity (1–10 nM) for WB-4101. The atypical α_1-adrenoceptor in the vascular smooth muscle of the guinea-pig nasal mucosa, in guinea-pig ileum and in the rat parotid gland is also insensitive to inactivation by CEC. In all three tissues, the atypical receptor has a low affinity (1–4 nM) for prazosin and low affinity (5–18 nM) for WB-4101. These similarities in pharmacological characteristics and the identification of both the α_{1L}- and the atypical α_1-adrenoceptor in vascular tissue from the guinea-pig suggest that these two receptor types may actually represent the same α_1-adrenoceptor subtype.

ACKNOWLEDGMENT

Supported by N.I.H. grant DE 06115 and the American Heart Association, Nebraska Affiliate.

References

Alonso-Llamazares, A., Zamanillo, D., Fernandez, A., Chinchetru, M.A. and Calvo, P. (1993) Differential expression of the α_{1c} adrenergic receptor subtype in rat tissues. *NeuroReport*, **4,** 1266–1268.

Cottecchia, S., Schwinn, D.A., Randall, R.R., Lefkowitz, R.J., Caron, M.G. and Kobilka, B.K. (1988) Molecular cloning and expression of the cDNA from the hamster alpha$_1$-adrenergic receptor. *Proc. Natl. Acad. Sci.*, **85,** 7159–7163.

Faure, C., Pimoule, C., Arbilla, S., Langer, S.Z. and Graham, D. (1994) Expression of α_1-adrenoceptor subtypes in rat tissues:implications for α_1-adrenoceptor classification. *Europ. J. Pharmacol.*, **268,** 141–149.

Feng, F., Abel, P.W., Scofield, M., Liu, F., Wolff, D.W. and Jeffries, W.B. (1993) Heterogeneous expression of α_1-adrenoceptor subtypes among rat nephron segments. *J. Pharmacol. Exptl. Ther.*, **44,** 926–933.

Flavahan, N.A. and Vanhoutte, P.M. (1986) α-Adrenoceptor subclassification in vascular smooth muscle. *Trends Pharmacol. Sci.*, **7,** 347–349.

Ford, A.P.D.W., Williams, T.J., Blue, D.R. and Clarke, D.E. (1994) α_1-Adrenoceptor classification: sharpening Occam's razor. *Trends Pharmacol. Sci.*, **15,** 167–170.

Forray, C., Bard, J.A., Wetzel, J.M., Chiu, G., Shapiro, E., Tang, R., Lepor, H., Hartig, P.R., Weinshank, R.L., Branchek, T.A. and Gluchowski, C. (1994) The α_1-adrenergic receptor that mediates smooth muscle contraction in human prostate has the pharmacological properties of the cloned human α_{1c} subtype. *Mol. Pharmacol.*, **45,** 703–708.

Han, C., Abel, P.W. and Minneman, K.P. (1987a) α_1-Adrenoceptor subtypes linked to different mechanisms for increasing intracellular Ca^{2+} in smooth muscle. *Nature*, **329,** 333–335.

Han, C., Abel, P.W. and Minneman, K.P. (1987b) Heterogeneity of α_1-adrenergic receptors revealed by chloroethylclonidine. *Mol. Pharmacol.*, **32,** 505–510.

Lomasney, J.W., Cotecchia, S., Lorenz, W., Leung, W.-Y., Schwinn, D.A., Yang-Feng, T.L., Brownstein, M., Lefkowitz, R.J. and Caron, M. (1991) Molecular cloning and expression of the cDNA for the α_{1a}-adrenergic receptor. *J. Biol. Chem.*, **266,** 6365–6369.

Minneman, K.P., Fox, A.W. and Abel, P.W. (1982) Occupancy of alpha$_1$-adrenergic receptors and contraction of rat vas deferens *Mol. Pharmacol.*, **23,** 359–368.

Morrow, A.L. and Creese, I. (1986) Characterization of alpha$_1$-adrenergic receptor subtypes in rat brain: a reevaluation of ^{3}H-WB-4101 and ^{3}H-prazosin binding. *Mol. Pharmacol.*, **29,** 321–330.

Muramatsu, I., Ohmura, T., Kigoshi, S., Hashimoto, S. and Oshita, M. (1990) Pharmacological subclassification of α_1-adrenoceptors in vascular smooth muscle. *Br. J. Pharmacol.*, **99,** 197–201.

Perez, D.M., Piascik, M.T. and Graham, R.M. (1991) Solution-phase library screening for the identification of rare clones: Isolation of an α_{1d}-adrenergic receptor cDNA. *Mol. Pharmacol.*, **40,** 876–883.

Piatak, M. Jr., Ka-Cheung, L., Williams, B. and Lifson, J.D. (1993) Quantitative competitive polymerase chain reaction for accurate quantitation of HIV DNA and RNA species. *BioTechniques*, **14,** 70–80.

Porter, J.E., Dowd, F.J. and Abel, P.W. (1992) Atypical alpha-1 adrenergic receptors on the rat parotid gland acinar cell. *J. Pharmacol. Exptl. Ther.*, **263,** 1062–1067.

Rokosh, D.G., Bailey, B.A., Stewart, A.F.R., Karns, L.R., Long, C.S. and Simpson, P.C. (1994) Distribution of α1C-adrenergic receptor mRNA in adult rat tissues by RNase assay and comparison with α1B and α1D. *Biochem. Biophys. Res. Comm.*, **200,** 1177–1184.

Schwinn, D.A., Lomasney, J.W., Lorenz, W., Szklut, P.J., Fremeau, R.T. Jr., Yang-Feng, T.L., Caron, M.G., Lefkowitz, R.J. and Cotecchia, S. (1990) Molecular cloning and expression of the cDNA for a novel alpha$_1$-adrenergic receptor subtype. *J. Biol. Chem.*, **265,** 8183–8189.

Schwinn, D.A. and Lomasney, J.W. (1992) Pharmacologic characterization of cloned α_1-adrenoceptor subtypes:selective antagonists suggest the existence of a fourth subtype. *Europ. J. Pharmacol.*, **227,** 433–436.

SIGNAL TRANSDUCTION BY α_1-ADRENERGIC RECEPTOR SUBTYPES IN CONTINUOUS CELL LINES

TIMOTHY A. ESBENSHADE and KENNETH P. MINNEMAN

Department of Pharmacology, Emory University School of Medicine, Atlanta, GA 30322

α_1-Adrenergic receptors (ARs) are a heterogeneous family of receptors, comprised of at least two natively expressed subtypes (α_{1A} and α_{1B}) and three cloned subtypes (α_{1B}, α_{1C}, $\alpha_{1A/D}$). α_1-ARs evoke cellular responses by increasing intracellular Ca^{2+} ($[Ca^{2+}]_i$). Initial studies in isolated tissues suggested that the α_{1A} subtype activated voltage-gated Ca^{2+} influx while the α_{1B} subtype stimulated formation of inositol phosphates and mobilization of stored intracellular Ca^{2+}. Further work has provided exceptions to this generalization, and the mechanisms by which different α_1-AR subtypes initiate signals is still unclear. All three cloned subtypes activate inositol phosphate (InsP) formation and release intracellular Ca^{2+}, although with different coupling efficiencies. We summarize studies on α_1-AR signalling in cell lines where molecular mechanisms are more accessible.

KEY WORDS: Norepinephrine, epinephrine, α_1-adrenergic receptors, inositol phosphates, calcium, diacylglycerol

INTRODUCTION

α_1-ARs regulate blood pressure, cell growth, and other processes. Activation of α_1-ARs increases $[Ca^{2+}]_i$, initiating signalling cascades affecting the metabolic or contractile state of cells. However, the mechanism(s) by which α_1-ARs increase $[Ca^{2+}]_i$ is not completely clear. Responses to α_1-AR activation in different tissues depend on different sources of Ca^{2+} and are associated with changes in several second messenger systems, suggesting that α_1-AR signal transduction may involve a number of different molecular mechanisms. Whether these different mechanisms are associated preferentially with the recently identified α_1-AR subtypes remains to be clarified.

α_1-ADRENERGIC RECEPTOR SUBTYPES

Two natively expressed α_1-AR subtypes have been described. α_{1A}-ARs have a high affinity for certain competitive antagonists and are not inactivated by chloroethylclonidine (CEC), while α_{1B}-ARs have a lower affinity for these antagonists and are inactivated by CEC (Minneman and Esbenshade, 1994). Three α_1-AR subtypes have been cloned from a variety of species, designated α_{1B}, α_{1C}, and $\alpha_{1A/D}$ (Cotecchia *et al.*, 1988; Schwinn *et al.*,

Correspondence to: Timothy A. Esbenshade, Department of Pharmacology, Emory University School of Medicine, Atlanta, GA 30322.

1990; Lomasney *et al.*, 1991; Perez *et al.*, 1991). Although the α_{1B}-AR cDNA clone appears to encode the natively expressed α_{1B} subtype, there is as yet no clear relationship between the α_{1C}- and $\alpha_{1A/D}$- clones and pharmacologically defined subtypes. Neither of these clones encodes a subtype with the exact properties of native α_{1A}-ARs, and it is possible that this subtype has not yet been cloned. However, some reports suggest that the $\alpha_{1A/D}$ (Garcia-Sainz and Romera-Avila, 1993; Kenny *et al.*, 1994) or α_{1C} clone encodes the native α_{1A} subtype (Laz *et al.*, 1993).

SIGNAL TRANSDUCTION MECHANISMS ACTIVATED BY α_1-ARs

Increases in $[Ca^{2+}]_i$ caused by α_1-AR activation are often due to stimulation of phospholipase C, hydrolysis of phosphatidylinositol (4,5) bisphosphate, and production of inositol (1,4,5) trisphosphate [Ins(1,4,5)P_3] and diacylglycerol (Berridge and Irvine, 1989). Ins(1,4,5)P_3 releases stored Ca^{2+} from the endoplasmic reticulum, while diacylglycerol activates protein kinase C. Release of stored Ca^{2+} by Ins(1,4,5)P_3 is often followed by Ca^{2+} influx, probably of a capacitative type, through channels which are not voltage-gated (Putney, 1990). In other tissues, however, α_1-ARs stimulate influx of extracellular Ca^{2+} through voltage-gated channels, which are potently blocked by dihydropyridines (Minneman and Esbenshade, 1994). The mechanism by which receptor activation increases voltage-gated Ca^{2+} influx is not well understood. α_1-ARs also activate other signalling pathways, such as stimulation of phospholipases A_2 and D, and potentiation of adenylate cyclase activity (Minneman and Esbenshade, 1994). The effect of these pathways on $[Ca^{2+}]_i$ is not yet clear.

The G-proteins involved in α_1-AR signalling appear to involve at least two families, (G_q and G_h) which can both activate phospholipase C. All three cloned α_1-AR subtypes (α_{1B},α_{1C},$\alpha_{1A/D}$) activate the β_1 isoform of phospholipase C through α-subunits of the G_q family in recombinant systems (Wu *et al.*, 1992). A recently described novel GTP-binding protein designated G_h possesses tranglutaminase activity but also has been shown to mediate α_{1B}-AR stimulation of phospholipase C (Nakaoka *et al.*, 1994) in recombinant systems. The nature of the G-proteins that mediate α_1-AR activation of voltage-gated Ca^{2+} influx is presently not well understood, but may involve a pertussis toxin sensitive G-protein(s) (Nichols *et al.*, 1989) unlike G_q and G_h.

DIFFERENTIAL SIGNALLING BY DIFFERENT SUBTYPES

The possibility that specific α_1-AR subtypes may be linked to different signalling pathways was suggested early on. Studies in isolated smooth muscles suggested that α_{1B}-ARs mobilize intracellular Ca^{2+} stores (probably through Ins(1,4,5)P_3), while α_{1A}-ARs activate voltage-gated Ca^{2+} influx (Minneman and Esbenshade, 1994). Later studies provided exceptions to this generalization (Schwietert *et al.*, 1992; Lepretre *et al.*, 1994). However, since most of the tissues studied contained mixtures of α_1-AR subtypes, analysis of signal transduction mechanisms was difficult.

All three cloned subtypes increase InsP formation and mobilize intracellular Ca^{2+} (Cotecchia *et al.*, 1988; Schwinn *et al.*, 1991; Perez *et al.*, 1993). However, the relative coupling efficiencies of the cloned subtypes appear to be different, with $\alpha_{1C} > \alpha_{1B} > \alpha_{1A/D}$ (Schwinn *et al.*, 1991; Perez *et al.*, 1993). There is also evidence for activation of arachidonic acid release by cloned subtypes, and some indirect evidence for coupling to dihydropyridine-sensitive channels through a pertussis toxin-sensitive G protein (Perez *et al.*, 1993). However, there is as yet little information available on signal transduction mechanisms activated by cloned α_1-AR subtypes.

α_1-AR SIGNALLING PATHWAYS IN CONTINUOUS CELL LINES

The use of continuous cell lines has greatly facilitated our understanding of the mechanisms by which receptor activation mediates signal transduction. A variety of cell lines have now been identified where α_1-ARs are expressed, although most of these cell lines express only the α_{1B}-subtype (see Table 1).

Cell lines expressing only α_{1B}-ARs

One of the first cell lines shown to express α_1-ARs was the BC_3H_1 cell line, which was isolated from mouse brain, possibly of smooth muscle origin. Activation of α_1-ARs mobilizes intracellular Ca^{2+}, probably through InsP metabolism in these cells (Ambler *et al.*, 1984). Pharmacological studies showed that these receptors are α_{1B} (Han *et al.*, 1992) and only α_{1B} mRNA transcripts are found (Esbenshade *et al.*, 1993).

In the DDT_1 cell line derived from hamster ductus deferens, α_1-AR activation also increases InsP formation (Nelemans *et al.*, 1990) and mobilizes intracellular Ca^{2+} (Reynolds and Dubyak, 1985). Influx of extracellular Ca^{2+} in response to α_1-AR activation was also observed, through channels which were not sensitive to dihydropyridines (Reynolds and Dubyak, 1985). This cell line served as the source for sequencing and cloning the first α_1-AR subtype (Cottechia *et al.*, 1988), and expression of this clone showed that it encoded the α_{1B} subtype. Later pharmacological studies showed that the α_1-ARs in this cell line were exclusively α_{1B} (Nelemans *et al.*, 1990; Han *et al.*, 1992) and only α_{1B} mRNA transcripts were found (Esbenshade *et al.*, 1993).

The MDCK canine kidney cell line also expresses α_1-ARs linked to phosphatidylinositol metabolism. α_1-AR activation also increases arachidonic acid release and PGE_2 formation in these cells, presumably by activation of phospholipase A_2 (Slivka and Insel, 1987).

TABLE 1

α_1-AR Subtypes in Continuous Cell Lines

Cell Line	*Source*	*Pharmacology*	*mRNA*	*Inositol Phosphate Response*	*Release Stored Ca^{2+}*	*Dihydropyridine-Sensitive Ca^{2+} influx*
BC3H1	mouse	α_{1B}	α_{1B}	YES	YES	NO
DDT1MF-2	hamster	α_{1B}	α_{1B}	YES	YES	NO
FRTL-5	rat	α_{1B}	α_{1B}	YES	YES	NO
MDCK	dog	α_{1B}	α_{1B}	YES	YES	NO
NB41A3	mouse	α_{1B}	α_{1B}	YES	YES	NO
rMTC 6-23	rat	α_{1B}	$\alpha_{1B}/\alpha_{1A/D}$	YES	NO	YES
SK-N-MC	human	α_{1A}/α_{1B}	$\alpha_{1B}/\alpha_{1C}/\alpha_{1A/D}$	YES	YES	NO

NE stimulates mobilization of intracellular Ca^{2+} and influx of extracellular Ca^{2+}, again through voltage-independent channels. The properties of the α_1-ARs in MDCK cells suggests that they are exclusively α_{1B}-(Blue *et al.*, 1994), and only α_{1B} mRNA transcripts are found (Esbenshade *et al.*, 1993).

α_1-ARs in the FRTL-5 rat thyroid cell line mobilize intracellular Ca^{2+} and stimulate iodide efflux. Interestingly, the activities of both phospholipase C and A_2 are increased by α_1-AR activation in these cells, and differential effects of pertussis toxin suggest the involvement of two separate G proteins (Burch *et al.*, 1986). Pharmacological characterization of the receptors in these cells suggest they are exclusively α_{1B}- (Shimura *et al.*, 1990).

Mouse NB41A3 neuroblastoma cells contain exclusively α_{1B}-ARs linked to InsP formation and mobilization of intracellular Ca^{2+}. Only α_{1B} mRNA transcripts are found in NB41A3 cells (Esbenshade *et al.*, 1993).

The reason why these cell lines express only the α_{1B}-AR subtype is not yet clear, but may indicate that expression of other α_1-AR subtypes are more precisely regulated, and lost under culture conditions. In all cell lines thus characterized where the α_{1B}-AR is the only subtype expressed, it couples to formation of InsPs and mobilization of intracellular Ca^{2+}. Although other signalling pathways are activated in some cell lines (phospholipase A_2, D, voltage-independent Ca^{2+} influx), these appear to be subsequent to phospholipase C activation. These observations are consistent with the hypothesis that signal transduction by the α_{1B} subtype is primarily through activation of phospholipase C.

SKNMC human neuroepithelioma cells—α_{1A}- and α_{1B}-ARs mobilize intracellular Ca^{2+}

Fisher and Landon (1991) first reported that human SKNMC neuroepithelioma cells expressed α_1-ARs linked to InsP formation. Preliminary pharmacological analysis suggested that these receptors were of the α_{1A} subtype, prompting Esbenshade *et al.* (1993) to conduct a more extensive examination. Studies on radioligand binding, InsP formation and Ca^{2+} showed that both pharmacologically defined α_{1A}- and α_{1B}-ARs coexist in this cell line.

NE stimulated both transient and sustained increases in $[Ca^{2+}]_i$ which were potently blocked by α_{1A}-selective antagonists. Chelation of extracellular Ca^{2+} eliminated the sustained component of the Ca^{2+} signal but not the major transient component, suggesting that this component was due to mobilization of stored Ca^{2+}. Nifedipine had no effect on either component. These observations showed clearly that the α_{1A}-ARs in SK-N-MC cells coupled to mobilization of intracellular Ca^{2+}.

Studies on the mRNA species expressed in SKNMC cells have been confusing. Initial Northern blots showed strong hybridization to the hamster α_{1B} cDNA, less hybridization to the rat $\alpha_{1A/D}$ cDNA, and no hybridization to the bovine α_{1C} cDNA (Esbenshade *et al.*, 1993). Ramarao *et al.* (1992) also showed the presence of the human α_{1B} transcript in these cells using RNase protection assays. However, more recent experiments using RNase protection assays with the human clones suggests the presence of α_{1C} and $\alpha_{1A/D}$ mRNA transcripts, but not α_{1B} (personal communication, D. Schwinn). The relationship of these transcripts to the pharmacologically expressed subtypes is still not clear.

rMTC 6-23 rat thyroid carcinoma—α_{1B}-ARs increase voltage-gated Ca^{2+} influx

The rMTC cell line was established from a rat medullary thyroid carcinoma (Zeytinoglu *et al.*, 1980) to study mechanisms of calcitonin secretion. Zink and Raue (1992) showed that

NE activated Ca^{2+} influx in rMTC 6-23 cells and that this response was blocked by verapamil, suggesting that it was due to activation of voltage-gated Ca^{2+} channels. Esbenshade *et al.* (1994) examined the AR subtype involved, leading to the first description of an α_1-AR coupled to activation of voltage-gated Ca^{2+} channels in a cell line.

The increases in $[Ca^{2+}]_i$ in rMTC 6-23 cells caused by NE consist of both a transient and sustained component. This effect is completely blocked by nifedipine but not by elimination of intracellular Ca^{2+} stores by thapsigargin, suggesting involvement of voltage-gated Ca^{2+} channels. Pharmacological studies showed that the effect of NE was mediated through α_1-ARs, apparently of the α_{1B}-subtype. Radioligand binding studies indicated an apparently homogeneous population of α_{1B}-like ARs existed in rMTC cells. Additionally, the pharmacological profile of NE-stimulated InsP formation was characteristic of the α_{1B} subtype. However, Northern blot analysis revealed rMTC 6-23 cells expressed mRNA transcripts for both the α_{1B} and $\alpha_{1A/D}$-ARs, suggesting that both subtypes coexist in these cells. The relatively homogeneous α_{1B} pharmacology observed is not surprising, since both α_{1B} and $\alpha_{1A/D}$ subtypes have almost identical "α_{1B}-like" pharmacology.

Addition of ATP, KCl, or the Ca^{2+} channel agonist BayK 8644 also increased $[Ca^{2+}]_i$ in rMTC cells, responses which were also blocked by nifedipine. However, unlike NE, these compounds did not increase InsP formation, suggesting a separation between these two signalling pathways. The effect of NE on Ca^{2+} influx in rMTC cells was not blocked by pretreatment with pertussis toxin (Zink and Raue, 1992), in contrast to α_1-AR regulation of voltage-gated Ca^{2+} channels in vascular smooth muscle (Nichols *et al.*, 1989). The phospholipase C inhibitor U-73122 decreased NE-stimulated InsP formation but not NE-stimulated increases in $[Ca^{2+}]_i$, and activation of protein kinase C by a phorbol ester increased $[Ca^{2+}]_i$ in a nifedipine-insensitive manner. These observations suggest that activation of voltage-gated Ca^{2+} channels by α_1-AR activation appears to be independent of release of intracellular Ca^{2+}, InsP formation, or activation of protein kinase C.

SUMMARY

Much data supports the coupling of α_{1B}-ARs to InsP formation and mobilization of intracellular Ca^{2+} and the coupling of α_{1A}-ARs to voltage-gated Ca^{2+} influx. However, it is now clear that α_{1A}-ARs can also couple to InsP formation and mobilization of intracellular Ca^{2+}, while α_{1B}-ARs can couple to voltage-gated Ca^{2+} influx. Our general impression of the weight

TABLE 2
Summary of Signalling by α_1-Adrenergic Receptor Subtypes

	Inositol Phosphate Formation	*Release of Stored Ca^{2+}*	*Ca^{2+} Influx (not DHP-sensitive)*	*DHP-sensitive Ca^{2+} influx*
Native Subtypes				
rat α_{1A}	+	+	+	+++
rat α_{1B}	+++	+++	+++	+
Transfected Clones				
hamster α_{1B}	++	++	++	+(?)
bovine α_{1C}	+++	++	++	?
rat $\alpha_{1A/D}$	+	+	+	+(?)

of available data is summarized in Table 2, supporting the hypothesis that coupling of α_1-AR subtypes to signalling pathways may be dependent on cell type, and the availability of G-proteins and other intracellular signalling molecules. If all subtypes can promiscuously activate the same signalling mechanisms, it is likely that the coupling efficiencies of each subtype may be substantially different. The correlations which have been observed between specific subtypes and intracellular signalling mechanisms, however, suggests a preferential coupling of particular subtypes to specific molecular transduction processes. Quantitation and comparison of the relative "coupling efficiencies" of these closely related subtypes in activating different second messenger systems presents an interesting challenge.

ACKNOWLEDGMENTS

This work was supported by NS21325. TAE is a recipient of NRSA HL08448.

References

Ambler, S.K., Brown, R.D. and Taylor, P. (1985) The relationship between phosphatidyl-inositol metabolism and mobilization of intracellular calcium elicited by α_1-adrenergic receptor stimulation in BC3H-1 muscle cells. *Mol. Pharmacol.*, **26,** 405–413.

Berridge, M.J. and Irvine, R.F. (1989) Inositol phosphates and cell signalling. *Nature*, **341,** 197–205.

Blue, D.R., Jr., Craig, D.A., Ransom, J.T., Camacho, J.A., Insel, P.A. and Clarke, D.E. (1994) Characterization of the α_1-adrenoceptor subtype mediating ^{3}H-arachidonic acid release and calcium mobilization in Madin-Darby Canine Kidney cells. *J. Pharmacol. Exp. Ther.*, **268,** 1588–1596.

Burch, R.M., Luini, L.A. and Axelrod, J. (1986) Phospholipase A_2 and phospholipase C are activated by distinct GTP binding proteins in response to α_1-adrenergic stimulation in FRTL5 thyroid cells. *Proc. Nat. Acad. Sci.*, **83,** 7201–7205.

Cotecchia, S., Schwinn, D.A., Randall, R.R., Lefkowitz, R.J., Caron, M.G. and Kobilka, B.K. (1988) Molecular cloning and expression of the cDNA for the hamster α_1-adrenergic receptor. *Proc. Natl. Acad. Sci. USA*, **85,** 7159–7163.

Esbenshade, T.A., Han, C., Murphy, T.J. and Minneman, K.P. (1993) Comparison of α_1-adrenergic receptor subtypes and signal transduction in SK-N-MC and NB41A3 neuronal cell lines. *Mol. Pharmacol.*, **44,** 76–86.

Esbenshade, T.A., Theroux, T.L. and Minneman, K.P. (1994) Increased voltage-dependent calcium influx produced by α_{1B}-adrenergic receptor activation in rat medullary thyroid carcinoma 6-23 cells. *Mol. Pharmacol.*, **45,** 591–598.

Fisher, S.K. and Landon, R.E. (1991) Identification of multiple phosphoinositide-linked receptors on human SK-N-MC neuroepithelioma cells. *J. Neurochem.*, **57,** 1599–1608.

Garcia-Sainz, J.A. and Romero-Avila, M.T. (1993) Characterization of the α_{1A}-adrenoceptors of guinea pig liver membranes using 5-^{3}H-methylurapidil. *Mol. Pharmacol.*, **44,** 589–594.

Han, C., Esbenshade, T.A. and Minneman, K.P. (1992) Subtypes of α_1-adrenoceptors in DDT_1 MF-2 and BC3H-1 clonal cell lines. *Eur. J. Pharmacol.*, **226,** 141–148.

Kenny, B.A., Naylor, A.M., Greengrass, P.M., Russell, M.J., Friend, S.J., Read, A.M. and Wyllie, M.G. (1994) Pharmacological properties of the cloned $\alpha_{1A/D}$-adrenoceptor subtype are consistent with the α_{1A}-adrenoceptor characterized in rat cerebral cortex and vas deferens. *Brit. J. Pharmacol.*, **111,** 1003–1008.

Laz, T.M., Forray, C., Smith, K.E., Vaysse, P., Hartig, P., Gluchowski, C., Branchek, T.A. and Weinshank, R.L. (1993) Cloned rat homolog of the bovine α_{1C}-adrenergic receptor exhibits an α_{1A}-like receptor pharmacology. *Neuroscience Abstracts*, 733.2.

Lepetre, N., Mironneau, J., Arnaudeau, S., Tanfin, Z., Harbon, S., Guillon, G. and Ibarrondo, J. (1994) Activation of α_{1A}-adrenoceptors mobilizes calcium from the intracellular stores in myocytes from rat portal vein. *J. Pharmacol. Exp. Ther.*, **268,** 167–174.

Lomasney, J.W., Cotecchia, S., Lorenz, W., Leung, W.Y., Schwinn, D.A., Yang-Feng, T.L., Caron, M.G., Lefkowitz, R.J. and Caron, M.G. (1991) Molecular cloning and expression of the cDNA for the α_{1A}-adrenergic receptor, the gene for which is located on human chromosome 5. *J. Biol. Chem.*, **266,** 6365–6369.

Minneman, K.P. and Esbenshade. T.A. (1994) α_1-Adrenergic receptor subtypes. *Ann. Rev. Pharmacol. Toxicol.*, **34,** 117–133.

Nakaoka, H., Perez, D.M., Baek, K.J., Das, T., Husain, A., Misono, K., Im, M.-J. and Graham, R.M. (1994) G_h: A GTP-binding protein with transglutaminase activity and receptor signalling function. *Science*, **264,** 1593–1596.

Nelemans, A., Hoiting, B., Molleman, A., Duin, M. and den Hertog, A. (1990) α-Adrenoceptor regulation of inositol phosphates, internal calcium and membrane current in DDT_1 MF-2 smooth muscle cells. *Eur. J. Pharmacol. Mol. Pharmacol. Sect.*, **189,** 41–49.

Nichols A.J., Motley, E.D. and Ruffolo, R.R. (1989) Effect of pertussis toxin treatment on postjunctional α_1- and α_2-adrenoceptor function in the cardiovascular system of the pithed rat. *J. Pharmacol. Exp. Ther.*, **249,** 203–209.

Perez, D.M., Piascik, M.T. and Graham, R.M. (1991) Solution-phase library screening for the identification of rare clones: isolation of an α_{1D}-adrenergic receptor cDNA. *Mol. Pharmacol.*, **40,** 876–883.

Perez, D.M., DeYoung, M.B. and Graham, R.M. (1993) Coupling of expressed α_{1B}- and α_{1D}-adrenergic receptors to multiple signalling pathways is both G-protein and cell type specific. *Mol. Pharmacol.*, **44,** 784–795.

Putney, J.W. Jr. (1990) Receptor-regulated calcium entry. *Pharmacol. Ther.*, **48,** 427–434.

Ramarao, C.S., Kincade-Denker, J.M., Perez, D.M., Gaivin, R.J., Riek, R.P. and Graham, R.M. (1992) Genomic organization of the human α_{1B}-adrenergic receptor. *J. Biol. Chem.*, **267,** 21936–21945.

Reynolds, E.E. and Dubyak, G.R. (1985) Activation of calcium mobilization and calcium influx by α_1-adrenergic receptors in a smooth muscle cell line. *Biochem. Biophys. Res. Comm.*, **130,** 627–632.

Schwietert, H.R., Mathy, M.-J., Wilhelm, D., Wilffert, B., Pfaffendorf, M. and van Zwieten, P.A. (1992) α_1-Adrenoceptor-mediated Ca^{2+} entry from the extracellular fluid and Ca^{2+} release from intracellular stores. *J. Autonom. Pharmacol.*, **12,** 125–136.

Schwinn, D.A., Lomasney, J.W., Lorenz, W., Szklut, P.J., Fremeau, Jr., R.T., Yang-Feng, T.L., Caron, M.G., Lefkowitz, R.J. and Cotecchia, S. (1990) Molecular cloning and expression of the cDNA for a novel α_1-adrenergic receptor subtype. *J. Biol. Chem.*, **265,** 8183–8189.

Schwinn, D.A., Page, S.O., Middleton, J.P., Lorenz, W., Liggett, S.B., Yamamoto, K., Lapetina, E.G., Caron, M.G., Lefkowitz, R.J. and Cotecchia, S. (1991) The α_{1C}-adrenergic receptor: characterization of signal transduction pathways and mammalian tissue heterogeneity. *Mol. Pharmacol.*, **40,** 619–626.

Shimura, H., Endo, T., Tsujimoto, G., Watanabe, K., Hashimoto, K. and Onaya, T. (1990) Characterization of α_1-adrenergic receptor subtypes linked to iodide efflux in FRTL cells. *J. Endocrinol.*, **124,** 433–441.

Slivka, S.R. and Insel, P.A. (1987) α_1-Adrenergic receptor mediated phosphoinositide hydrolysis and prostaglandin E_2 formation in Madin-Darby kidney cells. *J. Biol. Chem.*, **262,** 4200–4207.

Wu, D., Katz, A., Lee, C.H. and Simon, M.I. (1992) Activation of phospholipase C by α_1-adrenergic receptors is mediated by the α-subunits of the G_q-family. *J. Biol. Chem.*, **267,** 25798–25802.

Zeytinoglu, F.N., DeLellis, R.A., Gagel, R.F., Wolfe, H.J. and Tashjian, A.H. (1980) Establishment of a calcitonin-producing rat medullary thyroid carcinoma cell line. *Endocrinology*, **107,** 509–515.

Zink, A. and Raue, F. (1992) Somatostatin inhibits the norepinephrine-activated calcium channels in rMTC 6-23 cells: possible involvement of a pertussis toxin-sensitive G-protein. *Acta Endocrinol.*, **124,** 433–441.

REGULATION OF RESPONSES MEDIATED BY ALPHA$_1$ ADRENERGIC RECEPTORS IN SMOOTH MUSCLE CELLS

BRIAN B. HOFFMAN and ZHUO-WEI HU

Department of Medicine, Stanford University School of Medicine, and Geriatric Research, Educational, and Clinical Center, VA Medical Center, Palo Alto, California

Catecholamines activate alpha1 receptors in vascular smooth muscle which markedly enhances blood vessel contraction in most peripheral vascular beds. In addition, alpha1 adrenergic agonists may have prominent effects of smooth muscle growth. Relatively little is known about the regulation of these responses over time with drug therapy or in pathophysiological settings. However, these adaptations may have prominent effects on drug action in a number of important disease states. This article briefly reviews work from the authors' laboratory which illustrate some of the multiple mechanisms involved in regulating responses mediated by alpha adrenergic agonists.

KEY WORDS: Alpha$_1$ adrenergic receptors, smooth muscle, protein kinase C, aorta, gene expression

Catecholamines promote vascular smooth muscle contraction by activating alpha adrenergic receptors and relaxation via beta adrenergic receptors. Regulation of smooth muscle contraction and peripheral resistance is important in the maintenance and regulation of blood pressure in health and disease. Drugs which activate or block vascular alpha receptors have important therapeutic actions in a variety of clinical settings including cardiovascular failure and hypertension. The responsiveness of blood vessels to catecholamines may be modified in a variety of settings including diseases such as hypertension and altered hormonal states. Furthermore, prolonged exposure to agonists or antagonists may modify the sensitivity of a vessel to subsequent stimulation by sympathomimetic drugs. These alterations in sensitivity have important implications for pathophysiology and drug responsiveness.

After a cell or tissue has been exposed for a period of time to an agonist, the tissue often becomes less responsive to further stimulation by that agent. A variety of terms have been used to describe such processes including tolerance and desensitization. Desensitization may occur for a wide range of drugs used in a variety of clinical settings. For example, there may be loss in efficacy when nitroglycerin is infused chronically in patients with ischemic heart disease. The manifestations of desensitization of drug responses have clinical implications as well as shedding light on cellular adaptive and regulatory mechanisms. The mechanisms involved in desensitization of responses medi-

Correspondence to: Brian B. Hoffman, MD, VA Medical Center, 3801 Miranda Avenue, Palo Alto, California 94304, USA.

ated by beta adrenergic receptors which activate adenylate cyclase have been extensively studied and involve primarily changes in receptor coupling or sequestration. Prolonged exposure to beta adrenergic agonists leads to disappearance of these receptors from the cell (down-regulation).

There are so far at least three subtypes of $alpha_1$ receptor that have been identified by molecular cloning (Cotecchia *et al.*, 1988; Schwinn *et al.*, 1990; and Lomasney *et al.*, 1991). The relation between these cloned receptors and pharmacologically identified receptors is not entirely clear and is described in detail in other articles in this Symposium. Desensitization of responses mediated by alpha1 adrenergic receptors is not well understood. Desensitization may develop in response to sympathomimetic drugs used in the treatment of shock; also, patients with pheochromocytoma, tumors which typically secrete enormous quantities of catecholamines, have hypertension which may be much less than would be expected for the markedly elevated concentrations of plasma norepinephrine. These clinical observations likely reflect desensitization of alpha1 adrenergic receptor-mediated vascular smooth muscle contraction. Catecholamine-induced desensitization of alpha1 receptor-mediated responses has been observed in several studies.

Multiple complex mechanisms are involved in desensitization of alpha1 receptor-mediated vascular smooth muscle contraction. As might be expected based on analogies with the beta adrenergic receptor, a number of biochemical investigations have examined post-translational changes in alpha1 adrenergic receptors induced by catecholamines or phorbol esters which activate protein kinase C. Phorbol esters rapidly diminish the ability of a agonists to activate inositol phospholipid breakdown and these changes are associated with phosphorylation of alpha1 receptors (Leeb-Lundberg *et al.*, 1987). These changes occurred within minutes. Catecholamine-induced desensitization of smooth muscle contraction typically develops much more slowly, taking several hours to become evident. We have found evidence suggesting that a number of other changes occur in vascular smooth muscle leading to desensitization; consequently, it is very important to take a broader than conventional view in investigating the desensitization of alpha1 responses in smooth muscle, including potential adaptations in endothelial cells lining the vessel and in the smooth muscle itself (Hoffman, 1991).

In the absence of endothelium, agonists such as acetylcholine and bradykinin no longer cause smooth muscle relaxation. These endothelial cell-dependent mechanisms of relaxation are important in a variety of vascular beds including the coronary circulation; activation of these receptors on endothelial cells releases endothelial dependent relaxing factor (EDRF), which likely is nitric oxide (Moncada *et al.*, 1991). EDRF induces relaxation of vascular smooth muscle by activating guanylate cyclase which increases cGMP concentrations. It is clear that alterations in release or action of EDRF may have importance for a number of major clinical situations. Decreased activity of endothelial dependent relaxation has been found in several models of hypertension as well as with diabetes and atherosclerosis including coronary artery disease. Moreover, arteries from humans with atherosclerosis or hypercholesterolemia are relatively unresponsive to endothelium-dependent vasodilators. On the other hand, there is evidence suggesting that the hypotension of endotoxemia or that induced by cytokines is mediated at least in part by induction of enhanced release of EDRF (Moncada *et al.*, 1991). There are multiple forms of the nitric oxide synthase. One isoform is thought to be constitutive, releasing nitric oxide for short periods in response to receptor stimulation; another form is induced after activation of endothelial cells or macrophages by cytokines and once expressed, synthesizes nitric oxide for long periods of time.

We have found that in blood vessels taken from rats harboring pheochromocytoma (exposed in vivo to catecholamines for weeks) or aorta incubated in vitro for several hours with catecholamines have markedly blunted contractile responses to catecholamines (Rosenbaum *et al.*, 1986; Hiremath *et al.*, 1991). Interestingly, much of the desensitization of contraction in these vessels is reversed when the endothelium is mechanically removed (Hiremath *et al.*, 1991). Also, the desensitization is partially recovered when nitric oxide synthase is inhibited, suggesting an important role of either increased EDRF release or action. Using sandwich bioassays of vascular strips we have found that prolonged exposure to alpha adrenergic agonists enhances release of EDRF from endothelial cells with a delay of several hours raising the possibility that catecholamines induce either the expression or activity of a form of nitric oxide synthase (Hu *et al.*, 1994).

However, the desensitization of smooth muscle contraction is only partially explained by enhanced EDRF release from endothelial cells. There are also inherent changes in smooth muscle contraction. As indicated above, in several cell culture models, there is evidence suggesting diminished activation of polyphosphoinositide metabolism after exposure to catecholamines or phorbol esters. We have found similar changes in rat aorta (Hu *et al.*, 1994). However, there are other molecular adaptations which occur as a consequence of prolonged exposure to catecholamines which contribute to heterologous desensitization of contraction. For example, there is enhanced capacity of guanylate cyclase to synthesize cGMP in blood vessels exposed to catecholamines for several hours; this enhances relaxation responses to nitric oxide donors such as nitroprusside (Hu *et al.*, 1992a). Also, prolonged activation of alpha adrenergic receptors in rat aorta leads to down-regulation of expression of protein kinase C which may heterologously contribute to desensitization (Hu *et al.*, 1992b).

To summarize, there are multiple mechanisms of desensitization of smooth muscle contraction and these are not fully understood. Regulatory changes occur at the level of the alpha adrenergic receptors themselves in the smooth muscle leading to diminished coupling the receptors. In addition, there are heterologous adaptations which enhance responses to nitric oxide and diminish protein kinase C mediated contractile responses. Furthermore, there are adaptations in endothelial cells which enhance basal EDRF release. Consequently, desensitization of alpha1 adrenergic responses is complex, involving at least several biochemical adaptations which contribute to impaired contractile responses.

Relatively little is understood about the regulation of expression of alpha1 adrenergic receptors. There is growing recognition that a variety of hormones may regulate expression of these receptors which could have pathophysiological significance in a number of settings. The expression of eukaryotic genes is generally controlled through *cis*-acting DNA elements surrounding a gene's coding region. *Cis*-acting sequences provide binding sites for proteins (*trans*-acting factors) that are important for cell-specific expression and regulating expression in response to extra-cellular signals. These *cis*-acting sequences are frequently located in regions upstream (5′) of transcription initiation sites. There is strong evidence that the expression of adrenergic receptor genes are regulated by a variety of hormones (Collins *et al.*, 1991). For example, cAMP regulates expression of beta and alpha2 adrenergic receptors (Collins *et al.*, 1989; Sakaue and Hoffman, 1991b). Also, glucocorticoids modify expression of beta adrenergic receptor genes (Hadcock *et al.*, 1989).

Steroid hormones modulate physiological processes by a number of mechanisms including regulation of gene expression. We wondered if glucocorticoids might induce expression of alpha1 adrenergic receptors which could contribute to the increased sensitivity of vascular smooth muscle to catecholamines that may occur with glucocorticoid excess. We examined the effects of dexamethasone on the expression of the alpha1B adrenergic receptor gene in DDT 1 MF-2 smooth muscle cells (Sakaue and Hoffman, 1991a). It was found that dexamethasone produced an almost two fold increase in expression of these receptors. Steady state values of alpha1B adrenergic receptor mRNA, analyzed by Northern blotting, increased almost 3 fold after 48 hours exposure to dexamethasone. Alpha1B receptor mRNA abundance was also increased by testosterone and aldosterone whereas beta-estradiol and progesterone had no effect. The alpha1B receptor gene transcription rate, determined in nuclear run-off assays, was increased several fold in cells treated with dexamethasone whereas the half-life of the alpha1B receptor mRNA was unchanged by dexamethasone. These data indicate that glucocorticoids regulate expression of alpha1B receptors by increasing the rate of transcription of this gene.

Since protein kinase C regulates many cellular functions; we wondered whether this enzyme regulates expression of the alpha1B adrenergic receptor gene. We found that exposure of DDT_1 MF 2 cells to 12-myristate 13-acetate (PMA) (10 nM) increased values of alpha1B receptor mRNA with peak changes found at 4 h, about a 3 fold increase (Hu *et al.*, 1993). There was a dose-response relationship for PMA concentrations between 10 pM and 100 nM. Also, PMA induced a time-dependent increase in the number of alpha1B receptors at the protein level. Induction of the alpha1B receptor mRNA was suppressed by the protein kinase C inhibitors H7 and staurosporine; the inactive phorbol ester 4 alpha-phorbol didecanoate and calcium ionophores A23187 and Bay K8644 did not increase alpha1B receptor mRNA. The half-life of the alpha1B receptor mRNA was unchanged by PMA in these cells; however, nuclear run-off assays demonstrated that PMA produced a marked increase in the transcription rate of alpha1B receptor gene. These results suggest that the induction of alpha1B adrenergic receptor expression is mediated by a protein kinase C-dependent mechanism that occurs at the level of transcription. On the other hand, Izzo *et al.* (1994) have recently found that phorbol esters markedly destabilize alpha1B mRNA in cultured rabbit aortic smooth muscle cells leading to a large fall in mRNA abundance. Consequently, results in these cells are quite different from results of similar experiments in DDT1 cells (Hu *et al.*, 1993). Whether these differences relate to differences in *cis* or *trans* factors in these two different cell types from different species is unknown.

Nonetheless, effects of protein kinase C on expression of alpha1 receptors may have considerable pathophysiological significance. For example, we have found that angiotensin II via a protein kinase C dependent mechanism activates expression of alpha1A/D and alpha1B receptors in cultured rat aortic smooth muscle cells (Hu *et al.*, in press). Angiotensin II powerfully induces vascular smooth muscle contraction and functions as a growth factor stimulating vascular smooth muscle cell hypertrophy. There is some evidence suggesting that catecholamines may contribute to the effects of angiotensin

II on vascular smooth muscle cell growth (van Kleef *et al.*, 1992). Consequently, "cross-talk" between these two signaling systems wherein angiotensin II may enhance adrenergic responses may have implications for vascular remodeling in settings such as hypertension. Also, catecholamines activate proto-oncogene expression in smooth muscle (Majesky *et al.*, 1990; Okazaki *et al.*, 1994) which could serve as potential overlap between these various growth stimulating agonists in blood vessels.

References

Collins, S., Bouvier, M., Bolanowski, M.A., Caron, M.G. and Lefkowitz, R.J. (1989) Cyclic AMP stimulates transcription of the beta2-adrenergic receptor gene in response to short term agonist exposure. *Proc. Natl. Acad. Sci.*, **86,** 4853–4857.

Collins, S., Caron, M.G. and Lefkowitz, R.J. (1991) Regulation of adrenergic receptor responsiveness through modulation of receptor gene expression. *Ann. Rev. Physiol.*, **53,** 497–508.

Cotecchia, S., Schwinn, D.A., Randall, R.R., Lefkowitz, R.J., Caron, M.G. and Kobilka, B.K. (1988) Molecular cloning and expression of the cDNA for the hamster alpha1-adrenergic receptor. *Proc. Natl. Acad. Sci.*, **85,** 7159–7163.

Hadcock, J.R., Wang, H. and Malbon, C.C. (1989) Agonist-induced destablization of beta adrenergic receptor mRNA. Attenuation of glucocorticoid induced up-regulation of beta adrenergic receptors. *J. Biol. Chem.*, **264,** 19928–19933.

Hiremath, A., Hu, Z. and Hoffman, B.B. (1991) Desensitization of a adrenergic receptor-mediated smooth muscle contraction: role of the endothelium. *J. Cardiovasc. Pharmacol.*, **18,** 151–157.

Hoffman, B.B. (1991) Unpopular mechanisms of desensitization of responses mediated by catecholamines. In *Adrenoceptors: Structure, Mechanisms, Function: Advances in Pharmacological Sciences.* (Edited by E Szabadi and CM Bradshaw). Birkhauser Verlag (Basel): 189–199.

Hu, Z.-W., Honda, H., Murad, F. and Hoffman, B.B. (1992a) Prolonged exposure to catecholamines enhances sensitivity of smooth muscle relaxation induced by sodium nitroprusside and atriopeptin. *J. Pharmacol. Exp. Ther.*, **260,** 756–761.

Hu, Z.-W., Azhar, S. and Hoffman, B.B. (1992b) Prolonged activation of alpha1 adrenergic receptors induces down-regulation of protein kinase C in vascular smooth muscle. *J. Cardiovasc. Pharmacol.*, **20,** 982–989.

Hu, Z-W., Miller, J.W. and Hoffman, B.B. (1994) Induction of enhanced release of endothelium-derived relaxing factor following prolonged exposure to alpha adrenergic agonists: role in desensitization of smooth muscle contraction. *J. Cardiovasc. Pharmacol.*, **23,** 337–43.

Hu, Z-W., Shi, X.Y., Sakaue, M. and Hoffman, B.B. (1993) Prolonged activation of protein kinase C induces transcription and expression of the alpha1B adrenergic receptor gene in DDT$_1$ MF-2 cells. *J. Biol. Chem.*, **268,** 3610–3615.

Hu, Z.-W., Shi, X.-Y., Okazaki, M. and Hoffman, B.B. Angiotensin II induces transcription and expression of α_1 adrenergic receptors in cultured rat vascular smooth muscle cells. *Am. J. Physiol.*, in press.

Izzo, N.J., Tulenko, T.N. and Colucci, W.S. (1994) Phorbol esters and norepinephrine destabilize alpha1B-adrenergic receptor mRNA in vascular smooth muscle cells. *J. Biol. Chem.*, **269,** 1705–1710.

Leeb-Lundberg, L.M., Cotecchia, S., Deblasi, A., Caron, M.G. and Lefkowitz, R.J. (1987) Regulation of adrenergic receptor function by phosphorylation. I. Agonist-promoted desensitization and phosphorylation of alpha1-adrenergic receptors coupled to inositol phospholipid metabolism in DDT$_1$ MF-2 smooth muscle cells. *J. Biol. Chem.*, **262,** 3098–3105.

Lomasney, J.W., Cotecchia, S., Lorenz, W., Leung, W.-Y., Schwinn, D.A., Yang-Feng, T.L., Brownstein, M., Lefkowitz, R.J. and Caron, M.G. (1991) Molecular cloning and expression of the cDNA for the alpha1A-adrenergic receptor. *J. Biol. Chem.*, **266,** 6365–6369.

Majesky, M.W., Daemen, M.J.A.P. and Schwartz, S.M. (1990) Alpha1-adrenergic stimulation of platelet-derived growth factor A-chain gene expression in rat aorta. *J. Biol. Chem.*, **265**, 1082–1088.

Moncada, S., Palmer, R.M.J. and Higgs, E.A. (1991) Nitric oxide: Physiology, pathophysiology and pharmacology. *Pharmacol. Rev.*, **43,** 109–142.

Okazaki, M., Hu, Z.-W., Fujinaga, M. and Hoffman, B.B. (1994) Alpha1 adrenergic receptor-induced *c-fos* gene expression in rat aorta and cultured vascular smooth muscle cells. *J. Clin. Invest.*, **94,** 210–218.

Rosenbaum, J.S., Zera, P., Umans, V.A., Ginsburg, R. and Hoffman, B.B. (1986) Desensitization of aortic smooth muscle contraction in rats harboring pheochromocytoma. *J. Pharmacol. Exp. Ther.*, **238,** 396–400.

Sakaue, M. and Hoffman, B.B. (1991a) Glucocorticoids induce transcription and expression of the alpha1B adrenergic receptor gene in DTT1 MF-2 smooth muscle cells. *J. Clin. Invest.*, **88,** 385–389.

Sakaue, M. and Hoffman, B.B. (1991b) cAMP regulates transcription of the alpha2A adrenergic receptor gene in HT29 cells. *J. Biol. Chem.*, **266,** 5743–5749.

Schwinn, D.A., Lomasney, J.W., Lorenz, W., Szklut, P.J., Fremeau, R.T., Yang-Feng, T.L., Caron, M.G., Lefkowitz, R.J. and Cotecchia, S. (1990) Molecular cloning and expression of the cDNA for a novel alpha1-adrenergic receptor subtype. *J. Biol. Chem.*, **265,** 8183–8189.

van Kleef, E.M., Smits, J.F.M., de Mey, J.G.R., Cleutjens, J.P.M., Lombardi, D.M., Schwartz, S.M. and Daemen, M.J.A.P. (1992) Alpha1-adrenoceptor blockade reduces the angiotensin II-induced vascular smooth muscle cell DNA synthesis in the rat thoracic aorta and carotid artery. *Circ. Res.*, **70,** 1122–1127.

SPECIES HETEROGENEITY OF LIVER α_1-ADRENOCEPTORS: SUBTYPES, SIGNAL TRANSDUCTION AND REGULATION

J. ADOLFO GARCÍA-SÁINZ and MARINA MACÍAS-SILVA

Departamento de Bioenergética, Instituto de Fisiología Celular, Universidad Nacional Autónoma de México. Apartado Postal 70-248, 04510 México D. F.

The α_1-adrenoceptor subtype present in the liver varies considerably among species. Thus, it was observed that rats, hamsters, mice, chicken, and fishes express predominantly the α_{1B} subtype. Livers from rabbits, dogs and humans express predominantly the α_{1C} subtype. The α_1-adrenoceptor of guinea pig liver has the pharmacological characteristics of the "classical" α_{1A} subtype but Northern analysis indicates hybridization with the $\alpha_{1A/D}$ probe.

Using hepatocytes from guinea pigs ($\alpha_{1A(1A/D)}$), rats (α_{1B}), and rabbits (α_{1C}) it was observed that activation of α_1-adrenoceptors increases the production [^{3}H]IP$_3$ and increases $[Ca^{2+}]_i$. The increase in cytosol calcium was observed both in the absence or presence of extracellular calcium. This is consistent with the idea that these receptors are coupled to the phosphoinositide turnover/calcium mobilization signal transduction process. Some differences in the kinetics of [^{3}H]IP$_3$ production and $[Ca^{2+}]_i$ were observed in hepatocytes from these species, but they seem to be more the result of dissimilarities in the cells themselves rather than in the receptor subtypes they express.

Activation of protein kinase C with active phorbol esters blocks the α_1-adrenergic action in these hepatocytes. However, hepatocytes from rats were considerably more sensitive to phorbol esters than cells from guinea pigs or rabbits.

KEY WORDS: α_1-adrenoceptors, liver, Signal transduction, Inositol tris-phosphate

INTRODUCTION

The major pharmacological characteristics of α_1-adrenoceptors were defined more than fifteen years ago (Berthelsen and Pettinger 1977). It was later observed that these adrenoceptors mediate those actions which involve inositol lipid turnover, release of intracellular calcium and entry of this cation from the extracellular space (Fain and García-Sáinz, 1980).

One of the major advances was the realization that α_1-adrenoceptors are not an homogeneous entity but rather constitute an heterogeneous family of receptors. Initially, the existence of two subtypes, the α_{1A}- and α_{1B}-adrenoceptors, was suggested on pharmacological basis (Morrow and Creese, 1986; Minneman, 1988). Later, molecular biological techniques allowed the cloning and expression of the α_{1B}-adrenoceptor (Cotecchia *et al.*, 1988). However, cloning of the pharmacologically defined α_{1A}-adrenoceptor has not been reported. Nevertheless, two other receptors of this family have been cloned, i. e. the α_{1C}- (Schwinn *et al.*, 1990) and the $\alpha_{1A/D}$-adrenoceptors (Lomasney *et al.*, 1991; Perez *et al.*, 1991; Schwinn and Lomasney, 1992). When these receptors were expressed, some pharmacological dif-

Correspondence to: J. Adolfo García-Sáinz, Instituto de Fisiología Celular, UNAM, Apdo. Postal 70-248, 04510 México D. F., México.

ferences with those expected for the "classical" α_{1A}-adrenoceptor were evidenced which suggested the existence of a fourth receptor subtype (Lomasney *et al.*, 1991; Perez *et al.*, 1991; Schwinn and Lomasney, 1992). However, very recently, some groups have proposed that the $\alpha_{1A/D}$- (Kenny *et al.*, 1994) or the α_{1C}-adrenoceptor (Forray *et al.*, 1994; Rokosh *et al.*, 1994) may in fact encode the "classical" α_{1A} subtype. Therefore, at this moment, the number and identity of the α_1-adrenoceptor subtypes is not completely defined.

Liver adrenoceptors

It is well known that the natural adrenergic amines are among the major regulators of liver metabolism. The actions of catecholamines are mainly mediated through α_1- and β-adrenoceptors; α_2-adrenoceptors are also present in liver but their physiological role is far from understood. Activation of these receptors modulates glycogenolysis, gluconeogenesis, ureogenesis, fatty acid biosynthesis, synthesis and secretion of proteins.

The relative density and physiological importance of α_1 and β-adrenoceptors, in the liver of different species, varies significantly (Sulakhe *et al.*, 1988). Nevertheless, the β-adrenoceptor subtype that is expressed seems to be the same, i. e., β_2-adrenoceptors. This has been observed in rats, guinea pigs, rabbits, humans, and fishes. Only in some pathological conditions, such as hepatomas, β_1-adrenoceptors seem to be present in liver cells.

Species heterogeneity of liver α_1-adrenoceptors

A very different situation exists regarding liver α_1-adrenoceptors. We have observed that there is considerable variation in the α_1-adrenoceptor subtype present in hepatocytes from diverse species. Thus, rats (García-Sáinz *et al.*, 1992b; García-Sáinz *et al.*, 1994), mice, hamsters (García-Sáinz *et al.*, 1994), chickens (Gutiérrez-Venegas and García-Sáinz, 1993) and fishes (manuscript in preparation) express α_{1B}-adrenoceptors in this organ. Liver cells from rabbits (García-Sáinz *et al.*, 1992b; Taddei *et al.*, 1993), dogs (manuscript in preparation) and humans (manuscript in preparation) express α_{1C}-adrenoceptors. Guinea pig liver expresses a receptor with the binding properties of the α_{1A}-subtype but Northern analysis indicated hybridization to the $\alpha_{1A/D}$ cDNA probe (García-Sáinz *et al.*, 1992b); therefore it will be referred to in this manuscript as $\alpha_{1A(1A/D)}$-adrenoceptor.

This information is summarized in Table 1.

Signal transduction

Hepatocytes and liver membranes from guinea pigs, rats and rabbits can be considered as attractive models to test drug selectivity and receptor properties.

In hepatocytes isolated from these animals activation of α_1-adrenergic receptors stimulates the phosphoinositide turnover/calcium mobilization signal transduction process (García-Sáinz *et al.*, 1992b). The time-courses for the production of $[^3H]$inositol tris-phosphate ($[^3H]IP_3$) are presented in Fig. 1. One minute after addition of an adrenergic agonist, a small but reproducible increase in $[^3H]IP_3$ is observed in hepatocytes from all three species. However, the time-courses show some differences. In rat liver cells, the maximum increase in $[^3H]IP_3$ was obtained five minutes after agonist addition; this was followed by a slow de-

TABLE 1
Predominant α_1-adrenoceptor subtype expressed in the liver of different species

$\alpha_{1A(1A/D)}$	α_{1B}	α_{1C}
Guinea pig (*C. porcellus*)	Rat (*R. norvegicus*)	Rabbit (*O. cuniculus*)
	Hamster (*M. auratus*)	Dog (*C. familiaris*)
	Mouse (*M. musculus*)	Human (*H. sapiens*)
	Chicken (*G. domesticus*)	
	Fish (*I. punctatus*)	

crease in the amount of this labeled second messenger. In rabbit hepatocytes, after an initial rapid increase (0–5 min), $[^3H]IP_3$ continued increasing but much more slowly. In guinea pig hepatocytes, the maximum was obtained between 15 and 30 min (Fig. 1). These differences in kinetics seem to be (due to the repertoire of inositol phosphate metabolizing enzymes) within the cells rather than to the α_1-adrenoceptor subtype present in them, since similar time-courses were observed for other non-adrenergic agonists in hepatocytes from these species (data not shown).

Dose-response curves to norepinephrine (in the presence of 10 μM propranolol to block its β-adrenergic activity) are presented in Fig. 2. It is surprising that in rat liver cells, which

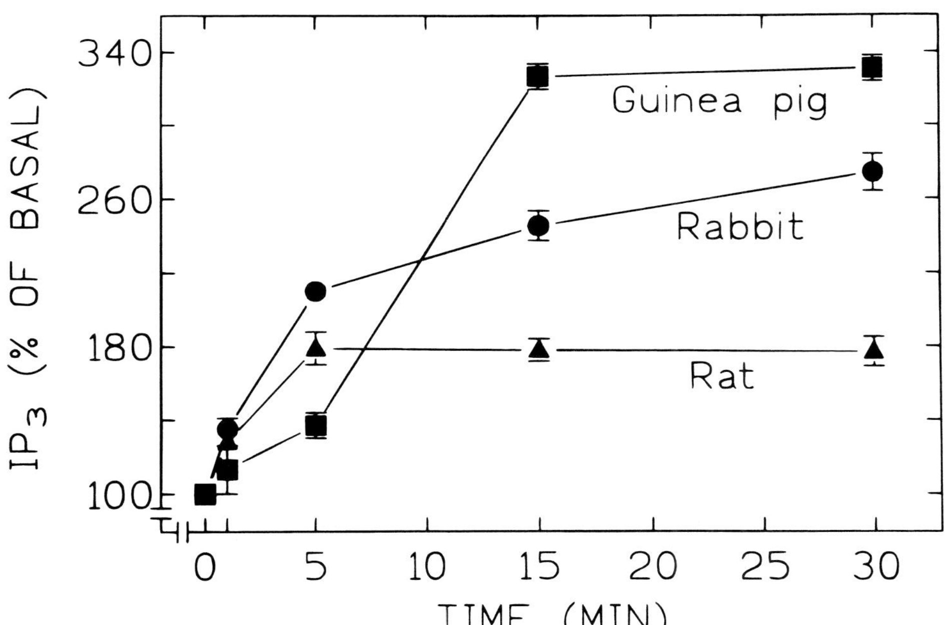

FIGURE 1 Time-courses of norepinephrine-stimulated mediated $[^3H]IP_3$ production. Cells were pre-labeled with [³H]inositol and washed, after this they were incubated in the absence or presence of 10 μM norepinephrine (in the presence of 10 μM propranolol). Time indicates minutes after norepinephrine addition. Cell isolation, labeling, washing and incubation were performed as described (García-Sáinz *et al.*, 1992b). Plotted are the means and vertical lines represent the S.E.M. of 3 different determinations in each case.

have a much bigger number of α_1-adrenoceptors than hepatocytes from the other species (García-Sáinz *et al.*, 1992b) the effect of noradrenaline was very small. Vasopressin, in contrast, can increase [^{3}H]IP$_3$ production up to 10-fold in rat hepatocytes.

The effects of norepinephrine (plus 10 μM propranolol) on cytosol calcium $[Ca^{++}]_i$ are presented in Fig. 3. In cells from the three species incubated in the presence of extracellular calcium, the concentration of calcium in the cytosol rapidly increased reaching its maximum almost immediately after the addition of the agonist. After this maximum increase $[Ca^{++}]_i$ gradually returned to near basal values; this decrease was much more rapid in rabbit hepatocytes that in cells from the other species. In the absence of extracellular calcium, the increases in $[Ca^{++}]_i$, were clearly of smaller magnitudes, reached their maximal values less rapidly, and returned to near basal values more rapidly. These data indicate that intracellular calcium stores are sufficient to initiate α_1-adrenergic actions in these cells but certainly insufficient to sustain the responses; therefore, gating of extracellular calcium seems to be required.

We would like to mention that oscillation of $[Ca^{++}]_i$ in response to α_1-adrenergic activations has been observed in isolated rat liver cells (Woods *et al.*, 1986). Interestingly, in suspensions of fish hepatocytes we have also noted such oscillatory pattern of response, which suggests that in these cells, synchronism of the $[Ca^{++}]_i$ response is clearly observed (manuscript in preparation).

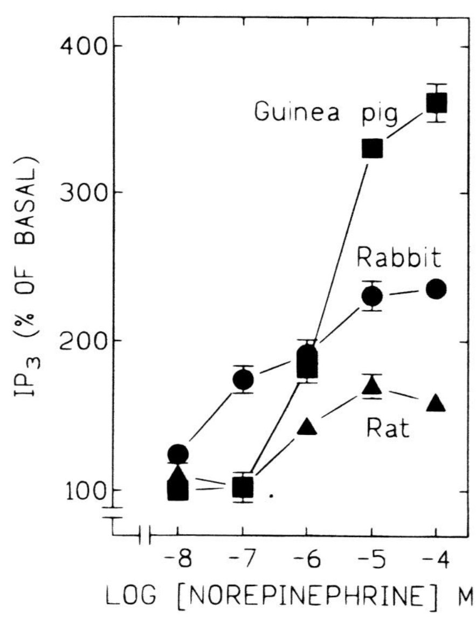

FIGURE 2 Dose-response curves to norepinephrine. Hepatocytes from the 3 different species were obtained and incubated as indicated in figure 1. Hepatocytes were incubated with the agonist for 30 minutes (guinea pig), 15 minutes (rabbit) or 5 minutes (rat). Plotted are the means and vertical lines represent the S.E.M. of 3 different determinations in each case.

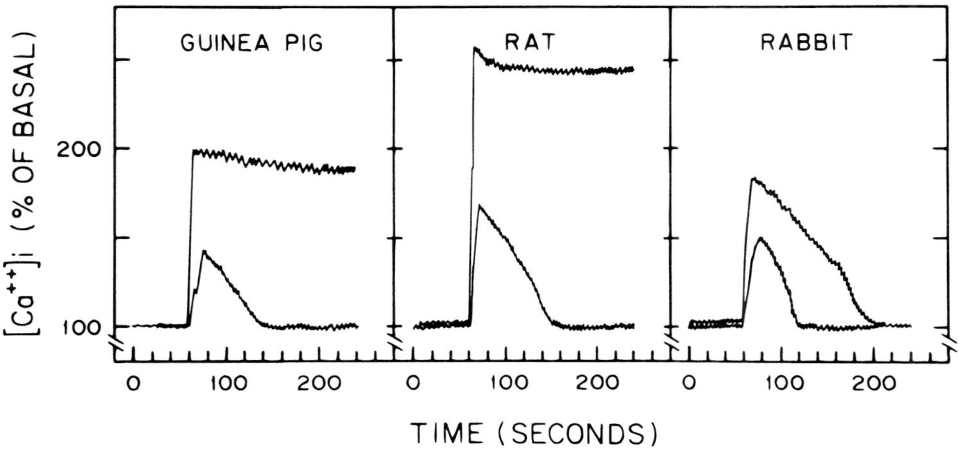

FIGURE 3 Effect of Norepinephrine on $[Ca^{2+}]_i$. Cytosol calcium was determined as described (García-Sáinz *et al.*, 1992a). Propranolol 10 μM did not affect $[Ca^{2+}]_i$ but addition of norepinephrine resulted in an immediate increase. For each species a representative tracing is presented. Incubations in regular buffer (bigger response) or in buffer without calcium and containing 25 μM EGTA (smaller response). Basal $[Ca^{2+}]_i$ were: 118 ± 11 nM (guinea pig hepatocytes n = 7), 83 ± 9 nM (rat hepatocytes n = 10), 295 ± 68 nM (rabbit hepatocytes n = 3).

Receptor regulation

Activation of protein kinase C blocks α_1-adrenergic responses in isolated rat hepatocytes (Corvera and García-Sáinz, 1984; Corvera *et al.*, 1985). This seems to be associated to receptor phosphorylation (Leeb-Lundberg *et al.*, 1985). We studied the effect of the active phorbol ester tetradecanoyl phorbol acetate (TPA), on norepinephrine-stimulated production of $[^3H]IP_3$ and observed that in the three species, this activator of protein kinase C inhibited dose-dependently the α_1-adrenergic action (Fig. 4). However, it is clear that rat liver cells were much more sensitive to this agent that those from rabbits or guinea pigs. Similar results were obtained when other parameters such as phosphatidylinositol labeling (resynthesis) or $[Ca^{++}]_i$ were studied (data not shown).

DISCUSSION

The work carried out in our laboratory during these last years clearly indicates that there is heterogeneity in the α_1-adrenoceptors expressed in liver cells among different species. To identify and characterize the α_1-adrenoceptors present in liver of diverse species several approaches have been employed, when it has been possible, i. e., molecular biological techniques, such as Northern analysis, binding in membranes and functional studies using whole cells. These approaches are complementary. Detection of an α_1-adrenoceptor subtype mRNA, does not necessarily indicates that such receptor is expressed and even less that it mediates an specific action. Direct detection and cuantitation of the α_1-adrenoceptors subtypes by binding studies is at the present difficult since available ligands are only partially subtype selective. Functional experiments are obviously of major importance if

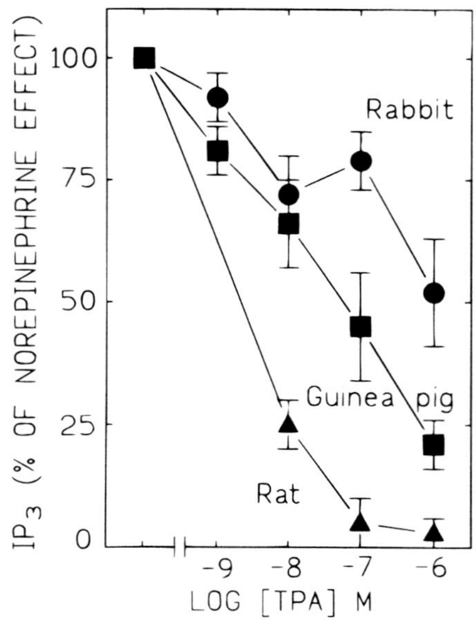

FIGURE 4 Effect of Tetradecanoyl Phorbol Acetate (TPA) on Norepinephrine-stimulated $[^3H]IP_3$ Production. Cells were incubated in the presence of different concentrations of TPA and 10 μM propranolol plus 10 μM norepinephrine. Results are presented as percentage of norepinephrine effect. Plotted are the means and vertical lines represent the S.E.M. of 3 different determinations in each case.

the receptor that mediates an action is to be defined. Important implications of the heterogeneity we have detected, are that data can not be extrapolated from one species to another and that, at least, for α_1-adrenergic subtypes no "tissue-specific" type of expression seems to exist. Obviously, this indicates the need to study directly the tissues of interest.

Regarding signal transduction, it is clear that all the subtypes, that we have been able to detect in liver cells, are coupled to the phosphoinositide/calcium mobilization signal transduction process. This does not indicate that all α_1-adrenergic actions in liver cells are necessarily mediated through this pathway. There is evidence that these receptors can be coupled to multiple signal transduction processes (Cotecchia *et al.*, 1990; García-Sáinz, 1993; Perez *et al.*, 1993). As mentioned before the differences observed in the production of IP_3 or in $[Ca^{++}]_i$, seem to be more the result of dissimilarities in the cells themselves, rather than in the receptor subtypes they express. In this regard, we would like to comment the elegant study of Perez *et al.*, (1993) who showed that cloned α_{1B} and $\alpha_{1A/D}$ can be coupled to multiple signal transduction processes but that such couplings depend on the cell type in which the receptors are expressed.

Finally, our data on the ability of activated protein kinase C to block α_1-adrenergic-mediated actions seem to be provocative. However, at this point, it is not clear if such variations in sensitivity are properties of the receptor subtypes present in the cells (i.e., their susceptibility to be substrates of the protein kinase(s) involved) or caused by other components of the signalling pathways, i. e. the spectrum of protein kinase C isoforms present in the different hepatocytes, accessibility to the receptors, role(s) of other kinase(s) or phosphatase(s), etc. Certainly this is an aspect of interest that deserves further attention; specially considering that protein kinase C seems to play a very important role in desensitization of α_1-adrenoceptors.

ACKNOWLEDGMENTS

This research was partially supported by Grant from CONACyT (4711-N9406) and DGAPA (IN 200193). The expert secretarial experience of Ms. Guadalupe Ramírez is gratefully acknowledged.

References

Berthelsen, S. and Pettinger W.A. (1977) A functional basis for the classification of α-adrenergic receptors. *Life Sci.*, **21,** 595–606.

Corvera, S. and García-Sáinz, J.A. (1984) Phorbol esters inhibit α1-adrenergic regulation of hepatocyte metabolism. *Biochem. Biophys. Res. Commun.*, **119,** 1128–1133.

Corvera, S., Schwarz, K.R., Graham, R.M. and García-Sáinz, J.A. (1986) Phorbol esters inhibit α_1-adrenergic effects and decreases the affinity of liver α_1-adrenergic receptors for epinephrine. *J. Biol. Chem.*, **261,** 520–526.

Cotecchia, S., Schwinn, D.A., Randall, R.R., Lefkowitz, R.J., Caron, M.G. and Kobilka, B.K. (1988) Molecular cloning and expression of the cDNA for the hamster α_1-adrenergic receptor. *Proc. Natl. Acad. Sci. USA*, **85,** 7159–7163.

Cotecchia, S., Kobilka, B.K., Daniel, K.W., Nolan, R.D., Lapetina, E.Y., Caron, M.G., Lefkowitz, R.J. and Regan, J.W. (1990) Multiple second messenger pathways of α-adrenergic receptor subtypes expressed in eukaryotic cells. *J. Biol. Chem.*, **265,** 63–69.

Fain, J.N. and García-Sáinz, J.A. (1980) Role of phosphatidylinositol turnover in alpha1 and of adenylate cyclase inhibition in alpha$_2$ effects of catecholamines. *Life Sci.*, **26,** 1183–1194.

Forray, C., Bard, J.A., Wetzel, J.M., Chiu, G., Shapiro, E., Tang, R., Lepor, H., Hartig, P.R., Weinshank, R.L., Branchek, T.A. and Gluchowski, C. (1994) The α_1-adrenergic receptor that mediates smooth muscle contraction in human prostate has the pharmacological poperties of the cloned human α_{1C} subtype. *Mol. Pharmacol.*, **45,** 703–708.

García-Sáinz, J.A. (1993) α_1-Adrenergic action: receptor subtypes, signal transduction and regulation. *Cell. Signalling*, **5,** 539–547.

García-Sáinz, J.A., Macías-Silva, M., Olivares-Reyes, J.A. and Romero-Avila, M.T. (1992a) Histamine activates phosphorylase and inositol phosphate production in guinea pig hepatocytes. *Eur. J. Pharmacol.*, **227,** 325–331.

García-Sáinz, J.A., Romero-Avila, M.T., Alcántara-Hernández, R., Macías-Silva, M., Olivares-Reyes, J.A. and González-Espinosa, C. (1992b) Species heterogeneity of hepatic α1-adrenoceptors: α1A, α1B and α1C subtypes. *Biochem. Biophys. Res. Commun.*, **186,** 760–767.

García-Sáinz, J.A., Casas-González, P., Romero-Avila, M.T. and González-Espinosa, C. (1994) Characterization of the Hepatic α_{1B}-adrenoceptors of rats, mice and hamsters. *Life Sci.*, **54,** 1995–2003.

Gutiérrez-Venegas, G. and García-Sáinz J.A. (1993) Characterization of the α_{1B}-adrenergic receptors of chicken hepatocytes. *Comp. Biochem. Physiol.*, **106C,** 797–803.

Kenny, B.A., Naylor, A.M., Greengrass, P.M., Russell, M.J., Friend, S.J., Read, A.M. and Wyllie, M.G. (1994) Pharmacological properties of the cloned $\alpha_{1A/D}$-adrenoceptor subtype are consistent with the α_{1A}-adrenoceptor characterized in rat cerebral cortex and vas deferens. *Brit. J. Pharmacol.*, **111,** 1003–1008.

Leeb-Lundberg, L.M.F., Cotecchia, S., Lomasney, J.W., DeBernardis, J.F., Lefkowitz, R.J. and Caron, M.G. (1985) Phorbol esters promote α_1-adrenergic receptor phosphorylation and receptor uncoupling from inositol phospholipid metabolism. *Proc. Natl. Acad. Sci. USA*, **82,** 5651–5655.

Lomasney, J.W., Cotecchia, S., Lorenz, W., Leung, W.-Y., Schwinn, D.A., Yang-Feng, T.L., Brownstein, M., Lefkowitz, R.J. and Caron, M.G. (1991) Molecular cloning and expression of the cDNA for the α_{1A}-adrenergic receptor. The gene for which is located on human chromosome 5. *J. Biol. Chem.*, **266,** 6365–6369.

Minneman, K.P. (1988) α_1-Adrenergic receptor subtypes, inositol phosphates and sources of cell Ca^{2+}. *Pharmacol. Rev.*, **40,** 87–119.

Morrow, A.L. and Creese, I. (1986) Characterization of α_1-adrenergic receptor subtypes in rat brain: a reevaluation of [^{3}H]WB4101 and [^{3}H]prazosin binding. *Mol. Pharmacol.*, **29,** 321–330.

Perez, D.M., Piascik, M.T. and Graham, R.M. (1991) Solution-phase library screening for the identification of rare clones: isolation of an α_{1D}-adrenergic receptor cDNA. *Mol. Pharmacol.*, **40,** 876–883.
Perez, D.M., DeYoung, M.B. and Graham, R.M. (1993) Coupling of the expressed α_{1B}- and α_{1D}-adrenergic receptors to multiple signaling pathways is both G protein and cells type specific. *Mol. Pharmacol.*, **44,** 784–795.
Rokosh, D.G., Bailey, B.A., Stewart, A.F.R., Karns, L.R., Long, C.S. and Simpson, P.C. (1994) Distribution of α_{1C}-adrenergic receptor mRNA in adult rat tissues by RNase protection assay and comparison with α_{1B} and α_{1D}. *Biochem. Biophys. Res. Commun.*, **200,** 1177–1184.
Schwinn, D.A. and Lomasney, J.W. (1992) Pharmacological characterization of cloned α_1-adrenoceptor subtypes, selective antagonists suggests the existence of a fourth subtype. *Eur. J. Pharmacol.*, **227,** 433–436.
Schwinn, D.A., Lomasney, J.W., Lorenz, W., Szklut, P.J., Frameau, R.T., Yang-Feng, T.L., Caron, M.G., Lefkowitz, R.J. and Cotecchia, S. (1990) Molecular cloning and expression of the cDNA for a novel α_1-adrenergic subtype. *J. Biol. Chem.*, **265,** 8183–8189.
Sulakhe, S.J., Pulga, V.B. and Tran, S. (1988) Hepatic α_1- and β-adrenergic receptors in various animal species. *Mol. Cell. Biochem.*, **83,** 81–88.
Taddei, C., Poggesi, E., Leonardi, A. and Testa, R. (1993) Affinity of different α_1-agonists and antagonists for the α_1-adrenoceptors of rabbit and rat liver membranes. *Life Sci.*, **53,** PL177-PL181.
Woods, N.M., Cuthbertson, K.S.R. and Cobbold, P.H. (1986) Repetitive transient rises in cytoplasmic free calcium in hormone-stimulated hepatocytes. *Nature*, **319,** 600–602.

MICROVASCULAR DISTRIBUTION OF SMOOTH MUSCLE α-ADRENOCEPTORS AND INTERACTION WITH METABOLIC CONTROL

JAMES E. FABER, JUN TATEISHI, CINDI J. LEECH, ANDREA D. ECKHART and MARY L. CLEMENTS

Department of Physiology, University of North Carolina, Chapel Hill, North Carolina 27599

Recent evidence indicates that vascular smooth muscle may express as many as five different α-adrenoceptor subtypes. This suprising heterogeneity has led to questions about the distribution of the subtypes among different vessels and whether the individual subtypes have unique contractile as well as non-contractile functions. In one species where the most information has been gathered (rat), different subtypes mediate adrenergic constriction of functionally distinct vessels: Large arteries rely on α1A/D and α1B, small arteries and arterioles (resistance vessels) on α1A/D and α2D, precapillary sphincters on predominantly α2D and venules/veins (capacitance vessels) on α1B and α2D. Moreover, subtype-specific constrictions of these vessel types vary greatly in their sensitivity to modulation by metabolic, myogenic, humoral and paracrine signals. This derives, at least in part, through interactions at the post-receptor level. Thus, expression of different subtypes may serve to target to specific vessel levels unique interactions between adrenergic and other vascular controls that permit optimal control of vascular resistance, capacitance and blood-tissue exchange.

KEY WORDS: Alpha-adrenergic receptor, vascular smooth muscle, hypoxia, acidosis, calcium channels, potassium channels, metabolic regulation, artery, vein, microcirculation

INTRODUCTION

Although characterization of α-adrenoceptor (AR) subtypes expressed by vascular smooth muscle is only just now beginning for the six cloned ARs [α1A/D, α1B, α1C, α2D (i.e., low rauwolscine-affinity α2A), α2B and α2C], evidence suggests that multiple subtypes may be expressed by vascular smooth muscle. mRNAs for at least four different subtypes may be expressed in rat blood vessels (Ping and Faber, 1993). Binding studies indicate the presence of both α1 and α2 ARs on large conduit vessels, and efforts have recently been made to characterize the functional (contractile) subtypes on macro- and microvessels (Han *et al.*, 1990; see Ping and Faber, 1993 for additional refs.). This brief report summarizes recent studies by our group which suggest that vessels with specific functions express different αAR subtypes, and that adrenergic constrictions mediated by the subtypes differ in their sensitivity to other vascular controls (e.g., metabolic and myogenic) with important consequences for peripheral vascular regulation.

Correspondence to: James E. Faber, PhD., Department of Physiology 7545, 474 Medical Research, University of North Carolina, Chapel Hill, NC 27599-7545.

FUNCTIONAL DISTRIBUTION OF SMOOTH MUSCLE α-ADRENOCEPTORS IN THE VASCULATURE

Constriction of Arterioles and Venules May Be Mediated by Different Subtypes

We have recently examined the specific subtypes of α1 and α2 ARs that signal constriction of arterioles and venules in the rat cremaster skeletal muscle microvasculature (Leech and Faber, 1993), having established earlier that the general αAR types (α1 and α2) are both present and activated by nerve stimulation (Faber, 1988; Ohyanagi, 1991). Arterioles were microdissected and cannulated with glass micropipets in a tissue bath to eliminate hemodynamic and parenchymal cell influences (Ikeoka *et al.*, 1992; Ikeoka and Faber, 1993). The vessel lumen and tissue bath were filled with a modified Krebs solution that was normally maintained at *in vivo* values of PO_2 70 mmHg, PCO_2 40 mmHg, pH 7.4 and 34°C. After development of resting tone with lumen pressure set to 70 mmHg (normal *in vivo* pressure), these "first order" arterioles have an average lumen diameter of approximately 100μm. Vessels were studied with the downstream pipet closed (except during wash periods) to eliminate flow-related mechanisms. Removal of endothelial cells and blockade of NO synthase and cyclo-oxygenase have indicated absence of basal release of NO or prostaglandins at vasoactive levels in these non-perfused vessels; these interventions have no effect on α1 or α2 AR constrictor responses or effects of acidosis and hypoxia (Ikeoka and Faber, 1993; Faber *et al.*, unpublished results). Smooth muscle contraction was measured as changes in lumen diameter with videomicroscopic image analysis. 10 pmolar angiotensin II (subthreshold for constriction) was present in the tissue bath to maintain α2AR sensitivity and efficacy at *in vivo* values (Ikeoka and Faber, 1993); 1μM propranolol was present to block β-ARs. All other agents were infused at 4-36 μl/min into a bath suffusion line that provided a non-recirculating flow of 4 ml/min into the 5ml tissue bath. Vessels and data were retained for analysis if: 1) vessels developed intrinsic tone sufficient to decrease their diameter by ≥ 20% from maximal dilation obtained with 0.1mM nitroprusside, 2) intrinsic tone remained constant throughout the experiment, and 3) myogenic reactivity to increases in pressure was normal and unchanged at the beginning and end of the experiment.

For venule studies, the single first-order venule that drains the cremaster and which has a lumen diameter of approximately 180μm was studied. Unlike arterioles, these vessels were not isolated because of their extreme delicacy. Instead, the cremaster muscle of an anesthetized rat was extended into a tissue bath maintained as above, and changes in venule lumen diameter for bath-added drugs were measured with videomicroscopy (Ohyanagi *et al.*, 1991). Two approaches were used to eliminate alterations in venule pressure and flow secondary to concomitant changes in diameter of vessels upstream from the first-order venule. A glass microoccluder was positioned on the venule several hundred microns upstream of the measurement site to stop flow and cause pressure below the occlusion to be determined by iliac vein pressure downstream of the exteriorized cremaster muscle. In a second approach vasopressin (3mU/ml) was present at all times in the bath to produce maximal arteriole constriction without changes in venules which lack vasopressin receptors (Muldowney and Faber, 1991); this served to "clamp" pressure and flow in the venule by eliminating additional arteriole constriction to other agents. In preliminary studies, identical dose-response curves and effects of antagonists were obtained with both approaches. Thus, results obtained using the vasopressin approach are presented below. None of the antagonists or vehicles had any effect on intrinsic tone. A single concentration-response curve for one agonist and antagonist was obtained in each vessel.

In α1AR subtype experiments (Leech and Faber, 1993) 1μM rauwolscine was present at all times to minimize α2-AR stimulation. Pretreatment for 20 min with the α1B-selective antagonist chlorethylclonidine (CEC; 35–70μM) had no effect on α1 constriction of arterioles with norepinephrine (NE). In contrast, 35uM CEC produced a non-parallel, 13-fold decrease in venule NE sensitivity and reduced maximal response by 35%. The α1A-selective antagonists WB 4101 (50nM) and 5-methylurapidil (350nM) induced parallel dextral shifts (35-fold and 6-fold, respectively) in the NE response curve for arterioles. However, in venules WB 4101 and 5-methylurapidil had no effect. Although examination of additional antagonists are required, these findings suggest that constriction of cremaster arterioles is mediated by α1A/D and venules by α1B ARs. A recent study of rat vena cava α1 ARs also concluded that constriction was mediated by the α1B (Sayet *et al.*, 1993). Han *et al.* (1990) suggested, on the basis of CEC, WB 4101 and nifedipine, that rat mesenteric artery and portal vein α1 constriction was mediated by the α1A/D and α1B. Renal artery and resistance vessels appear to utilize predominantly α1A/D (Han *et al.*, 1990; Elhawary *et al.*, 1992), while α1 subtype constriction of thoracic aorta (c.f. Ping and Faber, 1993 for refs.) appears to change with age (α1B in young, α1A/D in adult; Gurdal and Johnson, 1993). Although available antagonists make distinguishing between α1A/D and α1C ARs difficult, the greater potency of WB 4101 than 5-methylurapidil in our study is consistent with an α1A/D receptor. It is unclear whether rat vascular smooth muscle expresses α1C ARs (Ping and Faber, 1993).

α2 adrenoceptor studies (unpublished) were conducted in the presence of 10μM 5-methylurapidil to minimize α1 stimulation with the α2-selective agonist UK 14,304 (UK). For arterioles, the α2D-selective antagonist BRL 44408 (1μM) induced a 9-fold, parallel right shift in the UK response curve, while the α2B-selective antagonists ARC239 and BRL 41992 were without effect. BRL 44408 also significantly displaced the oxymetazoline response. Finally, a submaximal UK constiction was unaffected by multiple concentrations of ARC239 (0.03–3μM). Comparable studies in venules yielded similar results. Thus, these data suggest that constriction of both arterioles and venules in rat skeletal muscle is mediated by α2D ARs. A recent report indicated that α2 constriction of the human saphenous vein is mediated by α2B (Smith *et al.*, 1992), which is consistent with our evidence that α2B, and not α2D or α2C mRNAs are expressed by smooth muscle cells of this vessel (Ping and Faber, 1993).

FUNCTION OF MULTIPLE VASCULAR α-ADRENOCEPTORS

Selective Interaction With Metabolic and Myogenic Vascular Regulation

The above finding of α1A/D in arterioles and α1B in venules is intriguing in light of evidence that low oxygen (23 vs. 149 mmHg) greatly increases expression of α1B AR mRNA without effect on α1A/D or actin mRNAs (Eckhart *et al.*, 1993). Thus, the difference in oxygen tension in the smooth muscle environment of arteries and veins could influence this apparent differential expression of α1AR subtypes. This receptor heterogeneity and potential differences in subtype coupling to constriction may also underlie the sensitivity of α1A/D and especially α2D constriction of arterioles, but not α1B and α2D constriction of venules, to inhibition by acidosis (McGillivray- Anderson and Faber, 1990), non-hypotensive ischemia (McGillivray-Anderson and Faber, 1991), increased metabolic rate (Anderson and Faber, 1991), reduced pressure (Faber and Meininger, 1990; Ikeoka and Faber, 1992) and simultaneous ischemia and reduced pressure (Muldowney and Faber, 1991). This difference in metabolic and myogenic sensitivity between arterioles and venules may underlie the well-known observations that local control of resistance vessels supercedes adrenergic control to facilitate autoregulation of tissue oxygen delivery, while for venules, insuring the opposite heirarchy and maximizing adrenergic control of capacitance vessels for reflex control of venous return. Differences in adrenergic sensitivity to metabolic inhibition between pre- and post-capillary vessels has been recognized for many years although the underlying mechanism has not been known (c.f., Anderson and Faber, 1991 for refs.). We have proposed that this difference is conferred, in part, by expression on arterioles and venules of different αAR subtypes whose constrictions are selectively sensitive to metabolic inhibition (Anderson and Faber, 1991; Muldowney and Faber, 1991). In the studies described below we have sought to identify the cellular mechanisms that mediate the selective inhibition of α2D arteriolar constriction by hypoxia and acidosis.

Coupling Of αARs To Contractile Calcium In Microvessels

Considerable variation has been reported in the mechanisms by which α ARs induce increases in contractile calcium in different vessels. This may relate in part to differences, at the time unrecognized, in AR subtypes among vessels and species. Furthermore, little is known about coupling of αAR subtypes to calcium activation in microvessels-knowledge required to understand mechanisms underlying αAR interactions with metabolic and other vascular controls. We have examined AR coupling to calcium in isolated cremaster arterioles (Ikeoka *et al.*, 1992). Following removal of calcium from the vessel lumen and bath, stimulation of α1A/D ARs at an intermediate level produced transient constriction that was eliminated upon repeated stimulation. In contrast, α2D stimulation failed to induce constriction in zero calcium conditions. These results suggest that α1A/D but not α2D ARs couple to release of intracellular calcium, presumably by the phosphoinositide-sarcoplasmic reticulum pathway. When calcium was restored during continuous α1A/D AR stimulation, the resultant rapid constriction was not sensitive to blockade of dihydrophyridine calcium channels with up to 0.6μM nifedipine (NIF). In contrast, 0.3μM NIF attenuated α2D constriction by more than 50% using the same paradigm, and increasing NIF produced no additional inhibition of the remaining α2D constriction. Thus for these arterioles, α1A/D ARs induce release of intracellular calcium and calcium influx via a DHP-insensitive receptor-operated calcium channel (ROC). In contrast, α2D ARs appear to activate both ROC and NIF-sensitive, presumably voltage-operated calcium channels (VOC). In these studies we also determined that hypoxia (PO_2 10 mmHg) and acidosis (pH 7.0) selectively inhibited α2D constriction with UK 14,304 and not α1A/D constriction by phenylephrine, confirming previous studies of the same vessels *in vivo* (McGillivray-Anderson and Faber, 1990, 1991). Furthermore, this occurred in association with complete blockade of the VOC component of the constriction. The sensitivity of this α2D inhibition is well within the physiological range, with IC_{50}s for hypoxia of 24 mmHg and acidosis of pH 7.1 (Tateishi and Faber, 1994).

Interactions of hypoxia and acidosis

To examine in more detail the mechanism by which hypoxia and acidosis interfere with α2D but not α1A/D constriction, we tested whether VOC, to which the α2D is selectively coupled, are themselves inhibited. Neither hypoxia (PO_2 10) nor acidosis (pH 7.O) inhibited intermediate constriction produced by direct activation of VOC with the DHP agonist, SDZ 202-791. Hypoxia and acidosis also did not affect dose-dependent constriction produced by indirect VOC activation with KCl depolarization or stimulation of α1A/D ARs with the partial agonist St-587 whose constriction resembles α2D in its predominant reliance on VOC activation (Tateishi and Faber, 1994). These results suggest that inhibition of α2D constriction by hypoxia and acidosis does not involve direct inhibition of VOC.

It is possible that the mechanism by which α2D ARs activate VOC, currently unknown, is inhibited by hypoxia and acidosis. Recent evidence indicates that smooth muscle cells of some vessels possess a class of tonically active potassium channels which are further activated by hypoxia and reduced cellular ATP (designated K_{ATP} channels) (Bolton and Beech, 1992). Increases and decreases in activity of these channels produce relaxation and contraction of vascular smooth muscle in accordance with changes in membrane potential and VOC activity. We have recently tested in isolated arterioles the hypothesis that the α2D AR selectively couples to closure of K_{ATP}, and that this action confers sensitivity of α2D constriction to inhibition by hypoxia (Tateishi and Faber, 1995). K_{ATP} antagonists, glibenclamide (GLB,.01-10μM) and disopyramide (.001-1mM), alone, produced concentration-dependent constriction, indicating tonic activity of K_{ATP}. This constriction was prevented during an intermediate level of α2D stimulation by UK 14,304, but unaffected during α1A/D stimulation with either phenylephrine or St-587. (In these studies AR stimulation was maintained but contraction reversed with nitroprusside or adenosine, to eliminate complications by baseline changes). In a second experiment, 0.5μM GLB abolished all dose-dependent α2D constriction, while α1A/D was unaffected. Finally, α2D constriction was more sensitive than α1A/D constriction to inhibition by the K_{ATP} agonist, cromakalim (-log molar EC_{50}: 7.0 vs. 6.3, $p < .05$). These data suggest that α2D but not α1A/D ARs are coupled to closure of K_{ATP}, depolarization and contraction. In other studies we confirmed that hypoxia activates K_{ATP} in this vessel: Hypoxic (6 mmHg PO_2) inhibition of intrinsic tone was reversed by 0.1μ GLB, GLB constriction was reversed by hypoxia, and dose-dependent GLB constriction was greatly enhanced during hypoxia. Lastly, the abolition of GLB constriction during α2D stimulation, presumably due to prior closure of K_{ATP} by α2D activation, was completely restored if α2D constriction was first reversed by opening of K_{ATP} with hypoxia. These findings suggest that the selective sensitivity of α2D AR constriction to hypoxia extends from "antagonistic coupling", ie, because the α2D AR stimulates and hypoxia inhibits K_{ATP}.

In conclusion, these studies have raised the possibility that different αAR subtypes mediate constriction of functionally specific segments in the microvasculature. Furthermore, heterogeneity in coupling mechanisms for the different subtypes (and possibly for the same subtype but different vessels—e.g., the arteriole versus venule α2D), may form the basis for segment-specific modulation of adrenergic regulation of resistance vessels, precapillary sphincters and capacitance vessels by metabolic and other vasoactive mechanisms.

ACKNOWLEDGMENT

This study was supported by National Institutes of Health Grants HL-38783 and HL-02377.

References

Anderson, K.M. and Faber, J.E. (1991) Differential sensitivity of arteriolar α1- and α2-adrenoceptor constriction to metabolic inhibition during skeletal muscle contraction. *Circulation Research*, **69,** 174–184.

Bolton, T.B. and Beech, D.J. (1992) Smooth muscle potassium channels: their electrophysiology and function. In: Weston AH, Hamilton TC, eds. *Potassium Channel Modulators*. London, England: Blackwell Scientific Publications; pp. 144–180.

Eckhart, A.D., Chen, L.Q. and Faber, J.E. (1993) Effect of oxygen tension on vascular smooth muscle cell growth and mRNA for α-actin, β-actin and α-adrenoceptor subtypes. *FASEB Journal*, **7(4),** A884.

Elhawary, A.M., Pettinger, W.A. and Wolff, D.W. (1992) Subtype-selective alpha-1 adrenoceptor alkylation in the rat kidney and its effect on the vascular pressor response. *Journal of Pharmacology and Experimental Therapeutics*, **260,** 709–713.

Faber, J.E. (1988) In situ analysis of alpha-adrenoceptors on arteriolar and venular smooth muscle in rat skeletal muscle microcirculation. *Circulation Research*, **62,** 37–50.

Faber, J.E. and Meininger, G.A. (1990) Selective interaction of alpha-adrenoceptors with myogenic regulation of microvascular smooth muscle. *American Journal of Physiology*, **259,** H1126–H1133.

Gurdal, H. and Johnson, M.D. (1993) Alpha1 receptor subtypes in the aging aorta. *FASEB Journal*, **7,** A261.

Han, C., Li J. and Minneman, K.P. (1990) Subtypes of α1-adrenoceptors in rat blood vessels. *European Journal of Pharmacology*, **190,** 97–104.

Ikeoka, K., Nishigaki, K., Ohyanagi, M. and Faber, J.E. (1992) In vitro analysis of α-adrenoceptor interactions with the myogenic response in resistance vessels. *Journal of Vascular Research*, **29,** 313–321.

Ikeoka, K., Yamamoto, K. and Faber, J.E. (1992) Alpha-adrenoceptor subtype coupling to calcium regulation in microvascular smooth muscle and effect of acidosis. *FASEB Journal*, **6,** A2076.

Ikeoka, K. and Faber, J.E. (1993) Selective interference with alpha2 adrenoceptors after in vitro isolation of arteriole smooth muscle: restoration by angiotensin. *American Journal of Physiology*, **265,** H1988–1995.

Leech, C.J. and Faber, J.E. (1993) Pharmacological characterization of α1-adrenoceptor subtypes mediating contraction of resistance vessel vascular smooth muscle. *FASEB Journal*, **7(4),** A892.

McGillivray-Anderson, K.M. and Faber, J.E. (1990) Effect of acidosis on contraction of microvascular smooth muscle by alpha1 and alpha2 adrenoceptors: Implications for neural and metabolic regulation. *Circulation Research*, **66,** 1643–1657.

McGillivray-Anderson, K.M. and Faber, J.E. (1991) Effect of reduced flow on alpha1 and alpha2 adrenoceptor constriction of skeletal muscle microvessels. *Circulation Research*, **69,** 65–173.

Muldowney, S.M. and Faber, J.E. (1991) Venular but not arteriolar smooth muscle reactivity is preserved during reduced tissue perfusion. *Circulation Research*, **69,** 1215–1225.

Ohyanagi, M., Faber, J.E. and Nishigaki, K. (1991) Differential activation of alpha1 and alpha2 adrenoceptors on microvascular smooth muscle during sympathetic nerve stimulation. *Circulation Research*, **68,** 232–244.

Ping, P. and Faber, J.E. (1993) Characterization of alpha-adrenoceptor gene expression in arterial and venous smooth muscle. *American Journal of Physiology*, **265,** H1501–H1509.

Sayet, I., Neuilly, G., Rakotoarisoa, L., Mironneau, J. and Mironneau, C. (1993) Rat vena cava α1B-adrenoceptors: characterization by [3H]prazosin binding and contraction experiments. *European Journal of Pharmacology*, **246,** 275–281.

Smith, K. Connaughton, S. and Docherty, J.R. (1992) Investigations of the subtype of α2-adrenoceptor mediating contractions of human saphenous vein. *British Journal of Pharmacology*, **106,** 447–451.

Tateishi, J. Faber, J.E. (1994) Inhibition of arteriole α2A-Adrenoceptor constriction by hypoxia and acidosis does not involve calcium channel blockade. *FASEB Journal*, **8,** A34

Tateishi, J. Faber, J.E. (1995) ATP-Sensitive K^+ channels mediate α2-adrenergic receptor contraction of arteriolar smooth muscle and reversal of contraction by hypoxia. Circulation Res. **76**: in press, Jan.

CHARACTERIZATION AND REGULATION OF ALPHA-1 AND BETA ADRENOCEPTORS OF INTACT CELLS

SI-JIA ZHU and MYRON L. TOEWS

Department of Pharmacology, University of Nebraska Medical Center, Omaha, NE 68198-6260

KEY WORDS: Alpha$_1$-adrenoceptors, beta-adrenoceptors, receptor regulation, receptor desensitization, receptor internalization

Radioligand binding assays for adrenergic receptors on intact cells are advantageous in certain situations, as recently discussed (Bylund and Toews, 1993). However, agonist-induced changes in intact cell receptor binding properties complicate these assays. The literature on early studies of these agonist-induced changes has been reviewed in a previous article (Toews *et al.*, 1991) and is summarized briefly below.

Radioligand binding studies of beta adrenergic receptors (BARs) revealed that agonists, but not antagonists, exhibited markedly lower apparent affinities in competition binding assays with intact cells than in similar assays with isolated membrane preparations. Subsequent studies revealed that the apparent low affinity observed with intact cells was due to an agonist-induced conversion of the receptors from a native form with high affinity for agonists to the form exhibiting lower apparent affinity. The native high affinity form of the receptor, which exhibited agonist affinities similar to those observed in assays with isolated membrane preparations, could be detected in short-time competition binding assays (before conversion to low affinity occurred) or in assays at reduced temperatures (which prevent conversion to the low affinity form).

Our working hypothesis has been that the low apparent affinity for agonists observed in assays with intact cells results from receptor internalization. Exposure of cells to agonist for the time required to achieve "equilibrium" during competition binding assays would lead to receptor internalization. The relatively lipophilic antagonists used as radioligands, such as [^{125}I]iodopindolol, would cross cell membranes easily and thus would label both internalized and cell surface receptors. In contrast, BAR agonists such as epinephrine and isoproterenol are quite hydrophilic, and they would be expected to cross cell membranes only poorly. These agonists would have limited accessibility to the internalized receptors and would therefore inhibit binding to internalized receptors only at relatively high extracellular concentrations. Consequently, they would exhibit apparent low affinity binding under these assay conditions, even though the binding properties of the receptor *per se*

Correspondence to: Myron L. Toews, Ph.D., Department of Pharmacology, University of Nebraska Medical Center, 600 South 42nd Street, Box 986260, Omaha, NE 68198-6260.

might remain unaltered. Antagonists would not induce receptor internalization, and thus they would exhibit similar binding properties in both intact cell and broken cell binding assays.

Numerous lines of evidence support this hypothesis in the case of BARs. The time and temperature dependence for conversion of intact cell receptors to the low affinity form are consistent with involvement of internalization. Strong support for this hypothesis came from studies showing that agonists exhibited high affinity binding to intact cells in competition binding assays with the hydrophilic radioligand [^{3}H]CGP-12177, which should only label those receptors still exposed on the cell surface (Staehelin *et al.*, 1983). More recent evidence in support of the internalization hypothesis came from studies with inhibitors of internalization. We showed that permeabilization of cells with either saponin or digitonin prevented BAR internalization and that BARs on these permeabilized cells exhibited markedly higher apparent binding affinities for agonists than did those on intact cells (Zhu *et al.*, 1992). Depletion of cellular ATP by treatment with antimycin A inhibits BAR internalization, and we showed that antimycin A treatment significantly increased the apparent binding affinity of agonists for BARs on intact cells (Zhu and Toews, 1993a). In both cases, antagonist binding was unaltered.

In more recent studies we examined the binding properties of intact cells expressing BARs mutated in receptor internalization, receptor coupling to G_s, or receptor phosphorylation (Zhu and Toews, 1994). Cells expressing BARs defective only in coupling to G_s showed conversion to the low affinity form similar to that for cells expressing wild-type receptors. In contrast, cells expressing BARs defective in both internalization and G_s coupling showed only minimal conversion to the low affinity form. Cells expressing BARs defective in receptor phosphorylation by either the cyclic AMP-dependent protein kinase or by the beta adrenergic receptor kinase or by both kinases showed much smaller, but still significant, decreases in conversion to the low affinity form. These results provide further evidence that receptor internalization plays a major role in conversion of BARs to the low affinity form. Receptor phosphorylation may play a smaller role, whereas receptor coupling to G_s appears not to be involved.

Muscarinic acetylcholine receptors also exhibited lower apparent binding affinity for agonist in assays with intact cells than in broken cell assays when a lipophilic radioligand was used. We showed that higher agonist affinity was observed when a hydrophilic radioligand was used instead. Higher agonist affinity also was observed in cells treated with antimycin A, which prevented muscarinic receptor internalization. These differences in apparent affinities were not observed in assays with antagonists. These results are all analogous to those described above for BARs, and they suggest that similar processes are involved in regulating the binding properties for both BARs and muscarinic receptors on intact cells (Hoover and Toews, 1989).

Alpha-1 adrenergic receptors (A1ARs) also exhibit lower apparent binding affinities for agonists in assays with intact cells than in broken cell assays, and this discrepancy is not observed for antagonists. Again in this case, the radioligand [^{3}H]prazosin is relatively lipophilic whereas the agonists are quite hydrophilic. We directly compared conversion to the low affinity form and internalization for both BARs and A1ARs in DDT_1 MF-2 (DDT) cells, which express comparable levels of both receptors (Toews, 1987). Agonist-induced conversion of A1ARs to the low affinity form similar to that for BARs was observed, although the fraction of receptors in the low affinity form was somewhat lower for A1ARs than for BARs. Conversion to the low affinity form for BARs was accompanied by receptor internalization as assessed by sucrose density gradient centrifugation assays of receptor subcellular distribution, but surprisingly, conversion to the low affinity form for A1ARs was not accompanied by receptor internalization. These initial studies suggested the possibility that different mechanisms might be involved in conversion to low affinity for A1ARs than for BARs.

Subsequent studies with internalization inhibitors also suggested different mechanisms for these two receptors in DDT cells. Cell permeabilization markedly inhibited low affinity binding to BARs but did not alter low affinity binding to A1ARs (Zhu *et al.*, 1992). Similar results were obtained with antimycin A treatment, which inhibited low affinity binding to BARs without altering binding to A1ARs (Zhu and Toews, 1993a).

We now believe these differences may be related to differences in the internalization pathways for A1ARs and BARs in these cells. Although exposure of DDT cells to agonist alone did not lead to A1AR "internalization" as assessed by sucrose density gradient assays, agonist alone did induce a "sequestration" of A1ARs within the plasma membrane as assessed by changes in surface accessibility of binding sites in intact cell assays at reduced temperature or in assays of hydrophilic ligand competition for lipophilic radioligand binding to intact cells (Leeb-Lundberg *et al.*, 1987; Cowlen and Toews, 1988). Inclusion of a strong protein kinase C activator along with the agonist induced A1AR internalization as assessed by gradient assays and also increased the fraction of receptors inaccessible in the cell surface binding assays (Cowlen and Toews, 1988). Thus, for A1ARs in DDT cells, agonist alone appears to induce a sequestration of receptors within the plasma membrane but not endocytosis into the light vesicle fraction detected in the gradient assays. Stronger activation of protein kinase C appears to be required to promote endocytosis into the light vesicle fraction. In contrast, for BARs, agonist alone is apparently sufficient to induce both the sequestration and the endocytosis steps of the pathway, and protein kinase C activation has no further effect.

Based on this two-step model of A1AR internalization, we have modified our hypothesis for intact cell low affinity binding to propose that either sequestration within the plasma membrane or endocytosis into intracellular vesicles will result in low affinity binding. Several experimental observations support this modified hypothesis. Protein kinase C activation increases the total fraction of sequestered plus internalized receptors, and it also increases the fraction of A1ARs in the low affinity form in intact cell assays (Cowlen and Toews, 1987). More recently, we have tested the effects of intracellular potassium depletion and of hypertonic sucrose on low affinity binding (Hatcher *et al.*, 1994). These two treatments have been shown to inhibit receptor internalization at an early step in the overall pathway (Ilondo *et al.*, 1991), which we hypothesized might be analagous to the sequestration step induced by agonist alone for A1ARs. Consistent with our hypothesis, both potassium depletion and hypertonic sucrose were able to decrease the extent of low affinity binding of agonists to both BARs and A1ARs. Again, neither of these treatments altered antagonist binding.

Based on all of our results, it appears that either sequestration of receptors within the plasma membrane or internalization of receptors within endocytotic vesicles is sufficient to convert receptors to a form exhibiting low apparent affinity for hydrophilic ligands, presumably due to receptor inaccessibility. At least in DDT cells, internalization into endocytotic vesicles appears to play a major role in the case of BARs, whereas sequestration within the plasma membrane may be more important in the case of A1ARs.

Our previous studies of A1AR regulation focused on DDT cells, where we could compare agonist-induced changes in A1ARs with those for BARs in the same cells under the same conditions. Two major differences in receptor regulation were observed (Toews, 1987; Cowlen and Toews, 1988). First, agonist alone induced both sequestration and internalization for BARs but only sequestration for A1ARs, and protein kinase C activation modulated these processes for A1ARs but not for BARs. Second, agonist-induced down-regulation of receptor binding sites during long-term agonist exposure occurred to a much greater extent for BARs (80%) than for A1ARs (20%). To address the extent to which these regulatory differences were inherent in the A1AR structure or were unique to the DDT cell background in which the receptors were expressed, we conducted studies of A1AR regulation in Chinese hamster ovary cells transfected to express the hamster alpha-1B adrenergic receptor (CHO-A1AR cells), which was originally cloned from the DDT cells (Cotecchia *et al.*, 1988).

Both the internalization and down-regulation properties of the receptor were different in the CHO cell background than in DDT cells (Zhu and Toews, 1993b and unpublished data). Agonist alone or phorbol ester alone were both able to induce internalization of A1ARs in CHO-A1AR cells, and the combination of phorbol ester and agonist did not further increase the extent of internalization. The protein kinase C inhibitor staurosporine inhibited internalization induced by agonist and by phorbol ester, providing further evidence of a role for protein kinase C in A1AR internalization. The fraction of A1ARs in the agonist low affinity form in intact cell binding assays was greater in CHO-A1AR cells than in DDT cells and was not further increased by phorbol ester. Staurosporine, which inhibited A1AR internalization, also inhibited conversion to the low affinity form. These results provide further evidence for the relationship between internalization and low affinity binding for A1ARs, since each factor that altered the extent of A1AR internalization had a corresponding effect on the extent of conversion of A1ARs to the low affinity form.

Long-term exposure of CHO-A1AR cells to agonist led to an increase in A1AR binding sites, in marked contrast to the modest down-regulation observed for A1ARs in DDT cells. This "up-regulation" was both time- and agonist concentration-dependent and was accompanied by an increase in A1AR agonist-stimulated phosphoinositide hydrolysis. The mechanisms involved remain to be determined. Our observations for both internalization and down-regulation in CHO- A1AR cells indicate the importance of considering cell-specific as well as receptor-specific determinants in studies of receptor regulation and of using caution in interpreting studies of receptor regulation conducted in transfected cell systems.

References

Bylund, D.B. and Toews, M.L. (1993) Radioligand binding methods: practical guide and tips. *American Journal of Physiology*, **265** (*Lung Cellular and Molecular Physiology* **9**), L421–L429.

Cotecchia, S., Schwinn, D.A., Randall, R.R., Lefkowitz, R.J., Caron, M.G. and Kobilka, B.K. (1988) Molecular cloning and expression of the cDNA for the hamster α_1-adrenergic receptor. *Proceedings of the National Academy of Sciences USA*, **85,** 7159–7163.

Cowlen, M.S. and Toews, M.L. (1987) Effects of agonist and phorbol ester on adrenergic receptors of DDT_1 MF-2 cells. *Journal of Pharmacology and Experimental Therapeutics*, **243,** 527–533.

Cowlen, M.S. and Toews, M.L. (1988) Evidence for alpha-l adrenergic receptor internalization in DDT_1 MF-2 cells following exposure to agonists plus protein kinase C activators. *Molecular Pharmacology*, **34,** 340–346.

Hatcher, L.I., Brown, J., Whittle, S.M., Zhu, S.-J. and Toews, M.L. (1994) Effects of internalization inhibitors on the binding properties of beta and alpha-1 adrenergic receptors measured on intact cells. *FASEB Journal*, **8,** A354.

Hoover, R.K. and Toews, M.L. (1989) Evidence for an agonist-induced, ATP-dependent change in muscarinic receptor binding properties measured in intact cells. *Journal of Pharmacology and Experimental Therapeutics*, **251,** 63–70.

Ilondo, M.M., Smal, J., De Meyts, P. and Courtoy, P.J. (1991) Comparison of the effects of hypertonic sucrose and intracellular potassium depletion on growth hormone receptor binding kinetics and down-regulation in IM-9 cells: Evidence for a sequential block of receptor-mediated endocytosis. *Endocrinology*, **128,** 1597–1602.

Leeb-Lundberg, L.M.F., Cotecchia, S., DeBlasi, A., Caron, M.G. and Lefkowitz, R.J. (1987) Regulation of adrenergic receptors by phosphorylation: Agonist-promoted desensitization and phosphorylation of α_1-adrenergic receptors coupled to inositol phospholipid metabolism in DDT_1 MF-2 smooth muscle cells. *Journal of Biological Chemistry*, **262,** 3098–3105.

Staehelin, M., Simons, P., Jaeggi, K. and Wigger, N. (1983) CGP-12177: a hydrophilic β-adrenergic receptor radioligand reveals high affinity binding of agonists to intact cells. *Journal of Biological Chemistry*, **258,** 3496–3502.

Toews, M.L. (1987) Comparison of agonist-induced changes in beta- and alpha-1 adrenergic receptors of DDT_1 MF-2 cells. *Molecular Pharmacology*, **31,** 58–68.

Toews, M.L., Shreve, P.E. and Bylund, D.B. (1991) Regulation of adrenergic receptors. In *Signaling Mechanisms in Secretory and Immune Cells* (Ed. J.R. Martinez, *et al.*), San Francisco Press, Inc., San Francisco, pp. 1–17.

Zhu, S.-J. and Toews, M.L. (1993a) Effects of antimycin A on the binding properties of beta and alpha-1 adrenergic receptors measured on intact cells. *Journal of Pharmacology and Experimental Therapeutics*, **267,** 123–127.

Zhu, S.-J. and Toews, M.L. (1993b) Role of protein kinase C in internalization and low affinity binding of agonists to Chinese hamster ovary cells expressing alpha-1 adrenergic receptors. *Society for Neuroscience Abstracts*, **19,** 500.

Zhu, S.-J. and Toews, M.L. (1994) Intact cell binding properties of cells expressing altered beta adrenergic receptors. *Molecular Pharmacology*, **45,** 255–261.

Zhu, S.-J., Yost, L.E. and Toews, M.L. (1992) High affinity binding of agonists to beta but not alpha-1 adrenergic receptors of permeabilized DDT_1 MF-2 hamster smooth muscle cells. *Pharmacology Communications*, **1,** 163–176.

USE OF ANTISENSE cDNA TO ASSESS α_1-ADRENERGIC RECEPTOR COUPLING TO G_α PROTEINS

PAUL A. INSEL, KAZUSHI URASAWA and NAOKI MOCHIZUKI

Department of Pharmacology, 0636, University of California San Diego, La Jolla, CA 92093

We have used transient and stable transfection of MDCK-D_1 cells with antisense cDNA constructs to define the G_α protein(s) linked to α_{1b}-adrenergic 5-receptor-promoted arachidonic acid (AA) release in these cells. For transient transfectants, we ligated the cDNAs of $G_{q\alpha}$ $G_{11\alpha}$ and $G_{i2\alpha}$ in an antisense orientation in the pCMV vector, which utilizes the cytomegalovirus promoter. For stable transfectants, we cloned antisense $G_{q\alpha}$ into the pCEP4 vector, which also utilizes the cytomegalovirus promoter. In transient transfectants, antisense $G_{q\alpha}$ or $G_{11\alpha}$, but not $G_{i2\alpha}$, selectively decreased epinephrine but not ATP-stimulated AA release. Stable transfectants prepared antisense for $G_{q\alpha}$ had blunted epinephrine-promoted AA release without a decrease in AA release promoted by ATP, bradykinin, phorbol myristate acetate or A23187. These preliminary results suggest that transfection with antisense constructs can be used to define G protein linkages of adrenergic receptors and that α_{1b}-adrenergic receptors in MDCK-D_1 cells utilize $G_{q/11}$ to transduce receptor occupancy to the enhancement of arachidonic acid release.

KEY WORDS: Antisense cDNA, α_{1b}-adrenergic receptors, G_q protein, ATP receptor, Prostaglandin E_2, Arachidonic acid

INTRODUCTION

Adrenergic receptors are members of the super family of 7-membrane-spanning receptors that are linked to GTP-binding (G) proteins. β-adrenergic and α_2-adrenergic receptors preferentially couple, respectively, to G_s and G_{i2} (Simonds *et al.*, 1989a, 1989b) whereas the preferred G protein for α_1 adrenergic-receptors is not adequately defined. Alpha$_1$-adrenergic receptors activate phosphoinositide hydrolysis, and because this activation generally is pertussis toxin-insensitive, it is presumed that α_1-adrenergic receptors activate one or more members of the G_q family of G proteins (e.g. G_q, G_{11}, G_{14}, G_{15}, G_{16}). Some, but not all, data obtained with receptors and G proteins expressed in heterologous systems suggest that $G_{q\alpha}$ family members link to these receptors (Wu *et al.*, 1992; Perez *et al.*, 1993). Only limited data are available regarding G protein linkage of α_1 receptors in native cell systems. Techniques that block G protein expression provide a means to do this. One approach is the use of antisense oligonucleotides to inhibit expression of a G_α protein (Wong, *et al.*, 1992; Hescheler and Schultz, 1993). An alternative approach is to transfect larger cDNA antisense constructs and to evaluate transient or stable transfectants for protein function. We

Correspondence to: Paul A. Insel, Department of Pharmacology, University of California San Diego, La Jolla, CA 92093. This work was supported by grants from the National Institutes of Health.

describe here our preliminary efforts to utilize this latter method to define the G_α protein linkage of α_1 adrenergic-receptors in MDCK-D_1 cells, a renal epithelial cell line.

MATERIALS AND METHODS

Construction of Antisense G_α Expression Vectors

For transient transfection studies, full-length cDNA corresponding to $G_{q\alpha}$, $G_{11\alpha}$ (both obtained from Dr. M. Simon, Caltech) and $G_{i2\alpha}$ (obtained from Dr. Y. Kaziro, DNAX) were purified from plasmid constructs by restriction digestion and agarose gel electrophoresis and cloned into the pCMV vector (Clonetech) in the reverse orientation. For stable antisense studies, we cloned the cDNA corresponding to $G_{q\alpha}$ into the pCEP4 vector (Invitrogen) in the reverse orientation. The orientation of inserts was tested by restriction enzyme digestions and by PCR sequencing using Taq DNA polymerase.

Transfection of MDCK D1 cells

MDCK-D_1 cells (1×10^6) were suspended in potassium phosphate buffer (137 mM KCl, 2.7 mM NaCl, 8.1 mM Na_2HPO_4, 1.5 mM KH_2 PO_4, 4mM $MgCl_2$, pH7.5) on ice. Cells were then pulsed electrically by use of an Electroporator (Invitrogen). Stable transfectants were isolated by growth in the presence of hygromycin B (200 units/ml). Transient transfectants were examined 48 hr after transfection.

Cell culture and arachidonic acid release

MDCK D_1 cells were grown in monolayer culture, as described previously (Slivka and Insel, 1987). Arachidonic acid release was assessed by labelling cells for 12 hr with [^{3}H]arachidonic acid (0.3 μCi/ml). Washed cells were then incubated with agonists for 30 min at 37°C. Radioactivity in extracellular medium and solubilized cell extract prepared (using 0.1% sodium dodecyl sulfate), was determined in order to calculate % release of the total labelled arachidonic acid pool.

RESULTS AND DISCUSSION

MDCK D_1 cells, a clonal isolate from MDCK cells, possess a pure population of α_{1b}-adrenergic receptors. Initial results had suggested that these cells might have atypical α_{1b}-adrenergic receptors (Klijn *et al.*, 1991), but more recent data indicate that these receptors are typical of α_{1b}-adrenergic receptors (Blue *et al.*, 1994). Other studies have demonstrated that these cells possess a variety of different G_α proteins, including $G_{s\alpha}$, $G_{q\alpha}$, $G_{11\alpha}$, $G_{12\alpha}$, and $G_{i2\alpha}$ (data not shown). Alpha$_{1b}$-adrenergic receptors in MDCK-D_1 cells activate phospholipase A_2 and phospholipase C in a parallel and independent manner, increase intracellular Ca^{++}, and activate protein kinase C (Slivka and Insel, 1987; Slivka *et al.*, 1988; Blue *et al.*, 1994 and data not shown). Phospholipase activation leads to generation of arachidonic acid, which is rapidly converted to prostaglandin E_2.

In the current studies we used transient and stable transfection of antisense cDNA to block expression of G_α proteins in order to define G protein linkage of α_{1b}-adrenergic receptors in MDCK D_1 cells. For transient transfection we defined conditions of electroporation for maximum efficiency of transfection (using a transient β-galactosidase expression vector, pCMV-β and assessing efficiency by staining with x-Gal) with minimal killing of cells. Six different conditions were tested: three voltages (150V, 200V, and 250V) and two capacitances (250 µF and 500 µF). The combination of 150V and 500 µF produced 40–50% efficiency of transfection with only modest (~25%) reduction in cell number.

We used those conditions to transfect MDCK-D_1 cells with antisense constructs for $G_{q\alpha}$, $G_{11\alpha}$, and $G_{i2\alpha}$ and two days later assessed epinephrine and ATP-stimulated arachidonic acid release (Fig. 1). Compared to CMV vector-alone transfected cells, cells transfected with antisense $G_{q\alpha}$, $G_{11\alpha}$, but not $G_{i2\alpha}$, had a decrease in epinephrine-stimulated (i.e. α_{1b}-adrenergic receptor-mediated) [^{3}H]arachidonic acid release. The decrease was statistically significant ($p < .01$) at 10 nM epinephrine, but at 10 µM epinephrine, the decrease was only significant at $p = .10$. ATP-stimulated arachidonic acid release was similar in vector-transfected and all the G_α-antisense-transfected cells.

In spite of our attempt to optimize conditions for transient transfection, there are several limitations of this technique in this type of study: Cells are killed by the electroporation procedure, it is difficult to know the healthiness of the viable cells, efficiency of transfection is not complete and since we are testing for a loss in function, this may require high level of expression of the antisense construct. In addition, we do not know the functional lifetime of $G_{q\alpha}$ in the membrane, even if synthesis is blocked.

We thus decided to obtain stable transfectants by incubating electroporated cells with the antibiotic hygromycin and thereby to select for clonal transfectants that were hygromycin-resistant. In the current studies, we utilized only $G_{q\alpha}$ antisense constructs (Fig. 2). Relative to vector-transfected control cells, $G_{q\alpha}$-antisense cells had decreased arachidonic acid release in response to epinephrine, while responses to ATP, bradykinin, phorbol myristate ac-

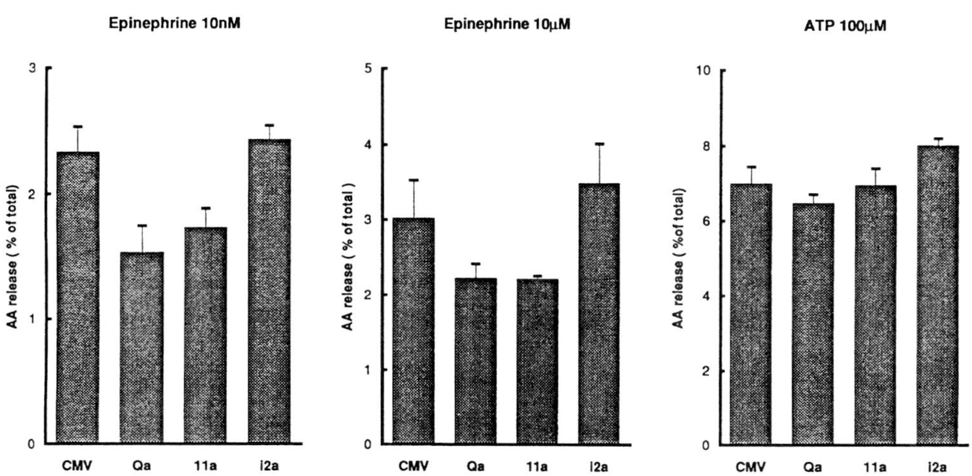

FIGURE 1 Epinephrine and ATP stimulated arachidonic acid release from transient transfectants. MDCK-D_1 cells were transfected using electroporation with antisense $G_{q\alpha}$, $G_{11\alpha}$, $G_{i2\alpha}$, or with vector control (CMV). Two days later, epinephrine (10nM, left panel; 10 µM, center panel) or ATP (100 µM, right panel) were used to stimulate [^{3}H] arachidonic acid release. The decreased responses observed with antisense $G_{q\alpha}$ (Qa) and $G_{11\alpha}$ (11a) were statistically significant ($P < 0.01$) at 10 nM but were $p = .10$ at 10 µM.

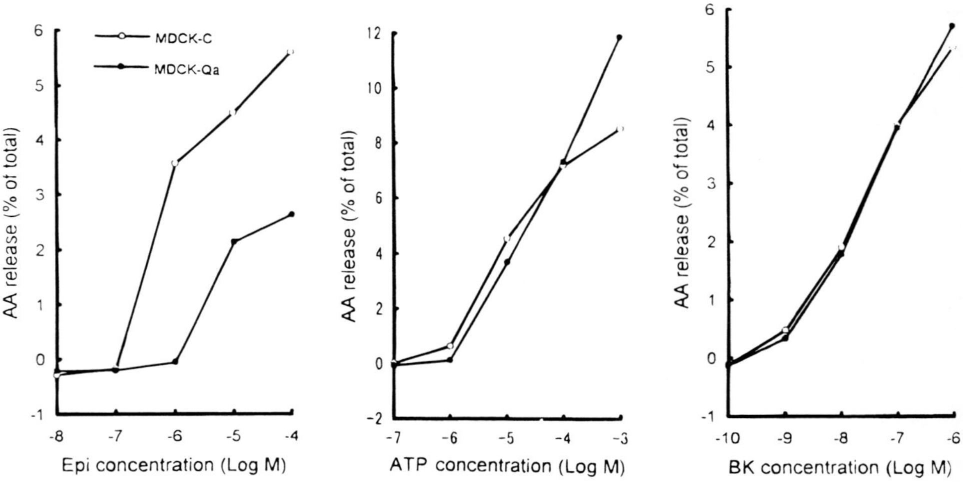

FIGURE 2 Agonist-stimulated arachidonic and release from stable transfectants. MDCK-D_1 cells were transfected with antisense $G_{q\alpha}$ in pCEP4 and selected for growth in hygromycin B. [^{3}H]-arachidonic acid release was assessed in clonal transfectants (MDCKQ_a) and vector control (MDCK-C) cells in response to varying concentrations of epinephrine (epi, left panel), ATP (middle panel) and bradykinin (BK, right panel).

etate (data not shown), and the calcium ionophore A23187 (data not shown) were similar in control and antisense-$G_{q\alpha}$ cells.

Taken together, these data suggest that antisense technology may prove useful to assess G protein linkage of particular classes of cellular receptors. Other investigators have utilized intranuclear injection of antisense oligonucleotides for different G_α, G_β and G_γ subunits to define subunits that link receptors to calcium channels in rat pituitary GH_3 cells (Heschler and Schultz, 1993). Stable transfectants have also been used to assess chemotactic factor receptor linkage to G proteins in HL60 cells (Goetzl et al, 1994).

The evidence that α_{1b}-adrenergic receptors of MDCK-D_1 cells link to $G_{q/11}$ is consistent with some, but not all, previous observations in these and other cell types (Pfeilschifter et al, 1991; Wu et al, 1992; Perez et al, 1993). Our previous work indicated that these receptors link to a pertussis toxin-insensitive G protein (Terman et al, 1987). Microinjection of G protein antibodies suggests that cardiac myocyte α_1 receptors link to members of the G_q family (LaMorte et al, 1994). Precise choices utilized by the receptors among members of this family will require additional study. Future work will also require more precise quantitative analysis of G protein expression and determination of stoichiometry between G proteins, receptors, and effector molecules.

ACKNOWLEDGMENT

This work was supported by grants from the National Institutes of Health.

References

Blue, D.R. Jr., Craig, D.A., Ransom, J.T., Camacho, J.A., Insel, P.A., and Clarke, D.E. (1994) Characterization of the α_1-adrenoceptor subtype mediating [^{3}H]arachidonic acid release and calcium mobilization in Madin-Darby canine kidney cells. *Journal of Pharmacology and Experimental Therapeutics*, **268,** 1588–1596.

Goetzl, E.J., Shames, R.S., Yang, J., Birke, F.W., Liu, Y.F., Albert, P.R. and An, S. (1994)Inhibition of human HL-60 cell responses to chemotactic factors by antisense messenger RNA depletion of G proteins. *Journal of Biological Chemistry*, **269,** 809–12.

Hescheler, J. and Schultz, G. (1993) G-proteins involved in the calcium channel signalling system. *Current Opinion in Neurobiology*, **3,** 360–367.

Klijn, K., Slivka, S.R., Bell, K. and Insel, P.A. (1991) Renal alpha$_1$-adrenergic receptor subtypes: MDCK-D$_1$ cells, but not rat cortical membranes possess a single population of receptors. *Molecular Pharmacology*, **39,** 407–413.

LaMorte, V.J., Thorburn, J., Absher, D., Spiegel, A., Brown, J.H., Chien, K.R. Feramisco, J.R. and Knowlton, K.U. (1994) Gq- and ras-dependent pathways mediate hypertrophy of neonatal rat ventricular myocytes following alpha 1-adrenergic stimulation. *Journal of Biological Chemistry*, **269,** 13490–13496.

Perez, D.M., DeYoung, M.D. and Graham, R.M. (1993) Coupling of expressed alpha$_{1B}$ and alpha$_{1D}$-adrenergic receptor to multiple signaling pathways is both G-protein and cell type specific. *Molecular Pharmacology*, **44,** 784–795.

Pfeilschifter, J., Paulmichl, M., Woll, E., Paulmichl, R. and Lang, F. (1991) Cellular mechanisms of adrenaline-induced hyperpolarization in renal epithelioid MDCK cells. *Biochemical Journal*, **274,** 243–248.

Simonds, W.F., Goldsmith, P.K., Codina, J., Unson, C.G., Spiegel, A.M. (1989a) G_{i2} mediates α_2-adrenergic inhibition of adenylyl cyclase in platelet membranes:in situ identification with G_α C-terminal antibodies. *Procedings of National Academy of Science USA*, **86(20),** 7809–13.

Simonds, W.F., Goldsmith, P.K., Woodard, C.J., Onson, C.G. and Spiegel, A.M. (1989b) Receptor and effector interactions of G_s. Functional studies with antibodies to the α_s carboxyl-terminal decapeptide. *FEBS Letters*, **249**(2), 189–194.

Slivka, S.R. and Insel, P.A. (1987) Alpha$_1$-adrenergic receptor-mediated phosphoinositide hydrolysis and prostaglandin E2 formation in Madin-Darby canine kidney cells. *Journal of Biological Chemistry*, **262,** 4200–7.

Slivka, S.R., Meier, K.E. and Insel, P.A. (1988) Alpha 1-adrenergic receptors promote phosphatidylcholine hydrolysis in MDCK-D1 cells. A mechanism for rapid activation of protein kinase C. *Journal of Biological Chemistry*, **263,** 12242–12246.

Terman, B.I., Slivka, S.R., Hughes, R.J. and Insel, P.A. (1987) Alpha$_1$-adrenergic receptor-linked guanine-nucleotide binding protein in muscle and kidney epithelial cells. *Molecular Pharmacology*, **311,** 12–20.

Wong, H.Y., Watkins, D.C. and Malbon, C.C. (1992) Antisense oligodeoxynucleotides to G_s protein α-subunit sequence accelerate differentiation of fibroblasts to adipocytes. *Nature*, **358,** 334–337.

Wu, D., Katz, A., Lee, C.H. and Simon, M.I. (1992) Activation of phospholipase C by α_1-adrenergic receptors is mediated by α subunits of G_q family. *Journal of Biological Chemistry*, **27,** 25798–802.

REC 15/2739 (SB 216469): A NOVEL PROSTATE SELECTIVE α_1-ADRENOCEPTOR ANTAGONIST

R. TESTA[1], C. TADDEI[1], E. POGGESI[1], C. DESTEFANI[1], S. COTECCHIA[2], J. P. HIEBLE[3], A. C. SULPIZIO[3], D. NASELSKY[3], D. BERGSMA[3], C. ELLIS[3], A. SWIFT[3], S. GANGULY[3], R. R. RUFFOLO, JR.[3] and A. LEONARDI[1]

[1]Research and Development Division-Recordati S.p.A. via M. Civitali 1, 21048 Milano-ITALY, [2]Institut de Pharmacologie-Universite de Lausanné rue du Bugnon 27, CH-1005 Lausanne-SWITZERLAND, and [3]Departments of Pharmacology, Molecular Genetics and Gene Expression Sciences, SmithKline Beecham Pharmaceuticals, 709 Swedeland Road, King of Prussia, Pennsylvania, 19406-USA

Rec 15/2739 (SB 216469) is a novel agent having potent α_1-adrenoceptor antagonist activity. Rec 15/2739 selectively inhibited [^{3}H] prazosin binding to tissue homogenates of native α_{1A}-and α_{1C}-adrenoceptors and to recombinant α_{1C}-adrenoceptors. Rec 15/2739 also produced potent inhibition of [^{3}H] prazosin binding to human prostate membranes (K_i = 2.0 nM). In functional studies using rabbit tissues, Rec 15/2739 was 39 fold more potent in prostatic strips (K_B = 2.7 nM) than in segments of ear artery (K_B = 106 nM). This degree of functional selectivity was not observed with any of the other α_1-adrenoceptor antagonists tested. The α_1-adrenoceptor antagonists currently utilized for the therapy of benign prostatic hyperplasia (BPH) are often associated with side-effects attributable to blockade of vascular α_1-adrenoceptors. Hence, Rec 15/2739 may offer a therapeutic advantage, since it may be possible to block prostatic α_1-adrenoceptors with this drug at a dose not influencing the vascular α_1-adrenoceptors.

INTRODUCTION

In recent years it has become clear that the α_1-adrenoceptor can be further subclassified. Based on functional, radioligand binding and molecular studies, the presence of three subtypes has been established. Receptors in native tissues are designated as α_{1A}-, α_{1B}- and α_{1C}-, and the recombinant receptors are commonly designated as $\alpha_{1A/D}$, α_{1B} and α_{1C}. The different designations reflect some confusion regarding the correspondence between native and recombinant α_1-adrenoceptors (see Review by Bylund *et al.*, 1994).

Prostatic smooth muscle from several species, including man, is responsive to α_1-adrenoceptor stimulation (Hieble *et al.*, 1986). In patients with benign prostatic hyperplasia (BPH), a component of the urinary obstruction induced by the enlarged gland is postulated to result from a "dynamic" component, due to activation of α_1-adrenoceptors on prostatic stromal smooth muscle via norepinephrine released from sympathetic nerve terminals (Caine, 1990). This provides a rationale for the clinical utility of α_1-adrenoceptor antagonists in BPH . Most of the α_1-adrenoceptor antagonists currently employed in BPH do not have subtype selectivity (Forray *et al.*, 1994a), and often produce side-effects attributable to

Correspondence to: R. Testa, Research and Development Division-Recordati S.p.A. via M. Civitali 1, 21048 Milano-Italy.

blockade of vascular α_1-adrenoceptors. Several of the α_1-adrenoceptor antagonists currently in clinical development for BPH, including tamsulosin (Yamada *et al.*, 1994) and SL 89-0591 have been reported to have selectivity for prostate versus the vasculature. A potential approach to the design of additional prostate selective α_1-adrenoceptor antagonists involves subtype selectivity, with the current focus on the α_{1C} adrenoceptor.

Rec 15/2739 (SB 216469), is a potent and selective α_1-adrenoceptor antagonist. In the present study we compare its affinity, in a variety of native tissue preparations and in cells expressing recombinant α_1-adrenoceptors, to a series of standard α_1-adrenoceptor antagonists, as well as several other drugs currently being developed for the treatment of BPH. In addition, the functional selectivity of several of these antagonists for rabbit prostate versus rabbit ear artery is determined.

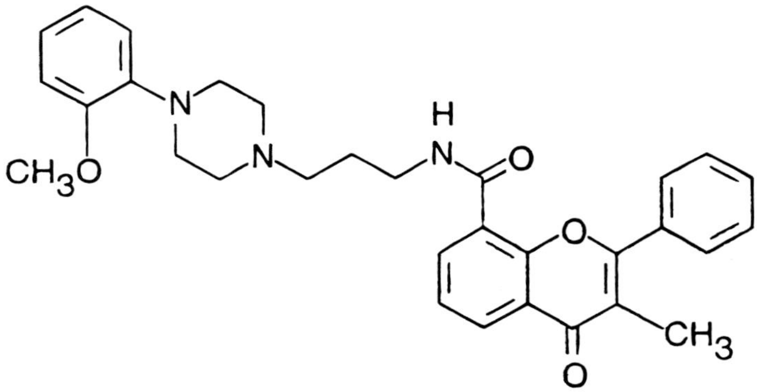

MATERIALS AND METHODS

Cloning of human α_1-adrenoceptors

α_{1A}-adrenoceptor The complete coding region of the human α_{1A}-adrenoceptor was cloned from two overlapping clones: A 1.1 kilobase (kb) clone encompassing appproximately 0.3 kb of 5′ untranslated region (UTR) and 0.8 kb of coding sequence was isolated from prostate cDNA library using a rat α_{1A}-adrenoceptor cDNA as a hybridization probe. A second 0.9 kb clone, encompassing the 3′ coding portion of the human α_{1A}-adrenoceptor (not including any 3′ UTR sequences) was isolated from a hippocampus cDNA library using a mouse serotonin $5HT_{5A}$ cDNA as a hybridization probe. This cDNA was spliced to the 1.1 kb clone at a common XhoI restriction site to form a 2.0 kb cDNA encompassing 0.3 kb of 5′UTR, and 1.7 kb of the complete coding region (576 amino acids) of the human α_{1A}-adrenoceptor. This cDNA clone was introduced into the Asp718/HindIII restriction sites of the vector pCDN for expression in mammalian cells.

α_{1B}-adrenoceptor A 2.3 kB cDNA clone encoding the human α_{1B}-adrenoceptor was isolated from a human hippocampus cDNA library using a rat α_{1B}-adrenoceptor cDNA as a hybridization probe. This clone contained 0.2 kb and 0.5 kb of 5′- and 3′-untranslated sequences, respectively, and 1.6 kb of coding sequence (encoding 517 amino acids). This cDNA clone was introduced into the Eco RI restriction site of the vector pCDN for expression in mammalian cells.

α_{1C}-adrenoceptor Two cDNA clones encoding portions of the human α_{1C}-adrenoceptor were isolated from a human hippocampus cDNA library using a bovine α_{1C}-adrenoceptor cDNA as a hybridization probe. One clone contained a region of DNA which encompassed only the first 80 N-terminal amino acids of the α_{1C}-adrenoceptor. The second cDNA clone (2.1 kb in length) encoded approximately 370 amino acids of the α_{1C}-adrenoceptor but lacked the region encoding approximately 100 amino acids of the N-terminus of the protein. A 375 base pair cDNA fragment encoding the missing 5′ coding region was derived from human brain mRNA by the polymerase chain reaction (PCR) utilizing primers corresponding to nucleotide sequences at the 5′ end, and 375 base pairs within, the coding region of the cDNA. This PCR DNA fragment was ligated to a 1.4 kb fragment (derived from the 2.1 kb cDNA by PCR) encompassing the remaining portion of the α_{1C}-adrenoceptor cDNA to form a complete coding region (466 amino acids). This cDNA fragment was subcloned into the Asp 718/Bam HI restriction sites of the vector pCDN for expression in mammalian cells.

Establishment of stable CHO clones expressing α_1-adrenoceptors

Chinese hamster overy cells, grown in suspension using MR1MOD3 medium (in house formulation), were transfected by electroporation with DNA containing the genes encoding each α_1-adrenoceptor subtype. Cells were then plated in 96 well plates at 3000 viable cells per well. After 48 hours, 0.4 mg/ml G418 (Gibco-BRL) was added to the plates and the cells were fed every 3 to 4 days until colonies appeared (usually after 10–14 days). Cells were expanded and tested for the presence of receptor expression by binding of [^{3}H] prazosin.

Sources of biological membranes

COS-7 membranes Transient expression of rat $\alpha_{1A/D}$-, hamster α_{1B}- and bovine α_{1C}-adrenoceptors in COS-7 cells was performed as previously described (Cotecchia *et al.*, 1988; Schwinn *et al.*, 1990). Transfected cells from culture flasks were scraped into 2 ml of 50 mM Tris-HCl, pH 7.4, containing 5 mM EDTA and lysed by sonication. The cell lysates were pelleted at 30,000 × g for 15 min at 4°C and washed three times with 10 ml ice-cold 50 mM Tris-HCl, pH 7.4. Membranes were then stored at −70°C until utilized.

Rat and rabbit membranes Membranes of rat hippocampus pretreated with chloroethylclonidine (CEC) (α_{1A}), rat liver (α_{1B}), and rabbit liver (α_{1C}) were prepared using previously described procedures (Testa *et al.*, 1993; Taddei *et al.*, 1993).

Human prostatic membranes: Prostatic adenoma specimens were obtained from patients with symptomatic BPH undergoing either open or trans-urethral resection of the prostate. The postatic tissues were frozen immediately on dry ice and stored at −70°C until utilized. Membranes were prepared from human prostatic tissue using a previously described procedure (Testa *et al.*, 1993).

Radioligand binding assays Membranes were incubated in 50 mM Tris-HCl, pH 7.4, containing 10 μM pargyline and 0.1% ascorbic acid, with 0.3–0.6 nM [^{3}H] prazosin in the absence or presence of the displacing drug (0.1 pM- 100 μM). Incubation volumes were 0.22 ml for the COS cell membranes and 2.0 ml for the other membrane preparations. Non-specific binding was determined in the presence of 10–100 μM phentolamine. The reaction mixture was incubated for 30 min at 25°C and then stopped by the addtion of ice-cold Tris-HCl buffer and rapid filtration through Whatman GF/B filters.

Data analysis

The saturation curves were analyzed by using the non-linear curve fitting program Ligand to determine tha apparent dissociation constant (K_d) and maximum number of binding sites (Bmax) for [^{3}H] prazosin. The ability of the drugs to displace specific [^{3}H] prazosin binding was estimated by the non-linear curve-fitting programs Acufit or Allfit.

Functional studies

Adult male, New Zealand White rabbits were sacrified by injection of pentobarbital into the marginal ear vein. Ring segments (3–4 mm long) were cut from the central ear artery and two tungsten wire hooks were passed through the lumen of each ring. The prostate was bisected and a length of 5-0 surgical silk was tied to each end of both sections. All tissues were suspended in organ baths between an isometric force-displacement transducer and a stationary tissue holder. The baths contained Krebs-Henseleit buffer (composition listed below) warmed to 38° C and aerated with 5% CO_2/95%O_2. A resting tension of 1 gram was applied to ear artery and prostate. The Krebs-Henseleit buffer was of the following composition (mmol/l): NaCl, 119; KCl, 4.7; KH_2PO_4, 1.2; $CaCl_2$, 2.5; $NaHCO_3$, 25 and glucose, 11. Cocaine (6 μM) was present in the buffer throughout the experiment to block neuronal uptake.

Concentration-response curves were constructed in each tissue by stepwise cumulative addition of norepinephrine. After completion of the concentration response curve, norepinephrine was washed from the bath by the overflow method. When the tension produced by each tissue returned to baseline, antagonist was added to the bath and remained in contact with the tissue for 30–45 minutes after which time the concentration-response curve to norepinephrine was repeated in the presence of the antagonist.

Two antagonist concentrations were usually tested for each antagonist, using separate tissue preparations for each concentration. The receptor dissociation constant (K_B) was calculated for each concentration, using the method of Furchgott (1972).

RESULTS

Table 1 compares the ability of Rec 15/2739 to inhibit the binding of [^{3}H] prazosin in a variety of α_1-adrenoceptor models, including native tissue models of α_{1A}- (CEC pretreated rat hippocampus), α_{1B}-(rat liver), and α_{1C}-(rabbit liver) adrenoceptors, human prostate, and cells expressing recombinant α_1-adrenoceptor subtypes from both animal and human sources. Rec 15/2739 shows highest affinity for the α_{1A} adrenoceptor in rat hippocampus, the human prostate α_1-adrenoceptor and for recombinant bovine and human α_{1C}-adrenoceptors (K_i = 0.3–2.0 nM). Correlation of affinity data for all of the antagonists shows the α_1-adrenoceptor of the human prostate to compare most closely with the recombinant bovine α_{1C}-adrenoceptor (r = 0.911, regression slope = 1.036) , the native α_{1A}-adrenoceptor of rat hippocampus (r = 0.921, regression slope = 0.981) and the native α_{1C}-adrenoceptor of rabbit liver (r = 0.953, regression slope = 1.025), which also correlated well with each other. Correlations of data from the native tissues was not correlated with that from the recombinant human α_1-adrenoceptors, since all of the antagonists were not examined on the human receptors. Data with (+) niguldipine was also not included in the correlations, since it has been reported that the apparent affinity of this highly lipophilic compound is dependent on the absolute amount of membrane homogenate used in the assay (Michel *et al.*, 1994).

TABLE 1

Affinity of α_1-Adrenoceptor Antagonists for Native and Recombinant α_1-Adrenoceptors. All values are expressed as pK_i ($-\log_{10} K_i$) for inhibition of the binding of [^{3}H] prazosin . Each value represents the mean of at least two duplicate determinations. Tissue sources were as follows: [1]Human prostate; [2]CEC pretreated rat hippocampus, [3]rat liver; [4]rabbit liver; [5]recombinant rat $\alpha_{1A/D}$ adrenoceptor expressed in COS-7 cells; [6]recombinant human $\alpha_{1A/D}$ adrenoceptor expressed in CHO cells; [7]recombinant hamster α_{1B} adrenoceptor expressed in COS-7 cells; [8]recombinant human α_{1B} adrenoceptor expressed in CHO cells; [9]recombinant bovine α_{1C} adrenoceptor expressed in COS-7 cells; [10]recombinant human α_{1C} adrenoceptor expressed in CHO cells. [11]Correlation coefficients for linear regressions comparing pKi values between different α_1-adrenoceptor assays.

Compound	α_1-*HP*[1]	α_{1A}[2]	α_{1B}[3]	α_{1C}[4]	$\alpha_{1A/D}$[5]	$\alpha_{1A/D}$[6]	α_{1B}[7]	α_{1B}[8]	α_{1C}[9]	α_{1C}[10]
Rec 15/2739	8.69	8.87	7.95	8.22	7.61	8.19	7.29	7.72	9.54	9.15
Abanoquil	10.05	9.60	9.85	9.38	10.21		9.89		10.10	
Alfuzosin	7.75	7.70	8.35	7.22	7.81	8.46	7.93	8.66	7.63	8.00
Indoramin	8.17	7.55	7.36	7.47	6.56	7.42	7.61	7.92	8.30	8.52
5-Methylurapidil	8.16	8.33	6.66	7.76	6.30	7.74	5.98	6.69	8.69	8.55
Naftopidil	7.94	7.13	6.99	6.61	7.39	8.24	6.92	7.59	7.78	8.22
(+)-Niguldipine	8.05	8.20	6.42	8.26	7.08		7.60		9.42	
Phentolamine	7.87	7.79	7.05	7.46	7.17	7.83	7.05	7.62	8.49	8.30
Prazosin	9.54	9.03	9.44	9.00	8.71	9.52	9.34	9.52	9.14	9.22
SL 89-0591	8.63	7.82	7.26	7.59	8.70	8.56	7.48	7.95	8.86	8.50
Spiperone	8.08	7.42	8.81	7.65	7.66		8.15	9.01	7.87	8.10
Tamsulosin	9.54	9.70	8.39	9.10	8.75	10.04	8.17	9.20	9.82	10.30
Terazosin	8.21	8.24	8.45	7.47	7.46	8.58	7.52	8.66	7.58	8.16
Urapidil	6.78	6.56	6.37	6.34	5.78		5.89	6.58	6.54	7.15
WB-4101	9.28	8.89	7.59	9.02	8.20	9.22	7.24	8.49	9.21	9.52
Correlation Values[11]										
Human Prostate		0.921	0.697	0.953	0.861		0.752		0.911	
Native α_{1A}			0.630	0.948	0.726		0.607		0.901	
Native α_{1B}				0.679	0.782		0.943		0.479	
Native α_{1C}					0.762		0.688		0.896	
Recombinant $\alpha_{1A/D}$							0.853		0.727	
Recombinant α_{1B}									0.558	

In functional studies measuring potency against norepinephrine induced contraction of rabbit tissues, Rec 15/2739 also shows prostate selectivity. Rec 15/2739 was 39 fold more potent in prostatic strips, compared to ear artery segments (Table 2). Indoramin, 5-methylurapidil, tamsulosin and SL 89-0591 showed 2.2–4.7 fold prostate selectivity in this assay, while other compounds showed either no selectivity (phentolamine), or were more potent in the ear artery (prazosin).

TABLE 2

Functional selectivity of α_1-adrenoceptor antagonists against norepinephrine-induced contraction of rabbit prostatic strips or rabbit ear artery segments

Compound	K_B *(Prostate)*	K_B *(Ear Artery)*	*Ratio*
Rec 15/2739	2.7 nM	106 nM	39
Indoramin	19 nM	89 nM	4.7
5-Methylurapidil	16 nM	61 nM	3.8
Phentolamine	31 nM	48 nM	1.5
Prazosin	13 nM	2.0 nM	0.15
SL 89-0591	18 nM	40 nM	2.2
Tamsulosin	0.24 nM	1.0 nM	4.2

DISCUSSION

Based on radioligand binding studies in native tissues from several species, Rec 15/2739 shows a small degree of selectivity for the α_{1A}- versus either the α_{1B}-(8.3 fold) or α_{1C}- (4.5 fold) adrenoceptor subtype. When tested for affinity to the recombinant α_1-adrenoceptor subtypes, Rec 15/2739 shows selectivity for the α_{1C}-adrenoceptor subtype. Moderate α_{1C}-adrenoceptor selectivity is observed when affinity for the human recombinant α_1-adrenoceptor subtypes is compared (9 and 27 fold versus $\alpha_{1A/D}$-and α_{1B}-adrenoceptors, respectively), while the compound appears to be highly α_{1C}-selective when affinities for the commonly studied recombinant animal α_1-adrenoceptor subtypes are compared (85 and 178 fold higher affinity for the bovine α_{1C}-adrenoceptor versus the rat $\alpha_{1A/D}$-and hamster α_{1B}-adrenoceptor, respectively). Data in Table 1 also show 5-methylurapidil to have substantially greater α_{1C}-adrenoceptor selectivity when affinity for recombinant animal α_1-subtypes is compared than when affinity for the three human α_1-adrenoceptors is determined. Potency differences have been reported for other α_1-adrenoceptor antagonists between species homologs of the α_{1C}-adrenoceptor (Forray *et al.*, 1994b), and these results suggest that, if possible, subtype affinity should be determined in recombinant receptors from a single species.

Correlation of the affinities for all of the antagonists tested between native and recombinant receptors shows the native α_{1A}-adrenoceptor in CEC treated rat hippocampus to have characteristics very similar to that of the recombinant bovine α_{1C}-adrenoceptor. This is based both on the correlation coefficient between pK_i values (0.901), and on the slope (0.962) and y-intercept (0.661) of the best-fit linear regression, which are close to the line of identity. This close correlation is consistent with the hypothesis that the recombinant α_{1C}-adrenoceptor may be the receptor pharmacologically defined as the α_{1A}-adrenoceptor in native tissues (Laz *et al.*, 1993; Perez *et al.*, 1994; Forray *et al.*, 1994a). The pK_i values in human prostate correlate best with those in either native α_{1A}-adrenoceptors (Table 1; Testa *et al.*, 1993) or recombinant bovine α_{1C}-adrenoceptors (Table 1), consistent with the observation that m-RNA for the α_{1C}-adrenoceptor is predominant in human prostate (Price *et al.*, 1993) and that the functional potency of α_1-adrenoceptor antagonists in isolated human prostate correlates with their affinity for the recombinant human α_{1C}-adrenoceptor (Forray *et al.*, 1994a; Forray *et al.*, 1994c). Data for the 13 compounds tested on both recombinant α_{1C}-adrenoceptors shows a slightly better correlation for pK_i values in human prostate with corresponding values in the recombinant human α_{1C}-adrenoceptor ($r = 0.934$, slope = 0.95) than for its bovine homolog ($r = 0.888$, slope = 1.05).

Rec 15/2739 shows a dramatic functional selectivity between rabbit prostate and ear artery (Table 2). Interestingly, the compounds showing α_{1C}-adrenoceptor selectivity on the recombinant human α_1-adrenoceptors all show at least some functional selectivity between rabbit prostate and ear artery. However, the magnitude of the functional selectivity observed with Rec 15/2739 is greater than that which would be predicted based on its selectivity between the recombinant receptors, suggesting either differences in pharmacological properties between human and rabbit α_1-adrenoceptors, or an additional factor contributing to its tissue selectivity between prostate and blood vessel. In contrast to most of the other compounds tested, prazosin is less potent in the prostate than ear artery.

As noted above, the predominant α_1-adrenoceptor in the human prostate appears to be the α_{1C}-adrenoceptor, and the design of selective α_{1C}-adrenoceptor antagonists is being pursued as an approach to more effective α_1-adrenoceptor antagonists for the treatment of BPH.

Although the α_1-adrenoceptor subtype mediating vascular contraction has not been clearly defined, several highly selective α_{1C}-adrenoceptor antagonists, although equipotent to prazosin against α_1-agonist induced contraction in human prostate (Forray *et al.*, 1994c; Hieble *et al.*, 1986; Chapple *et al.*, 1989) are much less active in a rat model of orthostatic hypotension (Gong *et al.*, 1994). This suggests that selectivity for the α_{1C}-adrenoceptor may indeed reduce the cardiovascular liability of an α_1-adrenoceptor antagonist in the therapy of BPH, and is consistent with our observation that α_{1C}-adrenoceptor selectivity does confer some prostate versus vascular selectivity in the rabbit tissues (Table 2).

Rec 15/2739 does possess moderate α_{1C}-adrenoceptor subtype selectivity, greater than observed with tamsulosin or SL 89.0591. However, perhaps more important is the functional selectivity of Rec 15/2739 for α_1-adrenoceptors on prostatic versus vascular smooth muscle, which is greater than would be predicted based only on its subtype selectivity, and substantially superior to any of the other antagonists tested. This may result in greater in vivo selectivity and the ability to produce more effective blockade of prostatic α_1-adrenoceptors without side-effects associated with blockade of vascular α_1-adrenoceptors.

References

Bylund, D.B., Eikenberg, D.C., Hieble, J.P., Langer, S.Z., Lefkowitz, R.J., Minneman, K.P., Molinoff, P.B., Ruffolo, R.R., Jr. and Trendelenburg, U. (1994) International Union of Pharmacology Nomenclature of Adrenoceptors. *Pharmacol. Rev.*, **46**, 121–136.

Caine, M (1990) Alpha-adrenergic blockers for the treatment of benign prostatic hyperplasia. *Urol. Clin. North Amer.*, **17**, 641–649.

Chapple, C.R., Aubry, M.L., James, S., Greengrass, P.M., Burnstock, G., Turner-Warwick, R.T., Milroy, E.J. and Davey, M.J. (1989) Characterisation of human prostatic adrenoceptors using pharmacology, receptor binding and localization. *Br. J. Urol.*, **63**, 487–496.

Cotecchia, S., Schwinn, D.A., Randall, R.R. and Lefkowitz, R.J. (1988) Molecular cloning and expression of the cDNA for the hamster α_1-adrenergic receptor. *Proc. Natl. Acad. Sci. USA*, **85**, 7159–7163.

Forray, C., Bard, J.A., Wetzel, J.M., Chiu, G., Shaprio, E., Tang, R., Lepor, H., Hartig, P.R., Weinshank, R.L., Branchek, T.A. and Gluchowski, C. (1994a) The α_1-adrenergic receptor that mediated smooth muscle contraction in human prostate has properties of the cloned human α_{1C} subtype. *Mol. Pharmacol.*, **45**, 703–708.

Forray, C., Bard, J.A., Laz, T.M., Smith, K.E., Vaysse, P.J.J., Weinshank, R.L., Gluchowski, C. and Branchek, T.A. (1994b) Comparison of the pharmacological properties of the cloned bovine, human and rat α_{1C}-adrenergic receptors. *FASEB J.*, **8**, A353.

Forray, C., Chiu, G., Wetzel, J.M., Bard, J.A., Weinshank, J.A., Branchek, T.A., Hartig, P.A. and Gluchowski, C. (1994c) Effects of novel alpha-1C adrenergic receptor antagoniss on the contraction of human prostatic smooth muscle. *J. Urology*, **151**, 267A.

Furchgott, R.F. (1972) The classification of adrenoceptors (adrenergic receptors). In: *Handbook of Experimental Pharmacology—New Series.* O. Eichler, A. Farah, H. Herken and A.D. Welch, Eds, Berlin: Springer Verlag, pp. 283–335.

Gong, G., Chiu, G., Gluchowski, C., Branchek, T.A., Hartig, P.R., Pettinger, W.A. and Forray, C. (1994) α_{1C}-Adrenergic antagonists and orthostatic hypotension. *FASEB J.*, **8**, A353.

Hieble, J.P., Boyce, A. and Caine, M. (1986) Comparison of the α-adrenoceptor characteristics in human and canine prostate. *Fed. Proc.*, **45**, 2609–2614.

Laz, T.M., Forray, C., Smith, K.E., Vaysse, P.J.J., Hartig, P.R., Gluchowski, C., Branchek, T.A. and Weinshank, R.L. (1993) Recombinant rat homolog of the bovine α_{1C}-adrenergic receptor exhibits α_{1A}-like receptor pharmacology *Soc. Neurosci. Abstracts*, **19**, 1788.

Michel, M.C., Hanft, G. and Gross, G. (1994) Radioligand binding studies of α_1-adrenoceptor subtypes in rat heart. *Brit. J. Pharmacol.*, **111**, 533–538.

Perez, D.M., Chen, J.L., Malik, N. and Graham, R.M. (1994) Is the α_{1C}-adrenergic receptor the α_{1A}-subtype? *FASEB J.*, **8**, A353.

Price, D.T., Schwinn, D.A., Lomasney, J.W., Allen, L.F., Caron, M.G. and Lefkowitz, R.J. (1993) Indentification, quantification and localization of mRNA for three distinct α_1-adrenergic receptor subtypes in human prostate. *J. Urol.*, **150**, 546–551.

Schwinn, D.A., Lomasney, J.W., Lorenz, W., Szklut, P.J., Fremeau, R.T., Yang-Feng, T.L., Caron, M.G., Lefkowitz, R.J. and Cotecchia, S. (1990) Molecular cloning and expression of the cDNA for a novel α_1-adrenergic receptor subtype. *J. Biol. Chem.*, **265,** 8183–8189.

Taddei, C., Poggesi, E., Leonardi, A. and Testa, R. (1993) Affinity of different α_1-agonists and antagonists for the α_1-adrenoceptors of rabbit and rat liver membranes. *Life Sci.*, **53,** 177–181.

Testa , R., Guarneri, L., Ibba, M., Strada, G., Poggesi, E., Taddei, C., Simonazzi, I. and Leonardi, A. (1993) Characterization of α_1-adrenoceptor subtypes in prostate and prostatic urethra of rat, rabbit, dog and man. *Europ. J. Pharmacol.*, **249,** 307–315.

Yamada, S., Tanaka, C., Kimura, R. and Kawabe, K. (1994) Alpha$_1$-adrenoceptors in human prostate: Characterization and binding characteristics of alpha$_1$-antagonists. *Life Sci.*, **54,** 1845–1854.

III. α_2-ADRENOCEPTORS

TABLE 1
Lack of Changes in mRNA Levels During Agonist Treatment

			Time of Incubation, hr					
			0.5	*1*	*2*	*4*	*7*	*24*
Cell Line	*Receptor Subtype*	*n*	*Ratio Control to Norepinephrine mRNA level*					
CHO-C10	A2AR-A	(2)	1.08	0.98	—	1.59	1.33	1.15
CHO-RNG	A2AR-B	(1)	0.81	0.85	—	1.10	1.90	0.71
OK	A2AR-C	(2)	0.94	0.84	1.17	1.47	0.85	1.43

Cells were treated with 10 μM norepinephrine for the indicated times. mRNA levels determined using a Phosphoimager (Molecular Dynamics) to quantitate an RNase protection assay of 30 μg of total RNA per time point.

OK cell line. The probe for alpha-2A adrenergic receptor is a 350 bp fragment, the alpha-2B probe a 584 bp fragment and the alpha-2C probe a 320 bp fragment. In preliminary experiments over a 24 hr time course with 10 μM norepinephrine treated cells no significant changes in mRNA levels were observed (Table 1). The autoradiogram for the CHO cells transfected with the rat RNG receptor (CHO-RNG) is shown in Figure 1. Because the transfected alpha-2A and alpha-2B receptors are under the control of a strong viral promoter, and since the 5′ and 3′ regulatory sequences of the gene are not present, it seems unlikely that mRNA levels would be altered by norepinephrine. Our preliminary data from RNase protection assays confirms that down-regulation both in the CHO cells transfected with the alpha-2A or the alpha-2B is not accompanied by a decrease in mRNA levels. More interestingly, in the OK cell (alpha-2C), where the normal regulatory mechanisms are intact, down-regulation is also not accompanied by a decrease in mRNA levels. Furthermore, down-regulation of the alpha-2A receptor is the same in the untransfected HT29 cell and in the transfected CHO cell. Thus, the predominant mechanism involved appears to be at the level of synthesis or degradation of the receptor rather than an effect

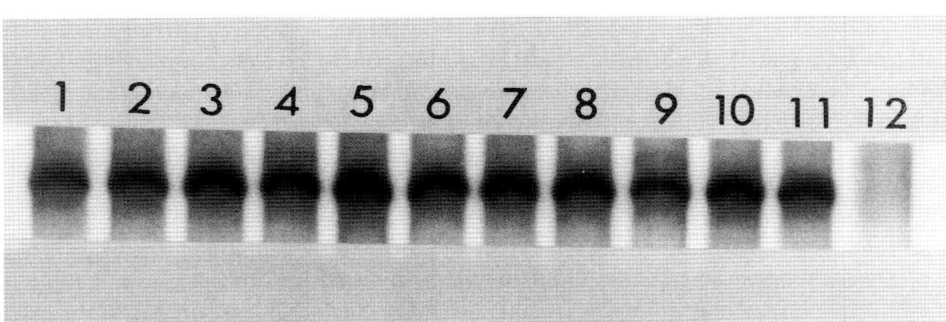

FIGURE 1 RNase protection assay for CHO cells stably transfected with the alpha-2B adrenergic receptor (rat RNG). Cells were treated with 10 μM norepinephrine for various times. The assay used 30 μg of total RNA at each time point. Lane 1, t = 0; Lanes 2 and 3, – and + norepinephrine at 0.5 hr; 4 and 5, – and + at 1 hr; 6 and 7, – and + at 4 hr; 8 and 9, – and + at 7 hr; 10 and 11, – and + at 24 hr; Lane 12, control yeast tRNA.

on mRNA levels. Our hypothesis is that a major mechanism of the decreased receptor density seen in homologous down-regulation of the alpha-2 adrenergic receptor is an increase in the rate of receptor degradation.

We have also utilized the RNase protection assay to study the distribution of the alpha-2A/D adrenergic receptor (RG20), the alpha-2B (RNG) and alpha-2C (RG10) in peripheral rat tissue (Blaxall *et al.*, 1994b). Alpha-2 adrenergic receptor mRNA was found in all tissues tested. Kidney, brain and spinal cord had transcripts for all three subtypes. Only one mRNA subtype was detected in several tissues. Aorta and spleen had only alpha-2A/D mRNA whereas, heart and liver had only alpha-2B mRNA. All other tissues had two alpha-2 adrenergic receptor subtype transcripts present. In contrast to the rat CNS, which contains predominately alpha-2A/D and alpha-2C mRNA with little alpha-2B mRNA, peripheral tissues contain predominately alpha-2A/D and alpha-2B mRNA with little alpha-2C mRNA. These results confirm those of a previous study (Handy *et al.*, 1993) in the cases where the same tissue was used (kidney, brain, spleen, liver, heart and adult lung) with the one exception that we found the presence of a low level of the alpha-2A/D adrenergic receptor transcript in addition to alpha-2B transcript in the neonatal lung.

A comparison of the relative distribution of mRNA transcripts for alpha-2 adrenergic receptor subtypes in human and rat tissues indicates a wide distribution of alpha-2 adrenergic subtype mRNA in both species. Kidneys from both have mRNA for all three subtypes. Adult human and rat lung and skeletal muscle have both alpha-2A/D and alpha-2B mRNA, but human also has alpha-2C mRNA which is absent in adult rat lung and skeletal muscle. The most striking species difference is that alpha-2C mRNA is widely distributed in human peripheral tissues but is only found in the rat periphery in the kidney. The knowledge for the distribution of alpha-2 adrenergic receptor subtypes should facilitate the elucidation of the physiological roles of these subtypes and the development of subtype specific agonists and antagonists for clinical intervention.

References

Blaxall, H.S., Cerutis, D.R., Hass, N.A., Iversen, L.J. and Bylund, D.B. (1994a) Cloning and expression of the α_{2C}-adrenergic receptor from the OK cell line. *Mol. Pharmacol.*, **45,** 176–181.

Blaxall, H.W., Hass, N.A. and Bylund, D.B. (1994b) Expression of alpha-2 adrenergic receptor genes in rat tissues. *Receptor*, **4**, 191–199.

Bylund, D.B. (1988) Subtypes of α_2-adrenoceptors: pharmacological and molecular biological evidence converge. *Trends Pharmacol. Sci.*, **9,** 356–361.

Bylund, D.B. (1992) Subtypes of alpha-1 and alpha-2 adrenergic receptors. *FASEB J.*, **6,** 832–839.

Bylund, D.B., Eikenberg, D.C., Hieble, J.P., Langer, S.Z., Lefkowitz, R.J., Minneman, K.P., Molinoff, P.B., Ruffolo, R.R., Jr. and Trendelenburg, U. (1994) International union of pharmacology nomenclature of adrenoceptors. *Pharmacol. Rev.*, **46,** 121–136.

Handy, D.E., Flordellis, C.S., Bogdanova, N.N., Bresnahan, M.R. and Gavras, H. (1993) Diverse tissue expression of rat α_2-adrenergic receptor genes. *Hypertension*, **21,** 861–865.

Jones, S.B., Leone, S.L. and Bylund, D.B. (1990) Desensitization of the alpha-2 adrenergic receptor in HT29 and opossum kidney cell lines. *J. Pharmacol. Exp. Ther.*, **254,** 294–300.

Lomasney, J.W., Cotecchia, S., Lefkowitz, R.J. and Caron, M.G. (1991) Molecular biology of α-adrenergic receptors: implications for receptor classification and for structure-function relationships. *Biochem. Biophys. Acta*, **1095,** 127–139.

Pleus, R.C., Shreve, P.E., Toews, M.L. and Bylund, D.B. (1993) Down-regulation of alpha-2 adrenoceptor subtypes. *Eur. J. Pharmacol.*, **244,** 181–185.

Shreve, P.E., Toews, M.L. and Bylund, D.B. (1991) α_{2A}- and α_{2C}-Adrenoceptor subtypes are differentially down-regulated by norepinephrine. *Eur. J. Pharmacol.*, **207,** 275–276.

FUNCTIONAL SUBCLASSIFICATION OF α_2-ADRENOCEPTORS

J. PAUL HIEBLE, ROBERT R. RUFFOLO, JR., ANTHONY C. SULPIZIO, DIANE P. NASELSKY, THERESA M. CONWAY, CATHY ELLIS, ANN M. SWIFT, SUBINAY GANGULY and DERK J. BERGSMA

Departments of Pharmacology, Molecular Genetics and Gene Expression Sciences, SmithKline Beecham Pharmaceuticals, 709 Swedeland Road, King of Prussia, PA, 19406, USA

Based on correlations of radioligand binding affinity, and on the primary sequences of recombinant receptors, the further subdivision of the α_2-adrenoceptor into four subtypes, designated as α_{2A}, α_{2B}, α_{2C} and α_{2D}, has been established. However, in addition to, and apparently independent of, this subclassification scheme, α_2-adrenoceptors can be functionally subclassified, based on the ability of certain antagonists to block some, but not all α_2-adrenoceptor mediated responses. These antagonists are found among several structural series', including the 6,9-disubstituted 3-benzazepines (*e.g.*, SK&F 104078), thieno-3-benzazepines (*e.g.*, SK&F 104856), furo-3-benzazepines (*e.g.*, SK&F 105854), benzisoindoles (*e.g.*, Abbott-65265), and arylpiperazines (*e.g.*, naftopidil). While it was initially postulated that these antagonists produced selective blockade of pre- versus postjunctional α_2-adrenoceptors, based on their affinity for those α_2-adrenoceptors mediating vascular contraction, but not for those mediating inhibition of atrial neurotransmission. However, it now seems as if certain prejunctional responses (*e.g.*, inhibition of neurotransmission in rat vas deferens) are sensitive, while some presumably postjunctional responses (*e.g.*, inhibition of lipolysis in the rat) are insensitive. In some cases the sensitivity of a response to these functionally selective antagonists is dependent on the specific agonist used to stimulate the α_2-adrenoceptors. Most of these novel antagonists also block the α_1-adrenoceptor, but SK&F 105854 produces potent blockade of α_2-adrenoceptors on canine saphenous vein with little or no affinity for atrial prejunctional receptors or vascular α_1-adrenoceptors. The demonstration of functional α_2-adrenoceptor selectivity suggests that additional subtypes of this receptor remain to be identified via molecular techniques, and may lead to additional therapeutic opportunities.

KEY WORDS: SK+F 104078, SK+F 104856, SK+F 105854, Naftopidil, Abbott-65265, Guinea Pig Atrium, Canine Saphenous Vein, Rat Vas Deferens

INTRODUCTION

The α_2-adrenoceptors were initially subdivided into two classes, designated as α_{2A} and α_{2B} adrenoceptors, based on the ability of several antagonists such as prazosin and ARC-239, previously thought to be selective for the α_1-adrenoceptor, to produce relatively potent inhibition of the binding of [^{3}H] rauwolscine to certain α_2-adrenoceptor populations, such as those of neonatal rat lung and a neuroblastoma-glioma cell line (NG-108) (Bylund *et al.*, 1985; Bylund *et al.*, 1987). These novel α_2-adrenoceptors were designated as α_{2B} adrenoceptors. By correlation of the ability of an extensive series of α-adrenoceptor antagonists to inhibit the binding of

Correspondence to: J. Paul Hieble, Ph.D., Division of Pharmacological Sciences, UW 2510, SmithKline Beecham Pharmaceuticals, 709 Swedeland Road, King of Prussia, PA, 19406, USA.

[^{3}H] rauwolscine or other α_2-adrenoceptor radioligands to a variety of tissue homogenates, two additional α_2-adrenoceptor subtypes have been identified, and designated as α_{2C} and α_{2D} adrenoceptors (Murphy and Bylund, 1988; Simonneaux *et al.*, 1991). The α_{2C} adrenoceptor has similar characteristics to the α_{2B} adrenoceptor, while the α_{2D} adrenoceptor resembles the α_{2A}, differing primarily by its relatively low affinity for yohimbine and rauwolscine.

This subclassification scheme has been supported by the cloning and expression of α_2-adrenoceptor proteins. Three distinct recombinant α_2-adrenoceptors have been identified from human (Lomasney *et al.*, 1991), rat (Harrison *et al.*, 1990; Zeng *et al.*, 1990) and mouse (Link *et al.*, 1992; Chruschinski *et al.*, 1992) tissues. The three recombinant human α_2-adrenoceptors appear to have pharmacologic characteristics corresponding to those of the native α_{2A}, α_{2B} and α_{2C} adrenoceptors, while those from rat and mouse correspond to α_{2B}, α_{2C} and α_{2D}. The human clone designated as α_2-C10 (Lomasney *et al.*, 1991) appears to be a species variant of the rat RG20 (Harrison *et al.*, 1990) and mouse Mα_2-C10 (Link *et al.*, 1993). Although these three proteins have at least 90% amino acid identity, the human receptor has α_{2A} characteristics, and the rat and mouse α_{2D}.

Functional in vitro studies have shown certain antagonists, such as SK&F 104078 (Hieble *et al.*, 1986; Hieble *et al.*, 1988) and SK&F 104856 (Hieble *et al.*, 1991) to be able to discriminate between α_2-adrenoceptor populations. These antagonists (Figure 1) will block α_2-adrenoceptor mediated contraction of canine, human or rabbit saphenous vein, while having no effect on the neuroinhibitory action of α_2-adrenoceptor agonists in atrial or vascular preparations. Other antagonists having similar characteristics have been identified, including naftopidil (Muramatsu *et al.*, 1991) and Abbott-65265 (Buckner *et al.*, 1989) (Figure 1).

It was initially postulated that SK&F 104078 produced selective blockade of post- versus prejunctional α_2-adrenoceptors. However, prejunctional α_2-adrenoceptors in rat vas deferens were blocked by this antagonist (Hieble *et al.*, 1988; Connaughton *et al.*, 1989); the ability of SK&F 104078, but not rauwolscine, to block α_2-adrenoceptor mediated neurotransmission in this tissue appears to be dependent on the agonist employed (Akers *et al.*, 1991; Oriowo *et al.*, 1991). Agonist dependent effects on the ability of SK&F 104078 to block α_2-adrenoceptors were also observed in the pithed rat (Akeus *et al.*, 1991). SK&F 104856 will also block α_2-adrenoceptor mediated inhibition of neurotransmission in the rat vas deferens, with a potency intermediate between that on vascular α_2-adrenoceptors in canine saphenous vein and on prejunctional α_2-adrenoceptors in guinea pig atrium (Hieble *et al.*, 1991).

In other α_2-adrenoceptor models SK&F 104078 and SK&F 104856 do not mimic the effects of nonselective α_2-adrenoceptor antagonists such as rauwolscine or SK&F 86466. These include suppression of glucose-induced hyperglycemia in the conscious rat (Hieble *et al.*, 1992) or potentiation of the lipolytic action of epinephrine in the anesthetized rat (Hieble and Kolpak, unpublished data). It is likely that the α_2-adrenoceptors involved in these responses are located at postjunctional sites, hence, while these novel antagonists are capable of functionally differentiating α_2-adrenoceptor mediated responses, they do not appear to have "anatomic" selectivity for postjunctional α_2-adrenoceptors.

METHODS

Cloning of human α-adrenoceptors

$\alpha_{1A/D}$-adrenoceptor The complete coding region of the human $\alpha_{1A/D}$-adrenoceptor was cloned from two overlapping clones: A 1.1 kilobase (kb) clone encompassing approxi-

mately 0.3 kb of 5′ untranslated region (UTR) and 0.8 kb of coding sequence was isolated from prostate cDNA library using a rat $\alpha_{1A/D}$-adrenoceptor cDNA as a hybridization probe. A second 0.9 kb clone, encompassing the 3′ coding portion of the human $\alpha_{1A/D}$-adrenoceptor (not including any 3′ UTR sequences) was isolated from a hippocampus cDNA library using a mouse serotonin 5HT5A cDNA as a hybridization probe. This cDNA was spliced to the 1.1 kb clone at a common XhoI restriction site to form a 2.0 kb cDNA encompassing 0.3 kb of 5′ UTR, and 1.7 kb of the complete coding region (576 amino acids) of the human $\alpha_{1A/D}$-adrenoceptor. This cDNA clone was introduced into the Asp718/HindIII restriction sites of the vector pCDN for expression in mammalian cells.

α_{1B}-adrenoceptor A 2.3 kB cDNA clone encoding the human α_{1B}-adrenoceptor was isolated from a human hippocampus cDNA library using a rat α_{1B}-adrenoceptor cDNA as a hybridization probe. This clone contained 0.2 kb and 0.5 kb of 5′- and 3′-untranslated sequences, respectively, and 1.6 kb of coding sequence (encoding 517 amino acids). This cDNA clone was introduced into the Eco RI restriction site of the vector pCDN for expression in mammalian cells.

α_{1C}-adrenoceptor Two cDNA clones encoding portions of the human α_{1C}-adrenoceptor were isolated from a human hippocampus cDNA library using a bovine α_{1C}-adrenoceptor cDNA as a hybridization probe. One clone contained a region of DNA which encompassed only the first 80 N-terminal amino acids of the α_{1C}-adrenoceptor. The second cDNA clone (2.1 kb in length) encoded approximately 370 amino acids of the α_{1C}-adrenoceptor but lacked the region encoding approximately 100 amino acids of the N-terminus of the protein. A 375 base pair cDNA fragment encoding the missing 5′ coding region was derived from human brain mRNA by the polymerase chain reaction (PCR) utilizing primers corresponding to nucleotide sequences at the 5′ end, and 375 base pairs within, the coding region of the cDNA. This PCR DNA fragment was ligated to a 1.4 kb fragment (derived from the 2.1 kb cDNA by PCR) encompassing the remaining portion of the α_{1C}-adrenoceptor cDNA to form a complete coding region (466 amino acids). This cDNA fragment was subcloned into the Asp 718/Bam HI restriction sites of the vector pCDN for expression in mammalian cells.

α_{2A}- and α_{2C}-adrenoceptors The human α_{2A}- and α_{2C}-adrenoceptors were isolated from a hippocampus cDNA library using as hybridization probes portion of serotonin receptor cDNAs. cDNA fragments of 1.4 kb and 1.8 kb encompassing the entire coding region (450 or 461 amino acids, respectively) of the α_{2A}- and α_{2C}-adrenoceptor cDNA were enzymatically excised from the original cDNA clones and subcloned into the cloning cassette of the vector pCDN for mammalian cell expression.

α_{2B}-adrenoceptor A genomic clone encoding the complete α_{2B}-adrenoceptor was isolated form a placental lambda library using portions of serotonin receptors as hybridization probes. Synthetic oligonucleotide primers were made corresponding to the amino and carboxy termini and used to obtain a 1.4 kb DNA fragment encoding the entire α_{2B}-adrenoceptor (450 amino acids) utilizing the polymerase chain reaction. This cDNA fragment was subcloned into the vector pCDN for expression in mammalian cells.

Establishment of Stable CHO Clones Expressing α-Adrenoceptors

Chinese hamster ovary cells, grown in suspension using MR1MOD3 medium (in house formulation), were transfected by electroporation with DNA containing the genes encoding $\alpha_{1A/D}$, α_{1B}-, α_{1C}-, α_{2A}-, α_{2B}-, and α_{2C}-adrenoceptors. Cells were expanded and tested for the presence of receptor expression by binding to [^{3}H] prazosin (α_1-adrenoceptors) or [^{3}H] rauwolscine (α_2-adrenoceptors).

Membrane Preparation

Previously frozen cell pellets were defrosted and homogenized in a Brinkman Polytron with cold TME buffer (10 mM Tris, 2 mM EDTA, pH 7.4). The homogenate was centrifuged at 500 × g for 10 minutes at 4 °C. The supernatant was collected and centrifuged at 100,000 × g for 30 minutes at 4 °C. The membrane pellet was resuspended in 50 mM Tris buffer, pH 7.4.

Inhibition studies

Membranes were incubated with test antagonists at concentrations ranging from 0.1 nM to 100 μM. Assays were initiated by the addition of membrane protein, and incubated at 25 °C for 30 ([^{3}H] rauwolscine) or 45 ([^{3}H] prazosin) minutes. 50 mM Tris-EDTA buffer, pH 7.4, was used for [^{3}H] rauwolscine and 50 mM Tris-HCl, pH 7.4, was used for [^{3}H] prazosin. Radioligand was present at a concentration near its K_d. Non-specific binding was defined using 10 μM phentolamine. K_i values were calculated using the London software program for nonlinear regression.

Functional experiments

The procedures for determination of postjunctional α-adrenoceptor antagonist activity in rabbit aorta, canine saphenous vein, guinea pig atrium and rat vas deferens have been previously reported (Hieble *et al.*, 1988). α_1-Adrenoceptors were activated by norepinephrine in rabbit aorta, and α_2-Adrenoceptors were activated by B-HT 920 in the canine saphenous vein and guinea pig atrium, and by UK 14,304 in the rat vas deferens.

RESULTS

Table 1 compares the potency of SK&F 86466, and several of its structural analogs in functional assays for α-adrenoceptor antagonist activity. In contrast to SK&F 86466, SK&F 104078, SK&F 104856 and SK&F 105854 show preferential affinity for vascular α_2-adrenoceptors of canine saphenous vein, compared to prejunctional receptors on guinea pig atrium or rat vas deferens. A similar profile is observed for the structurally unrelated compound, naftopidil, and, based on literature data, Abbott 65265. SK&F 104078, SK&F 104856 and naftopidil all have α_1-adrenoceptor antagonist potency comparable to that observed at vascular α_2-adrenoceptors.

Table 2 compares the affinity of SK&F 86466, its functionally selective analogs, and naftopidil, for recombinant human α_1- and α_2-adrenoceptors, stably expressed in CHO cells.

DISCUSSION

From the data in Table 1, it is clear that functional selectivity can be achieved between α_2-adrenoceptors of canine saphenous vein and those of guinea pig atrium or rat vas deferens. As observed for SK&F 86466, most other α_2-adrenoceptor antagonists are essentially

TABLE 1
α-Adrenoceptor antagonist activity of novel antagonists, determined in functional in vitro assays.

Compound	*DSV*[a]	*GPA*[b]	*RVD*[c]	*RA*[d]
SK&F 86466	42	17	11	600
SK&F 104078	76	> 3000	1400	151
SK&F 104856	29	> 3000	1200	25
SK&F 105854	40	> 3000	4000	1700
NAFTOPIDIL	335[e]	> 3000	> 3000	30[f]
ABBOTT 65265	2.7[g]	ND	74[g]	87[g]

[a]K_B (nM) for blockade of B-HT 920 induced contraction in isolated dog saphenous vein.
[b]K_B (nM) for blockade of the neuroinhibitory action of B-HT 920 in isolated superfused guinea pig atrium.
[c]K_B (nM) for blockade of the neuroinhibitory action of UK 14,304 in isolated rat vas deferens.
[d]K_B (nM) for blockade of norepinephrine induced contraction in isolated rabbit aorta.
[e]Mean K_B value of that obtained in this study and that reported by Muramatsu *et al.*, 1991.
[f]Data from Muramatsu *et al.*, 1991.
[g]Data from Buckner *et al.*, 1989.

equipotent between saphenous vein and atrium, with potency ratios ≤ 5 (Hieble *et al.*, 1986).

This functional selectivity cannot be attributed to selective affinity for one of the α_2-adrenoceptor subtypes identified via molecular biology or binding affinity correlation. None of the antagonists shows pharmacologically significant selectivity between these subtypes (Table 2). While SK&F 104078 shows somewhat lower affinity for the α_{2D} adrenoceptor, the magnitude of the difference from the other α_2-adrenoceptor subtypes (<3 fold) is not consistent with that of the functional selectivity (>20 fold). Furthermore, α_{2D} adrenoceptors are presumably not present in human tissues, but SK&F 104078 shows functional α_2-adrenoceptor selectivity between vascular and neuronal α_2-adrenoceptors in human saphenous vein (Hieble, unpublished data; Hieble *et al.*, 1991).

Interestingly, the functionally selective antagonists derived from SK&F 86466 both show selectivity for the recombinant $\alpha_{1A/D}$-adrenoceptor versus either the α_{1B}- or α_{1C}-subtypes

TABLE 2
Affinity of Novel Antagonists for Recombinant α-Adrenoceptors

Compound	$\alpha_{1A/D}$[a]	α_{1B}[b]	α_{1C}[c]	α_{2A}[d]	α_{2B}[e]	α_{2C}[f]	α_{2D}[g]
SK&F 86466	126	485	449	9.4	16	20	5.2
SK&F 104078	59	87	33	114[h]	142[h]	64[h]	290
SK&F 104856	1.6	23	36	24	3.4	21	80
SK&F 105854	72	783	3300	54	14	28	ND
NAFTOPIDIL	5.7	26	5.9	43	63	37	ND

[a]K_i (nM) for inhibition of [^{3}H] prazosin binding to CHO cells expressing the human $\alpha_{1A/D}$-adrenoceptor.
[b]K_i (nM) for inhibition of [^{3}H] prazosin binding to CHO cells expressing the human α_{1B}-adrenoceptor.
[c]K_i (nM) for inhibition of [^{3}H] prazosin binding to CHO cells expressing the human α_{1C}-adrenoceptor.
[d]K_i (nM) for inhibition of [^{3}H] rauwolscine binding to CHO cells expressing the human α_{2A} adrenoceptor.
[e]K_i (nM) for inhibition of [^{3}H] rauwolscine binding to CHO cells expressing the human α_{2B} adrenoceptor.
[f]K_i (nM) for inhibition of [^{3}H] rauwolscine binding to CHO cells expressing the human α_{2C} adrenoceptor.
[g]K_i (nM) for inhibition of [^{3}H] rauwolscine binding to expressed recombinant α_{2D} adrenoceptor (rat RG20 clone) or to tissues containing pure populations of this receptor (rat sublingual gland or bovine pineal gland). Data from Gleason and Hieble (1992); Harrison *et al.*, (1991); Michel *et al.*, (1989); Simonneaux *et al.*, (1991), Lanier *et al.*, (1991) and O' Rourke *et al.*, (1994).
[h]Mean values of data obtained in human or rat (RNG, RG10 clones) receptors, as reported by O' Rourke *et al.*, (1994); Lanier *et al.*, (1991); Uhlen *et al.*, (1992); Marjamaki *et al.*, (1993); Lomasney *et al.*, (1990); Xia *et al.*, (1993) and Harrison *et al.*, (1991).

(Table 2). This subtype selectivity is of similar magnitude in SK&F 104856 and SK&F 105854, even though SK&F 104856 has substantially greater affinity for all of the α_1-adrenoceptor subtypes, consistent with its greater potency as a functional α_1-adrenoceptor antagonist (Table 1). The similar potency of these two analogs in α_2-adrenoceptor assays (Table 1), would suggest that α_1-adrenoceptor blockade does not contribute to the functional selectivity observed.

These data show that α_2-adrenoceptor agonists are able to produce some of their actions via activation of a receptor sensitive to most α_2-adrenoceptor antagonists, including SK&F 86466 and rauwolscine, but insensitive to the other antagonists shown in Figure 1. This receptor does not correspond to any of the known recombinant α_2-adrenoceptors, suggesting that additional α_2-adrenoceptor subtypes remain to be identified. This is perhaps analogous to the prazosin-insensitive α_1-adrenoceptors, which can be demonstrated in functional assays (Muramatsu *et al.*, 1990), but have not yet been identified via molecular biological techniques.

References

Akers, I., Coates, J., Drew, G.M. and Sullivan, A.T. (1991) α_2-Adrenoceptor blocking profile of SK&F 104078: further evidence for receptor subtypes. *Br. J. Pharmacol.*, **102,** 943–949.

Buckner, S.A., Oheim, K.W., Morse, P.A., Cain, J.C., Hancock, A.A. and DeBernardis, J.F. (1989) In vitro effects of Abbott-65265, a selective postsynaptic alpha-2 receptor antagonist. *FASEB J.*, **3,** A437.

Bylund, D.B. (1988) Subtypes of α_2-adrenoceptors: Pharmacological and molecular biological evidence converge. *Trends in Pharmacol. Sci.*, **9,** 356–361.

Bylund, D.B., Ray-Prenger, C. and Murphy, T.J. (1988) Alpha-2A and alpha-2B adrenergic receptor subtypes: antagonist binding in tissues and cell lines containing only one receptor subtype. *J. Pharmacol. Exp. Ther.*, **245,** 600–607.

Chruscinski, A.J., Link, R.E., Daunt, D.A., Barsh, G.S. and Kobilka, B.K. (1992) Cloning and expression of the mouse homolog of the human α_2-C2 adrenergic receptor. *Biochem. Biophys. Res. Comm.*, **15,** 49–57.

Connaughton, S. and Docherty, J.R. (1989) No evidence for differences between pre- and postjunctional α_2-adrenoceptors in the periphery. *Br. J. Pharmacol.*, **99,** 97–102.

Gleason, M.M. and Hieble, J.P. (1992) The α_2-adrenoceptors of the human retinoblastoma cell line (Y79) may represent an additional example of the α_{2C}-adrenoceptor. *Br. J. Pharmacol.*, **107,** 222–225.

Harrison, J.K., D'Angelo, D.D., Zeng, D. and Lynch, K.R. (1991) Pharmacological characterization of rat α_2-adrenergic receptors. *Mol. Pharmacol.*, **40,** 407–412.

Hieble, J.P., Sulpizio, A.C., Edwards, R., Chapman, H., Young, P., Roberts, S.P., Blackburn, T.P., Wood, M.D., Shah, D.H., DeMarinis, R.M. and Ruffolo, R.R., Jr. (1991) Additional evidence for functional subclassification of alpha-2 adrenoceptors based on a new selective antagonist, SK&F 104856. *J. Pharmacol. Exp. Ther.*, **259,** 643–652.

Hieble, J.P., Sulpizio, A.C., Nichols, A.J., DeMarinis, R.M., Pfeiffer, F.R., Lavanchy, P.G. and Ruffolo, R.R., Jr. (1986) Pharmacological differentiation of pre- and postjunctional alpha-2 adrenoceptors. *J. Hypertens.*, **4 (Suppl 6),** S189–S192.

Hieble, J.P., Sulpizio, A.C., Nichols, A.J., Willette, R. and Ruffolo, R.R., Jr. (1988) Pharmacologic characterization of SK&F 104078, a novel alpha-2 adrenoceptor antagonists which discriminates between pre- and postjunctional alpha-2 adrenoceptors. *J. Pharmacol. Exp. Ther.*, **247,** 645–652.

Hieble, J.P., Sulpizio, A.C. and Goldstein, R.S. (1992) Determination of the α-adrenoceptor subtype involved in modulation of the response to glucose challenge in the rat. *Pharmacol. Commun.*, **1,** 59–69.

Lanier, S.M., Downing, S., Duzic, E. and Homcy, C.J. (1991) Isolation of rat genomic clones encoding subtypes of the α_2-adrenergic receptor. *J. Biol. Chem.*, **266,** 10470–10478.

Link, R., Daunt, D., Barsh, G., Chruscinski, A. and Kobilka, B. (1992) Cloning of two mouse genes encoding α_2-adrenergic receptor subtypes and identification of a single amino acid in the mouse α_2-C10 homolog responsible for an interspecies variation in antagonist binding. *Mol. Pharmacol.*, **42,** 16–27.

Lomasney, J.W., Cotecchia, S., Lefkowitz, R.J. and Caron, M.G. (1991) Molecular biology of α-adrenergic receptors: Implications for receptor classification and for structure-function relationships. *Biochim. et Biophys. Acta*, **1095,** 127–139.

Lomasney, J.W., Lorenz, W., Allen, L.F., King, K., Regan, J.W., Yang-Feng, T.L., Caron, M.G. and Lefkowitz, R.J. (1990) Expansion of the α_2-adrenergic receptor family: cloning and characterization of a human α_2-

adrenergic receptor subtype, the gene for which is located on chromosome 2. *Proc. Natl. Acad. Sci.*, **87,** 5094–5098.

Marjamaki, A., Luomala, K., Ala-Uotila, S. and Scheinin, M. (1993) Use of recombinant human α_2-adrenoceptors to characterize subtype selectivity of antagonist binding. *Eur. J. Pharmacol. (Mol. Pharmacol.)*, **246,** 219–226.

Michel, A.D., Loury, D.N. and Whiting, R.L. (1989) Assessment of imiloxan as a selective α_{2B}-adrenoceptor antagonist. *Br. J. Pharmacol.*, **99,** 560–564.

Muramatsu, I., Ohmura, T., Kigoshi, S., Hashimoto, S. and Oshita, M. (1990) Pharmacological subclassification of α_1-adrenoceptors in vascular smooth muscle. *Br. J. Pharmacol.*, **99,** 197–201.

Muramatsu, I., Yamanaka, K. and Kigoshi, S. (1991) Pharmacological profile of the novel α-adrenoceptor antagonist KT-611 (Naftopidil). *Jpn. J. Pharmacol.*, **55,** 391–398.

Murphy, T.J. and Bylund, D.B. (1988) Characterization of alpha-2 adrenergic receptors in the OK cell, an opossum kidney cell line. *J. Pharmacol. Exp. Ther.*, **244,** 571–578.

O'Rourke, M.F., Blaxalll, H.S., Iversen, L.J. and Bylund, D.B. (1994) Characterization of [^{3}H]RX821002 binding to alpha-2 adrenergic receptor subtypes. *J. Pharmacol. Exp. Ther.*, **268,** 1362–1367.

Oriowo, M.A., Hieble, J.P. and Ruffolo, R.R., Jr. (1991) Evidence for heterogeneity of prejunctional alpha-2 adrenoceptors. *Pharmacology*, **43,** 1–13.

Simonneaux, V., Ebadi, M. and Bylund, D.B. (1991) Identification and characterization of α_2D-adrenergic receptors in bovine pineal gland. *Mol. Pharmacol.*, **40,** 235–241.

Uhlen, S., Xia, Y., Chhajlani, V., Felder, C.C. and Wikberg, J.E.S. (1992) [^{3}H]-MK912 binding delinates two α_2-adrenoceptor subtypes in rat CNS one of which is identical with the cloned pA2d α_2-adrenoceptor. *Br. J. Pharmacol.*, **106,** 986–995.

Xia, Y., Uhlen, S., Chhajlani, V., Lien, E.J. and Wikberg, J.E.S. (1993) Further evidence for the existence of two forms of α_{2B}-adrenoceptors in rat. *Pharmacol. Toxicol.*, **72,** 40–49.

Zeng, D., Harrison, J.K., D'Angelo, D.D., Barber, C.M., Tucker, A.L., Lu, Z. and Lynch, K.R. (1990) Molecular characterization of a rat α_{2B}-adrenergic receptor. *Proc. Natl. Acad. Sci.*, **87,** 3102–3106.

PRESYNAPTIC α_2-ADRENOCEPTORS: SUBTYPE DETERMINATION

K. STARKE, A. U. TRENDELENBURG and N. LIMBERGER

Pharmakologisches Institut, Hermann-Herder-Strasse 5, D-79104 Freiburg i.Br., Germany

A summary is presented of subtype determinations of nine presynaptic α_2-adrenoceptors in three species: *rat* α_2-autoreceptors in submaxillary gland, atria and brain cortex as well as α_2-receptors inhibiting release of serotonin (5-HT) in brain cortex; *rabbit* α_2-autoreceptors in kidney and brain cortex as well as α_2-receptors inhibiting release of 5-HT in brain cortex and of dopamine in caudate nucleus; and *guinea-pig* α_2-autoreceptors in brain cortex. The receptors were at least predominantly α_{2D} in rat and guinea-pig but were α_{2A} in the rabbit. It is suggested that mammalian presynaptic α_2-adrenoceptors mainly belong to the $\alpha_{2A/D}$ group.

KEY WORDS: α_2-adrenoceptors, presynaptic receptors, autoreceptors, species differences

INTRODUCTION

The discovery of presynaptic α-adrenoceptors at noradrenergic axons led to the detection of two pharmacologic classes of α-adrenoceptors. α-Adrenoceptor agonists acted with greatly different relative potencies at presynaptic as compared to postsynaptic α-adrenoceptors (Starke, 1972; Starke *et al.*, 1974). The same was found for antagonists (Dubocovich and Langer, 1974), and 20 years ago the distinction between α_1 and α_2 was proposed (Langer, 1974).

Studies on presynaptic α_2-autoreceptors, as they came to be called, also were the first to show non-homogeneity of the α_2 class. Doxey and Everitt (1977) concluded that there were "differences between the presynaptic α-adrenoceptors of different sympathetic nerve endings". A species difference between rat and rabbit α_2-autoreceptors was a consistent finding (Starke, 1981; Lattimer and Rhodes, 1985): phentolamine and idazoxan were less potent autoreceptor antagonists than yohimbine and rauwolscine in rabbits, but were at least as potent as yohimbine and rauwolscine in rats.

Notwithstanding the pioneer role of the autoreceptors, the present α_2 subclassification is based on radioligand and genetic data (see Bylund, 1992). These data indicate that there are at least three α_2-adrenoceptor genes per species and that, across the species, the three genes form three groups of homologues. One gene codes for the α_{2B}-adrenoceptor in each species examined; across species, α_{2B}-receptors are are at least 82% identical in amino acid sequence and so similar pharmacologically that species differences are not detected with

Correspondence to: Dr. Klaus Starke, Pharmakologisches Institut, Hermann-Herder-Strasse 5, D-79104 Freiburg i.Br., Federal Republic of Germany.

the drugs available; hence the common designation α_{2B}. The second gene codes for the α_{2C}-adrenoceptor; α_{2C}-receptors again are at least 88% identical in amino acid sequence and pharmacologically similar across species: hence the common designation α_{2C}. The third gene gives rise to a protein that is at least 89% sequence identical across species; here, however, the structural difference engenders a pharmacologic one: phentolamine and idazoxan are less potent than yohimbine and rauwolscine at the receptor encoded by the human gene but are at least as potent as yohimbine and rauwolscine at the homologous rat and mouse receptor (Lanier *et al.*, 1991; Link *et al.*, 1992); the human type will be called α_{2A} and the mouse and rat type α_{2D} in this paper following the original nomenclature (Simonneaux *et al.*, 1991).

To which subtype or subtypes do those receptors belong that opened up the whole field—presynaptic α_2-autoreceptors? The question seems all the more relevant since little is known about the *functions* the various subtypes serve (Shen *et al.*, 1990; Bylund, 1992). Early suggestions such as that the autoreceptors were α_{2B} were based on too limited data and stood in contrast to much of the literature (see Gobbi *et al.*, 1990; Limberger *et al.*, 1991; Schwartz and Malik, 1992). We summarize here some of our subtype determinations of α_2-autoreceptors and other presynaptic α_2-receptors.

METHODOLOGIC CONSIDERATIONS

Specific problems are encountered when presynaptic α_2-adrenoceptors are characterized. These receptors change stimulus-evoked transmitter release. Release of noradrenaline, however, brings the transmitter into the vicinity of the receptors. When exogenous ligands then are administered, they compete with released noradrenaline, and this distorts agonist EC_{50} and antagonist pK_B values (Starke *et al.*, 1974; Limberger *et al.*, 1989). One way out is to use such stimulation conditions that any effect of released noradrenaline is negligible. Brief trains of high-frequency pulses are one possibility (Singer, 1988).

Terminal axons may possess α_1- in addition to α_2-adrenoceptors. The existence of presynaptic α_1-autoceptors is disputed. The noradrenaline release-inhibiting effect of α_1-adrenoceptor agonists may be mediated by prostaglandins and adenosine coming from postsynaptic cells (Bohmann *et al.*, 1993a). Still, the necessity remains to rule out drug effects at α_1-adrenoceptors, especially in tissues where they have been suggested to occur presynaptically such as rat heart and in the case of drugs with high α_1 affinity such as prazosin. Some α_2 ligands may also act at imidazoline receptors to inhibit the release of noradrenaline (Göthert and Molderings, 1991).

Drugs may change the release of neurotransmitter at sites other than membrane receptors for endogenous chemical signals. One mechanism is interference with the noradrenaline storage vesicles, leading to an increase in the efflux of ^{3}H-3,4-dihydroxyphenylglycol in experiments with ^{3}H-noradrenaline. Many α-adrenoceptor antagonists produce this effect but prazosin does so at a concentration as low as 0.1 μM. Not only may calculations of the evoked overflow of tritiated compounds be invalidated, but the reserpine component may *per se* enhance stimulus-evoked noradrenaline release.

Correlation of antagonist affinities to an unclassified α_2-receptor with affinities to known $\alpha_{2A\text{-}D}$ prototypes is important for subtype determination. For this method to yield clear results, antagonists with highly different affinities to α_{2A}, α_{2B}, α_{2C} and α_{2D} should be included: otherwise a good correlation with any of the four subtypes is obtained.

Superfused tissue slices were used in the experiments presented here. Transmitter release was elicited by electric pulses. Noradrenaline and serotonin (5-HT) release was assessed as overflow of tritiated compounds after preincubation with ^{3}H-noradrenaline and ^{3}H-5-HT, respectively. Dopamine release was assessed by fast cyclic voltammetry. Presynaptic α_2-receptors including autoreceptors were kept free from activation by released noradrenaline as far as possible by using brief trains of pulses. Concentration-response curves for the α_2-selective imidazoline-derived agonist UK 14304 were determined in the absence and in the presence of antagonists (see original publications for chemical names of code number compounds). EC_{50} values and maximal effects of UK 14304 were calculated by logistic curve fitting, and pK_B values of antagonists were obtained from the rightward shift of the UK 14304 concentration-response curve. Drug effects at sites other than presynaptic α_2-adrenoceptors were ruled out where advisable. Antagonists included those that are able to differentiate between α_2 subtypes such as prazosin, ARC 239, phentolamine and idazoxan.

RESULTS

Nine presynaptic α_2-adrenoceptors from three species were examined: autoreceptors in *rat* submaxillary gland, atria and brain cortex as well as the receptors modulating the release of 5-HT in brain cortex; autoreceptors in *rabbit* kidney and brain cortex as well as the receptors modulating 5-HT release in brain cortex and dopamine release in caudate nucleus; and autoreceptors in *guinea-pig* brain cortex. The rat submaxillary gland part will be described in some detail (Limberger *et al.*, 1992).

Slices of the gland were stimulated 4 times at intervals of 40 min (S_1 to S_4). Each stimulation period consisted of 10 trains of 6 pulses/100 Hz. Re-uptake of released ^{3}H-noradrenaline was blocked by desipramine. These conditions ensured sizeable peaks of tritium overflow with little autoinhibition: the overflow peaks were only slightly higher in the presence of phentolamine (S_1 in lower panel of Figure 1) than in its absence (S_1 in upper panel of Figure 1). In all other tissues, autoinhibition was even weaker or absent. Increasing concentrations of UK 14304 progressively reduced the evoked overflow of tritium. Higher concentrations of UK 14304 were necessary to produce inhibition in the presence of phentolamine. The basal outflow of tritium was not changed (Figure 1). In the absence of phentolamine, the EC_{50} of UK 14304 was 6.2 nM and the maximal inhibition 92%. Phentolamine shifted the curve to the right. The pK_B value was 8.3.

Analogous experiments were carried out with rauwolscine, BRL 44408, BRL 41992, WB 4101, imiloxan, SKF 104078, prazosin and corynanthine. All increased the evoked overflow of tritium (S_1), thus confirming some autoinhibition. Prazosin 1 μM greatly accelerated basal tritium efflux: the reserpine-like effect mentioned. Phentolamine was the most potent antagonist. The potency of the other antagonists decreased in the above order.

Experiments were carried out to rule out non-α_2 actions. The α_1-selective agonist methoxamine reduced the evoked overflow of tritium, but only at high concentrations, EC_{50} 21 μM. The pK_B values of rauwolscine (α_2-selective), imiloxan (α_2-selective) and prazosin (α_1-selective) against methoxamine were close to those against UK 14304, indicating absence of release-inhibiting α_1-adrenoceptors and an effect of methoxamine at α_2-autoreceptors.

In some experiments slices of the submaxillary gland were stimulated 4 times by trains of 60 pulses/1 Hz, conditions under which marked autoinhibition developed. Concentration-

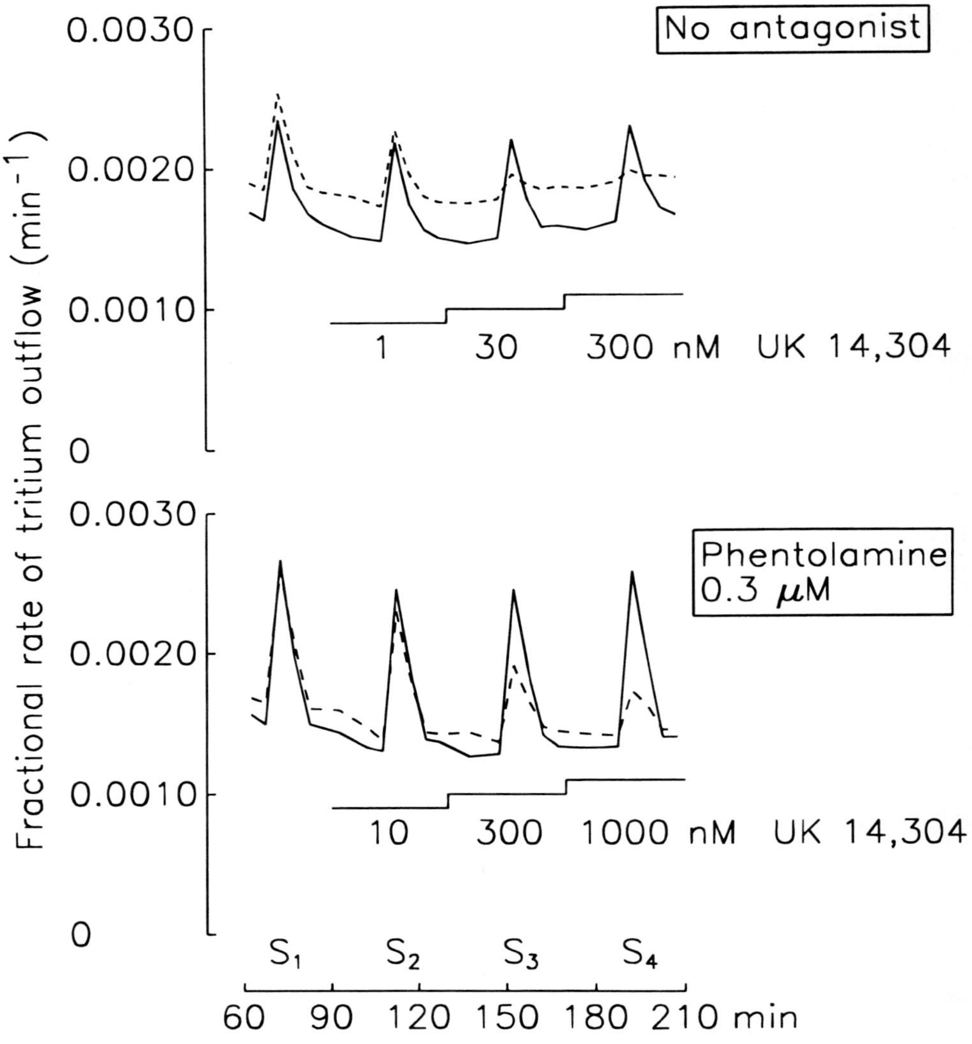

FIGURE 1 Tritium efflux from slices of rat submaxillary gland preincubated with ^{3}H-noradrenaline. Slices were superfused with medium containing desipramine 1 μM. In experiments of lower panel, phentolamine 0.3 μM was also present throughout superfusion. S_1 to S_4 each consisted of 10 trains of 6 pulses/100 Hz, train interval 30 s. Solid lines, experiments without UK 14304. Broken lines, experiments in which UK 14304 was added as indicated. Abscissae, minutes of superfusion. Means from 2 slices per group. Data from Limberger *et al.* (1992).

response curves for the release-enhancing effect of the 9 α antagonists were then determined and concentrations causing a 30% increase calculated (pEC_{30}). The pEC_{30} values correlated well with the pK_B values. The enhancement of release is due to antagonism against released noradrenaline. The correlation, hence, indicates that the antagonists acted in the same way against the imidazoline agonist UK 14304 and the endogenous non-imidazoline noradrenaline: UK 14304 did not act at an imidazoline receptor.

With this proof of an α_2 mode of action, the correlation of the autoreceptor pK_B values of the antagonists with pK_B values at prototypic α_{2A-D} radioligand binding sites was then calculated (Figure 2). The binding sites were membranes from the human HT29 cell line for α_{2A}, from neonatal rat lung for α_{2B}, from the opossum kidney OK cell line for α_{2C} (Blaxall *et al.*, 1991), and from rat submaxillary gland for α_{2D} (Michel *et al.*, 1989). The pK_B values at the submaxillary α_2-autoreceptors correlated only, and excellently, with pK_B values at the submaxillary α_2 binding sites, correlation coefficient 0.98 and slope of the regression line 0.91, indicating that the autoreceptors were not α_{2A-C} but α_{2D}.

Correlation data for the nine α_2-adrenoceptors examined are summarized in Table 1. The result is one of remarkable harmony. Three of the four *rat* receptors correlated only or best

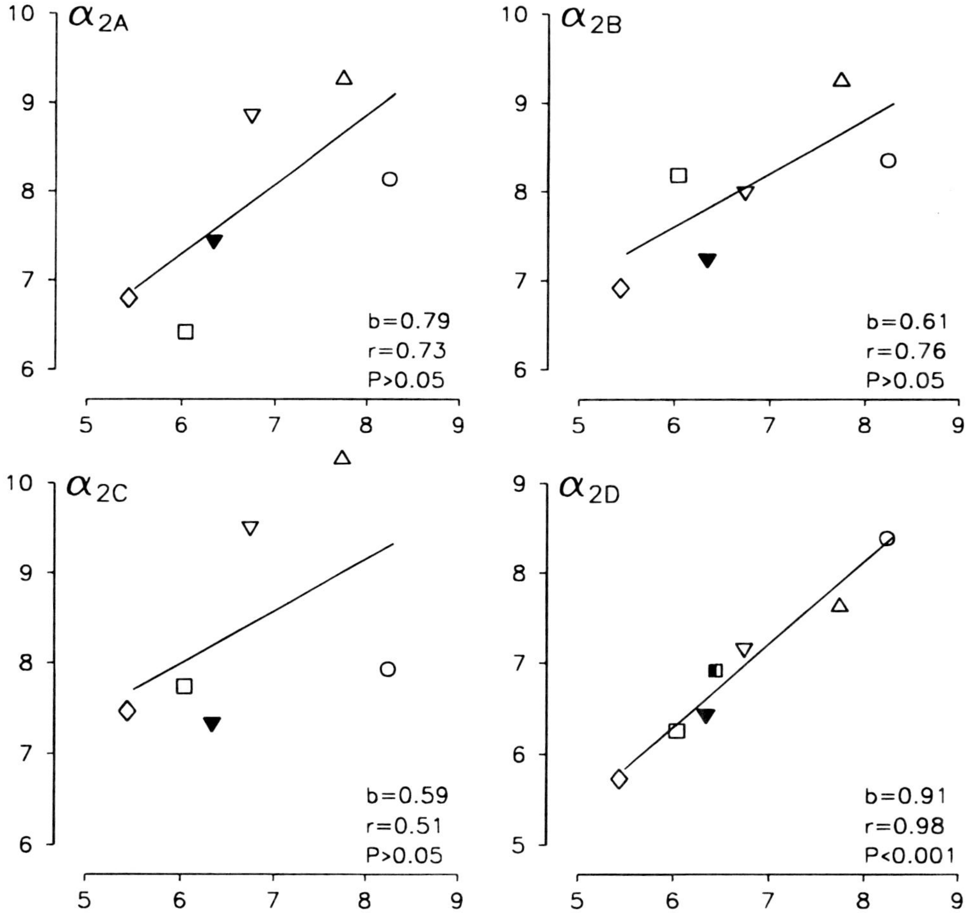

FIGURE 2 Correlation between qatnagonist affinities for the presynaptic α_2-autoreceptors of rat submaxillary gland and affinities for prototypic α_2 binding sites. Abscissae, autoreceptor pK_B values (present study). Ordinates: pK_B values at α_2 binding sites in **A** HT29 cells, **B** neonatal rat lung, **C** OK cells (Blaxall *et al.*, 1991) and D rat submaxillary gland (Michel *et al.*, 1989). Symbols: ❍ phentolamine; Δ rauwolscine; ∇ WB 4101; ■❑ imiloxan; ▼ SKF 104978; ❑ prazosin; ◊ corynanthine. Data from Limberger *et al.*, (1992).

TABLE 1

Correlation between antagonist affinities for presynaptic α_2-adrenoceptors and affinities for prototypic α_{2A}, α_{2B}, α_{2C} and α_{2D} binding sites

Binding sites	*Rat*				*Rabbit*				*Guinea pig*
	Submax. gland NA axons	*Atrium NA axons*	*Brain NA axons*	*Brain 5-HT axons*	*Kidney NA axons*	*Brain NA axons*	*Brain 5-HT axons*	*Brain DA axons*	*Brain NA axons*
					Correlation coefficient				
HT29 cells (α_{2A})	0.73	0.87*	0.72*	0.60	0.92**	0.95***	0.95**	0.97**	0.82*
Neonatal rat lung (α_{2B})	0.74	0.93**	0.63	0.74	0.55	0.33	0.92**	0.49	0.45
OK cells (α_{2C})	0.50	0.85*	0.45	0.35	0.81*	0.67*	0.79*	0.20	0.58
Rat submax. gland (α_{2D})	0.98***	0.79*	0.98***	0.97**	0.45	0.69	0.66	0.84*	0.96***
					Slope of regression line				
HT29 cells (α_{2A})	0.79	1.23	0.76	0.52	1.28	1.02	1.00	1.01	0.80
Neonatal rat lung (α_{2B})	0.60	0.98	0.50	0.56	0.59	0.26	0.85	0.37	0.32
OK cells (α_{2C})	0.58	0.98	0.59	0.44	1.47	0.82	1.23	0.14	0.53
Rat submax. gland (α_{2D})	0.92	0.91	0.90	0.87	0.54	0.71	0.72	0.74	0.83

The nine presynaptic α_2-adrenoceptors were autoreceptors (at noradrenaline = NA axons) or receptors at 5-HT or dopamine (DA) axons. Shown are correlation coefficients and slopes of the regression of "pK_B at α_2 binding site" on "pK_B at presynaptic receptor". Presynaptic receptor values from present study. pK_B values at α_2 binding sites from Blaxall *et al.* (1991) and Michel *et al.* (1989). Significant correlations: * $P < 0.05$; ** $P < 0.01$; *** $P < 0.001$.

with α_{2D}. The exception was atria where the receptors correlated best with α_{2B}. We therefore correlated the atrial pK_B values with pK_B values from the bovine pineal gland where the α_{2D} subtype was first described (Simonneaux *et al.*, 1991). The correlation coefficient was 0.95. Affinity *ratios* also indicated similarity to α_{2D}. We concluded that the atrial autoreceptors resemble, but are not identical with, α_{2D}; a mixed autoreceptor population also was not ruled out (Limberger *et al.*, 1992). All four *rabbit* receptors correlated best with α_{2A}. Finally, the autoreceptors in *guinea-pig* brain correlated best with α_{2D}.

DISCUSSION

Attempts to identify presynaptic α_2-adrenoceptors in terms of α_{2A-D} followed the growth of the classification system. Investigations of up to about 1991, for example, considered only the α_{2A-C} subtypes then known. It should also be noted that the α_{2D} designation for the rat and mouse homologue of the human α_{2A}-adrenoceptor is not used by all authors. However, despite these difficulties the "harmony" is confirmed by much of the available literature.

In the *rat*, Smith and Docherty (1992) studied submaxillary gland and vas deferens α_2-autoreceptors: both resembled most the α_2 binding sites in rat submaxillary gland (Michel *et al.*, 1989) and hence, in our nomenclature, are α_{2D}. The same holds true for the α_2-autoreceptors in rat kidney (Schwartz and Malik, 1992; Bohmann *et al.*, 1993b). The non-B-non-C character of the receptors at rat brain noradrenergic and 5-HT axons also agrees with recent papers (Gobbi *et al.*, 1990; Maura *et al.*, 1992). Even pre-subclassification studies retrospecticely support the "D" diagnosis. For example, prazosin has long been known to be very weak, and phentolamine to be more potent than yohimbine, at rat brain cortex α_2-autoreceptors and the receptors inhibiting 5-HT release (Taube *et al.*, 1977; Frankhuyzen and Mulder, 1982). Antagonist pK_B values at the α_2-autoreceptors of rat vas deferens were measured by Doxey and colleagues in 1983 and 1984. We correlated their pK_B values for idazoxan, phentolamine, rauwolscine, WB 4101, prazosin and corynanthine with the binding pK_Bs used in our Table 1: the correlation coefficient was 0.75 for α_{2A}, 0.61 for α_{2B} and 0.51 for α_{2C} but 0.99 for α_{2D}. Only the latter correlation was significant ($P < 0.001$). The receptor obviously is α_{2D}. The perfect agreement with later, post-subclassification results, we are tempted to say, is a token of the quality of (some) pharmacologic work. Smith and Docherty (1994) maintain that the autoreceptor in rat atria is α_{2B}. We agree that it differs from other rat α_2-autoreceptors (see above). Whether the difference is in fact due to expression of the B subtype in cardiac sympathetic neurones, or to other factors, remains to be seen (see Vidovic *et al.*, 1994, for a mRNA study arguing against relevant levels of α_{2B}).

In the *rabbit*, pulmonary artery α_2-autoreceptors were classified as α_{2A} (Molderings and Göthert, 1992), again in accord with early data (Table 2 of Starke, 1981).

In the *guinea pig* small intestine, α_2-autoceptors as well as the α_2-receptors at the cholinergic axons of the submucous plexus were described as α_{2A} (Shen *et al.*, 1990). In the neighbouring myenteric plexus, the receptors at the cholinergic neurones also were classified as α_{2A}, whereas α_2-autoreceptors seemed to be α_{2B} as judged from the release-enhancing effect of antagonists (Blandizzi *et al.*, 1993). Finally, Alberts (1992) classified the autoreceptors in guinea-pig urethra as α_{2A}. Perhaps re-appraisal will change some of the assignments to α_{2D}. In the case of the submucous plexus receptors this seems likely, idazoxan being more potent than yohimbine (Shen *et al.*, 1990). Preliminary pK_B values from our laboratory indicate that in the myenteric plexus the autoreceptors are α_{2D} rather than α_{2B}.

One marked discrepancy has to be mentioned. There is a pervading opinion in the literature that rat brain α_2-autoreceptors and the α_2-"heteroreceptors" modulating the release of 5-HT differ (Frankhuyzen and Mulder, 1982; Gobbi *et al.*, 1990; Maura *et al.*, 1992). The present findings show, however, that both are α_{2D}. In the rabbit brain autoreceptors and heteroreceptors also were the same, α_{2A}. Compounds previously suggested to distinguish auto- from heteroreceptors such as (–)-mianserin failed to do so when tested as antagonists against UK 14304. Possible reasons for the discrepancy have been discussed (Trendelenburg *et al.*, 1994a).

It is obvious that one of the early indications of non-homogeneity of α_2, the species difference rat *versus* rabbit α_2-autoreceptors, is borne out by the present data: the rat autoreceptors are at least predominantly α_{2D} and the rabbit receptors α_{2A}. A fascinating aspect is that the rat (and mouse) α_{2D}-receptor and the human α_{2A}-receptor are homologues, the A/D branch of the phylogenetic tree. Rabbit and guinea-pig α_2-adrenoceptors have not been cloned, but our findings suggest that one of their α_2-adrenoceptors belongs to the A/D branch genetically and structurally and that, in the rabbit, this receptor is very similar to the human α_{2A}- while in the guinea pig, it is very similar to the rat α_{2D}-receptor. This view implies, of course, that guinea-pig autoreceptors (unless some are α_{2B} or α_{2C}) must be α_{2D}, not α_{2A}. It also predicts that α_2-autoreceptors in the mouse, where they have not been examined, should be α_{2D} (again unless some are α_{2B} or α_{2C}).

Do species of the Cohort *Glires*, to which rats, mice, guinea pigs (i.e. rodents) and rabbits (i.e. lagomorphs) belong, in general use mainly the $\alpha_{2A/D}$-subtype to bring their neurones under α_2-adrenergic control? Does this perhaps even hold true for all mammals (Trendelenburg *et al.*, 1993)? Do $\alpha_{2A/D}$ receptors generally possess α_{2D} pharmacologic properties in rodents and α_{2A} pharmacologic properties in lagomorphs and primates? The findings presented suggest these speculations should be tested. Raiteri *et al.* (1992) classified the human brain cortex presynaptic α_2-autoreceptor as α_{2A} or α_{2D}, in accord with the speculation. In marked contrast, we recently found the human kidney presynaptic α_2-autoreceptor to be α_{2C} (Trendelenburg *et al.*, 1994b). The situation is exciting.

References

Alberts, P. (1992) Subtype classification of the presynaptic α-adrenoceptors which regulate [^{3}H]-noradrenaline secretion in guinea-pig isolated urethra. *British Journal of Pharmacology*, **105**, 142–146.

Blandizzi, C., Tarkovacs, G., Natale, G., del Tacca, M. and Vizi, E.S. (1993) Functional evidence that [^{3}H]acetylcholine and [^{3}H]noradrenaline release from guinea pig ileal myenteric plexus and noradrenergic terminals is modulated by different presynaptic alpha-2 adrenoceptor subtypes. *Journal of Pharmacology and Experimental Therapeutics*, **267**, 1054–1060.

Blaxall, H.S., Murphy, T.J., Baker, J.C., Ray, C. and Bylund, D.B. (1991) Characterization of the alpha-2C adrenergic receptor subtype in the opossum kidney and in the OK cell line. *Journal of Pharmacology and Experimental Therapeutics*, **259**, 323–329.

Bohmann, C., Schollmeyer, P. and Rump, L.C. (1993a) Methoxamine inhibits noradrenaline release through activation of α_1- and α_2-adrenoceptors in rat isolated kidney: involvement of purines and prostaglandins. *Naunyn-Schmiedeberg's Archives of Pharmacology*, **347**, 273–279.

Bohmann, C., Schollmeyer, P. and Rump, L.C. (1993b) α_2-Autoreceptor subclassification in rat isolated kidney by use of short trains of electrical stimulation. *British Journal of Pharmacology*, **108**, 262–268.

Bylund, D.B. (1992) Subtypes of α_1- and α_2-adrenergic receptors. *FASEB Journal*, **6**, 832–839.

Doxey, J.C. and Everitt, J. (1977) Inhibitory effects of clonidine on responses to sympathetic nerve stimulation in the pithed rat. *British Journal of Pharmacology*, **61**, 559–566.

Doxey, J.C., Lane, A.C., Roach, A.G. and Virdee, N.K. (1984) Comparison of the α-adrenoceptor antagonist profiles of idazoxan (RX 781094), yohimbine, rauwolscine and corynanthine. *Naunyn-Schmiedeberg's Archives of Pharmacology*, **325,** 136–144.

Doxey, J.C., Roach, A.G. and Smith, C.F.C. (1983) Studies on RX 781094: a selective, potent and specific antagonist of α_2-adrenoceptors. *British Journal of Pharmacology*, **78,** 489–505.

Dubocovich, M.L. and Langer, S.Z. (1974) Negative feed-back regulation of nonadrenaline release by nerve stimulation in the perfused cat's spleen: differences in potency of phenoxybenzamine in blocking the pre- and post-synaptic adrenergic receptors. *Journal of Physiology*, **237,** 505–519.

Frankhuyzen, A.L. and Mulder, A.H. (1982) Pharmacological characterization of presynaptic α-adrenoceptors modulating [^{3}H]noradrenaline and [^{3}H]5-hydroxytryptamine release from slices of the hippocampus of the rat. *European Journal of Pharmacology*, **81,** 97–106.

Gobbi, M., Frittoli, E. und Mennini, T. (1990) The modulation of [^{3}H]noradrenaline and [^{3}H]serotonin release from rat brain synaptosomes is not mediated by the α_{2B}-adrenoceptor subtype. *Naunyn-Schmiedeberg's Archives of Pharmacology*, **342,** 382–386.

Göthert, M. and Molderings, G.J. (1991) Involvement of presynaptic imidazoline receptors in the α_2-adrenoceptor-independent inhibition of noradrenaline release by imidazoline derivatives. *Naunyn-Schmiedeberg's Archives of Pharmacology*, **343,** 271–282.

Langer, S.Z. (1974) Presynaptic regulation of catecholamine release. *Biochemical Pharmacology*, **23,** 1793–1800.

Lanier, S.M., Downing, S., Duzic, E. and Homcy, C.J. (1991) Isolation of rat genomic clones encoding subtypes of the α_2-adrenergic receptor. *Journal of Biological Chemistry*, **266,** 10470–10478.

Lattimer, N. and Rhodes, K.F. (1985) A difference in the affinity of some selective α_2-adrenoceptor antagonists when compared on isolated vasa deferentia of rat and rabbit. *Naunyn-Schmiedeberg's Archives of Pharmacology*, **329,** 278–281.

Limberger, N., Mayer, A., Zier, G., Valenta, B., Starke, K. and Singer, E.A. (1989) Estimation of pA_2 values at presynaptic α_2-autoreceptors in rabbit and rat brain cortex in the absence of autoinhibition. *Naunyn-Schmiedeberg's Archives of Pharmacology*, **340,** 639–647.

Limberger, N., Späth, L. and Starke, K. (1991) Subclassification of the presynaptic α_2-autoreceptors in rabbit brain cortex. *British Journal of Pharmacology*, **103,** 1251–1255.

Limberger, N., Trendelenburg, A.U. and Starke, K. (1992) Pharmacological characterization of presynaptic α_2-autoreceptors in rat submaxillary gland and heart atrium. *British Journal of Pharmacology*, **107,** 246–255.

Link, R., Daunt, D., Barsh, G., Chruscinski, A. and Kobilka, B. (1992) Cloning of two mouse genes encoding α_2-adrenergic receptor subtypes and identification of a single amino acid in the mouse α_2-C10 homolog responsible for an interspecies variation in antagonist binding. *Molecular Pharmacology*, **42,** 16–27.

Maura, G., Bonanno, G. and Raiteri, M. (1992) Presynaptic α_2-adrenoceptors mediating inhibition of noradrenaline and 5-hydroxytryptamine release in rat cerebral cortex: further characterization as different α_2-adrenoceptor subtypes. *Naunyn-Schmiedeberg's Archives of Pharmacology*, **345,** 410–416.

Michel, A.D., Loury, D.N. and Whiting, R.L. (1989) Differences between the α_2-adrenoceptor in rat submaxillary gland and the α_{2A}- and α_{2B}-adrenoceptor subtypes. *British Journal of Pharmacology*, **98,** 890–897.

Molderings, G. and Göthert, M. (1992) Pharmacological characterization of the inhibitory α_2-autoreceptors on the sympathetic nerves of the rabbit pulmonary artery. *Naunyn-Schmiedeberg's Archives of Pharmacology*, **346,** R64.

Raiteri, M., Bonanno, G., Maura, G., Pende, M., Andrioli, G.C. and Ruelle, A. (1992) Subclassification of release-regulating α_2-autoreceptors in human brain cortex. *British Journal of Pharmacology*, **107,** 1146–1151.

Schwartz, D.D. and Malik, K.U. (1992) Characterization of prejunctional alpha-2 adrenergic receptors involved in modulation of adrenergic transmitter release in the isolated perfused rat kidney. *Journal of Pharmacology and Experimental Therapeutics*, **261,** 1050–1055.

Shen, K.Z., Barajas-Lopez, C. and Surprenant, A. (1990) Functional characterization of neuronal pre and postsynaptic α_2-adrenoceptor subtypes in guinea-pig submucosal plexus. *British Journal of Pharmacology*, **101,** 925–931.

Simonneaux, V., Ebadi, M. and Bylund, D.B. (1991) Identification and characterization of α_{2D}-adrenergic receptors in bovine pineal gland. *Molecular Pharmacology*, **40,** 235–241.

Singer, E.A. (1988) Transmitter release from brain slices elicited by single pulses: a powerful method to study presynaptic mechanisms. *Trends in Pharmacological Sciences*, **9,** 274–276.

Smith, K. and Docherty, J.R. (1992) Are the prejunctional α_2-adrenoceptors of the rat vas deferens and submandibular gland of the α_{2A}- or α_{2D}-subtype? *European Journal of Pharmacology*, **219,** 203–210.

Smith, K. and Docherty, J.R. (1994) Further investigations of the prejunctional α_2-adrenoceptors of rat atrium. *British Journal of Pharmacology*, **112,** 388P.

Starke, K. (1972) Alpha sympathomimetic inhibition of adrenergic and cholinergic transmission in the rabbit heart. *Naunyn-Schmiedeberg's Archives of Pharmacology*, **274,** 18–45.

Starke, K. (1981) α-Adrenoceptor subclassification. *Reviews of Physiology, Biochemistry and Pharmacology*, **88,** 199–236.

Starke, K., Montel, H., Gayk, W. and Merker, R. (1974) Comparison of the effects of clonidine on pre- and postsynaptic adrenoceptors in the rabbit pulmonary artery. *Naunyn-Schmiedeberg's Archives of Pharmacology*, **285,** 133–150.

Taube, H.D., Starke, K. and Borowski, E. (1977) Presynaptic receptor systems on the noradrenergic neurones of rat brain. *Naunyn-Schmiedeberg's Archives of Pharmacology*, **299,** 123–141.

Trendelenburg, A.U., Limberger, N. and Starke, K. (1993) Presynaptic α_2-autoreceptors in brain cortex: α_{2D} in the rat and α_{2A} in the rabbit. *Naunyn-Schmiedeberg's Archives of Pharmacology*, **348,** 35–45.

Trendelenburg, A.U., Trendelenburg, M., Starke, K. and Limberger, N. (1994a) Release-inhibiting α_2-adrenoceptors at serotonergic axons in rat and rabbit brain cortex: evidence for pharmacological identity with α_2-autoreceptors. *Naunyn-Schmiedeberg's Archives of Pharmacology*, **349,** 25–33.

Trendelenburg, A.U., Limberger, N. and Rump, L.C. (1994b) α_2-Adrenergic receptors of the α_{2C} subtype mediate inhibition of norepinephrine release in human kidney cortex. *Molecular Pharmacology*, **45,** M68–M76.

Vidovic, M., Cohen, D. and Hill, C.E. (1994) Identification of α_2 adrenergic receptor gene expression in sympathetic neurones using polymerase chain reaction and in situ hybridization. *Molecular Brain Research*, **22,** 49–56.

NEWER SUBTYPES OF THE α_2?

J. E. S. WIKBERG, G. TIGER*, Y. XIA, V. CHHAJLANI and S. UHLÉN

Department of Pharmaceutical Biosciences, Uppsala University and
**Department of Pharmacology, Umeå University, Sweden*

Adrenoceptors of the α_2 exist in at least three subtypes in mammals: α_{2A}, α_{2B} and α_{2C}. In earlier studies we found compelling evidence that the N-hydroxyguanidine guanoxabenz could further delineate α_{2A} and α_{2B} adrenoceptors into $\alpha_{2A1}/\alpha_{2A2}$ and $\alpha_{2B1}/\alpha_{2B2}$ subforms. However, newer results prompt a shift of the paradigm and indicates that the selectivity of guanoxabenz rather arises due to metabolic activation of the N-hydroxyguanidine in some tissues. A number of structural analogues to guanoxabenz appear to be metabolic inhibitors of the activation of guanoxabenz. Moreover, the tissue specific activation of guanoxabenz extends also to the functional level as the compound proved to show high α_2-adrenoceptor blocking activity on pre-junctional cholinergic neurones of the guinea pig ileum.

KEY WORDS: α_2-adrenoceptors, receptor classification, guanoxabenz, prejunctional adrenoceptors

INTRODUCTION

Overwhelming evidence suggest that there exist at least three subtypes of α_2-adrenoceptors (α_2-ARs) which are termed α_{2A}, α_{2B} and α_{2C}. Support for this idea comes from molecular cloning of three distinct genes encoding α_2-adrenoceptors in mammals and from pharmacological studies using both radioligand binding and physiological approaches. However, there is still some uncertainty whether or not there may be more subtypes of α_2-ARs present in mammals. It is clear that some of the earlier confusion was due to the presence of species differences in the pharmacology of species homologues of α_2-ARs (e.g. the α_{2A} versus the "α_{2D}"-ARs which arise due to slight difference in sequence of the human and rodent α_{2A}-adrenoceptor genes) and that some other of the confusion arose due to the difficulties to clearly delineate α_{2B} and α_{2C}-ARs pharmacologically. Nevertheless, in some recent studies we presented evidence that α_{2B}-ARs could be delineated further into α_{2B1} and α_{2B2}-types and α_{2A}-ARs into α_{2A1} and α_{2A2}-types (Uhlén and Wikberg 1991; Uhlén *et al.* 1993; Xia et al. 1993). The evidence for this suggestion came from the observations that the N-hydroxyguanidine guanoxabenz seemed to show high affinity for a population of α_{2B}-ARs termed "α_{2B1}" whereas it showed low affinity for another population termed "α_{2B2}". In a similar fashion guanoxabenz also showed high respectively low affinity for "α_{2A1}" and "α_{2A2}" populations of α_{2A}-ARs. However, in more recent studies we have obtained data that seem to disfavour the original hypothesis that α_{2A} and α_{2B}-ARs exist in further subforms. We will here give a short review of our previous studies regarding the delineation of α_2-ARs into subforms as well as present some new data pertaining to these studies.

Correspondence to: J.E.S. Wikberg, Pharmaceutical Biosciences, Uppsala University, S-75124 Uppsala, Sweden.

SUMMARY OF RESULTS AND DISCUSSION

Delineation of α_2-ARs into α_{2A}, α_{2B} and α_{2C} in the rat

By using the radioligands [^{3}H]RX821002 and [^{3}H]MK-912 we could delineate between three subtypes of α_2-ARs in the rat. In many of the rat tissues combinations of two of the three α_2-ARs are present, albeit often with big differences in their densities. To be able to detect these receptors we developed a multicurve approach where saturation and/or competition curves of drugs, obtained with and without masking concentrations of various subtype selective drugs, were analysed simultaneously in the computer (for details see Uhlén and Wikberg 1991). The radioligand used was also selected to match the densities of receptors in the tissues studied. In e.g. the kidney 13% of the receptors were of the α_{2A}-type and 87% of the α_{2B}-type. During these conditions the selection of an appropriate concentration of [^{3}H]RX821002 resulted in a signal of equal magnitude from both of the receptor types when applied in competition studies, because [^{3}H]RX821002 showed higher affinity for the α_{2A}-ARs ($K_d \sim 0.7$ nM) than for the α_{2B}-ARs ($K_d \sim 5$ nM), something which thus rendered the data optimal for computer analysis. A similar situation was present for α_{2A} and α_{2C}-ARs in the CNS. In e.g. the cerebral cortex 10% of the receptors were α_{2C} and 90% were α_{2A}. Still analysis of both subtypes was feasible since [^{3}H]MK-912 showed higher affinity for the α_{2C}-ARs ($K_d \sim 0.06$ nM) than for the α_{2A}-ARs ($K_d \sim 2$ nM) (Uhlén *et al.* 1992).

Using these approaches α_2-AR subtypes were detected in membranes prepared from various tissues of the rat (table 1). The K_d values for of a number of drugs determined for these receptors are given in table 2A, along with the corresponding values for the α_2-ARs expressed from either one of the three cloned rat α_2-AR genes. As can be seen from table 2B the correlations of log(K_d) values in between the same subtype of α_2-ARs were excellent ($r = 0.94 - 0.99$) whereas those of different subtypes did not correlate well. Among the most interesting compounds evaluated may be mentioned BRL44408 and guanfacine which are α_{2A}-selective, ARC-239 which is α_{2B}-selective and MK-912 which is α_{2C}-selective. Moreover, spiroxatrine was also found to be α_{2C}-selective (Uhlén and Wikberg 1993). WB4101 is also slightly α_{2C}-selective (table 2).

Do there exist further subforms of α_2?

The initial observation leading us to suggest such a possibility came from the finding that guanoxabenz (GOB) appeared to bind to two populations of α_{2B}-ARs in the kidney; one showing high and the other showing low affinity for GOB. Further studies in various tissues of the rat revealed that guanoxabenz also delineated two apparent populations of α_{2A}-

TABLE 1

Levels (in fmol/mg protein) of α_2-AR subtypes in membranes prepared from tissues of the rat as assayed by using [^{3}H]RX821002 or [^{3}H]MK-912 binding. (Data from Uhlén and Wikberg 1991a, 1992; Uhlén *et al.* 1992, 1993; Xia *et al.* 1993). The sign "—" denotes "not detectable".

	α_{2A}	α_{2B}	α_{2C}
Kidney	15	97	—
Cerebral cortex	300	—	33
Spinal cord	170	—	7
Cerebellum	110	—	19
Neonatal lung	—	230	—
Spleen	130	—	—

TABLE 2

K_d-values (in nM) of drugs determined from binding studies in rat tissues using [^{3}H]RX821002 (RX) or [^{3}H]MK-912 (MK). C. Ctx = Cerebral cortex, N. Lung = Neonatal lung. RG20, RNG and pA2d denotes the receptor expressed from respectively the α_2-ARs gene cloned by Lanier *et al.* 1991, Zeng *et al.* 1990 and Voigt *et al.* 1991 (pA2d gene is identical to the RG10 gene cloned by Lanier *et al.* 1991 and the RG20, RNG and pA2d are the rat homologues to the human α_2-C10, α_2-C2 and α_2-C4 genes cloned by Lefkowitz and co-workers). **B**. Correlation coefficients (r^2) for all possible correlations of the $\log_{10}(K_d)$ data in table 2A. (Table 2 adopted from Uhlén 1993, Umeå University Dissertation no. 374).

A *Ligand*	^{3}*H-RX*	^{3}*H-MK*	^{3}*H-RX*	^{3}*H-RX*	^{3}*H-MK*	^{3}*H-MK*	^{3}*H-MK*	^{3}*H-MK*	^{3}*H-MK*
Tissue	*Kidney*	*C.Ctx*	*Spleen*	*RG20*	*Kidney*	*N.Lung*	*RNG*	*C.Ctx*	*pA2d*
Drug/ Subtype	α_{2A}	α_{2A}	α_{2A}	α_{2A}	α_{2B}	α_{2B}	α_{2B}	α_{2C}	α_{2C}
MK912	2.5	1.8	1.5	—	1.3	0.77	0.79	0.075	0.046
guanfacine	31	20	30	35	1900	1600	1200	420	380
oxymetazoline	35	13	25	38	1700	860	790	120	99
yohimbine	58	50	49	72	16	10	11	2.8	2.4
rauwolscine	68	34	58	62	13	9.3	9.7	1.6	0.85
BRL41992	78	28	38	18	34	22	13	276	137
WB4101	180	140	150	150	53	26	25	4.5	3
prazosin	1400	490	520	930	52	19	56	30	32
ARC239	1500	760	830	810	14	8.8	7	46	47
(–) adrenaline	3100	3000	7100	5900	4700	3700	1200	2400	1600
(–) noradrenaline	13000	12000	17000	19000	3600	3000	1800	3300	3800
B									
RX/Kidney α_{2A}	1.00	0.98	0.97	0.94	0.25	0.24	0.25	0.46	0.49
MK/c.ctx α_{2A}		1.00	0.99	0.97	0.26	0.26	0.25	0.43	0.46
RX/Spleen α_{2A}			1.00	0.98	0.34	0.34	0.32	0.50	0.53
RX/RG20 α_{2A}				1.00	0.17	0.17	0.16	0.25	0.29
RX/kidney α_{2B}					1.00	0.99	0.98	0.77	0.77
MK/n. lung α_{2B}						1.00	0.97	0.77	0.74
MK/RNG α_{2B}							1.00	0.72	0.74
MK/c.ctx α_{2C}								1.00	0.99
MK/pA2d α_{2C}									1.00

ARs. For α_{2C}-ARs guanoxabenz showed only one low affinity (table 3). Several other GOB analogues were evaluated but only one, namely LT11 (1-[(3-methoxybenzylidene)-amino]-4-hydroxyguanidine) could delineate between apparent high and low affinity forms of α_{2B}-ARs (Xia *et al.* 1993). However, LT11 was not capable of delineating further α_{2A}-ARs (Uhlén *et al.* 1993). As can be seen from table 3 both the high and low affinity forms of the α_2-ARs appeared to coexist in some tissues, which rendered the competition curves of GOB distinctly biphasic. Examples of such competition curves for guanoxabenz in the rat spleen are shown in fig. 1. As may be seen from the computer drawn curves a two site model fitted the data excellently. Thus these experiments seemed to strongly suggest that two α_{2A}-receptor populations were present.

However, a number of reasons led us subsequently to doubt these interpretations. One of the major reasons being the inability to find genes that could account for the apparent high affinity forms of the α_{2A} and α_{2B}-ARs. As can be seen from table 3 the α_2-ARs expressed from the RG20 and RNG genes showed uniform low affinities for GOB. Repeated attempts to find extra α_2-AR genes using PCR were unsuccessful. Moreover, the long time that was elapsing without any reports appearing indicating that such extra genes were present also strongly indicated that in mammals there are only three distinct α_2-AR genes present. In order to further investigate the phenomenon we therefore designed the experiment shown in fig. 1A. In this experiment competition curves of GOB were obtained in the rat cerebral cortex and spleen membranes using [^{3}H]RX821002 as ligand (which are conditions for assays of only α_{2A}-ARs). As can be seen the competition curve of GOB in the spleen is biphasic whereas that in cerebral cortex it is uniphasic, as expected. Thus, in these particular experiments the calculations showed that 80% of α_{2A}-ARs had high affinity and 20% low affinity for GOB in the spleen. In the cerebral cortex 100% of the α_{2A}-ARs showed low affinity for GOB. In the subsequent step the membranes of cerebral cortex and spleen were mixed: 60% of α_{2A}-ARs from the cerebral cortex and 40% from the spleen. Still in these mixed membranes 80% of the GOB sites were of high affinity and 20% of low affinity despite the expected proportion 32% high and 68% low affinity. Thus, it appeared as if the presence of spleen membranes had "recruited" GOB high affinity α_{2A}-AR sites in the cerebral cortex membranes.

TABLE 3

K_d (in nM) of guanoxabenz (GOB) for apparent high and low affinity α_2-ARs in the rat determined in competition with[^{3}H]RX821002 or [^{3}H]MK-912. Proportion of sites (in %) given within parenthesis. (Table adopted from Uhlén 1993, Umeå University Dissertation no. 374).

Receptor/Tissue	*GOB high affinity*	*GOB low affinity*
α_{2A}-ARs		
Kidney	39 (100 %)	—
Spleen	35 (57 %)	8900 (43 %)
Cerebral cortex	—	4000 (100 %)
Spinal cord	—	3900 (100 %)
RG20 clone	—	5200 (100 %)
α_{2B}-ARs		
Kidney	193 (29 %)	25000 (71 %)
Neonatal lung	—	8400 (100 %)
RNG clone	—	6300 (100 %)
α_{2C}-ARs		
Cerebral cortex	—	18000 (100 %)
pA2d clone	—	18000 (100 %)

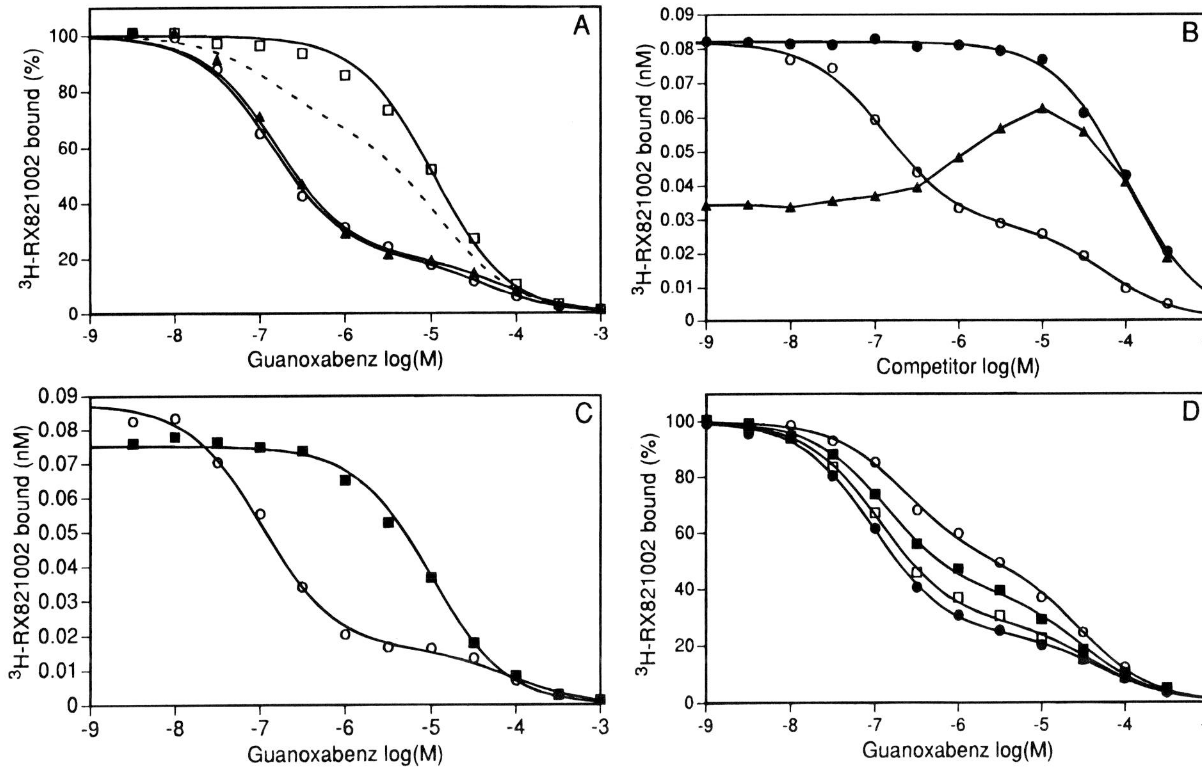

FIGURE 1A Competition curves of GOB using [^{3}H]RX821002 with membranes from the spleen (○), cerebral cortex (❑), and a 40/60% mixture (with respect to α_{2A}-AR density) of spleen and cerebral cortex (▲). Hatched line represent the theoretical competition curve of a 40/60% mixture of spleen and cortex assuming that no recruitment of GOB high affinity activity takes place **B**. Competition curves of GOB and LW03 using [^{3}H]RX821002 with membranes from the spleen. Shown are plain competition curves of GOB (○) and LW03 (●) and a competition curve of LW03 in the presence of 1.6 μM GOB (▲). **C**. Competition curves of GOB in the absence (○) and in the presence of 6 μM LW03 (■). **D**. Competition curves of GOB using [^{3}H]RX821002 with spleen membrane. The incubation time was 30 min (○), 60 min (■), 120 min (❑) and 180 min (●).

In further experiments we then found that a number of other N-hydroxyguanidines could block the formation of the apparent high affinity sites for GOB. As can be seen from fig. 1B the addition of LW03* 0.3 – 100 μM reversed the ability of 1.6 μM GOB to block the binding of [^{3}H]RX821002 to spleen membranes. Moreover, as can be seen from fig. 1C the addition of 6 μM LW03 completely eliminated the high affinity component of GOB binding to spleen membranes. It is also evident from fig. 1B that LW03 in itself could compete with [^{3}H]RX821002 for the α_{2A}-AR binding in the spleen membranes. In table 4 the effectiveness of some N-hydroxyguanidines to counteract the high affinity binding of GOB are compared with their α_{2A}-AR affinity. As may be seen from the table as simple compounds as hydroxyguanidine and hydroxyurea were capable of mediating the reversal of GOB high affinity binding, although many of the LW compounds were much more active. Morever, as seen in the table many of these compounds were distinctly more active as reversers of GOB binding than they were in binding to the α_{2A}-AR. Thus, their anti GOB activity showed a clear structure activity relationship that was separated from their α_2-AR binding activity. Moreover, the effect was related to N-hydroxyguanidines only, since classical α_2-AR agonists and agonists were not capable of inducing the reversal of GOB high affinity binding (data not shown).

What is the explanation for the GOB high affinity binding?

In view of these new results and for the reasons listed in the previous sections we have begun to strongly doubt that GOB high and low affinity forms of α_{2A} and α_{2B}-ARs are due to the presence of distinctly different subtypes of α_2-ARs. One possibility could be that certain α_2-ARs had become postranslationally modified so that they had acquired the ability to bind GOB with high affinity. In this model the other N-hydroxyguanidines would be capable of inducing further modification of the α_2-AR so that GOB lost its high affinity. However, this explanation seems quite far fetched. This is even more so in view of the experiments presented in fig. 1A, which makes it necessary to assume that the cerebral cortex α_{2A}-ARs were converted to GOB high affinity receptors during the incubation with the spleen membranes. We therefore favour a more simple explanation, namely that GOB is being metabolised to a compound that possesses higher affinity for the α_2-ARs than GOB itself. In this model it is necessary to assume that the conversion of GOB to its high affinity product becomes rate limited above ~ 1 μM. Higher concentrations of GOB would then not be capable of accumulating enough metabolite to completely prevent the binding of [^{3}H]RX821002 to e.g. the spleen membranes, thus yielding apparent biphasic competiton curves of GOB. Such behaviour would be present if GOB was metabolised via an enzymatic system that became saturated above 1 μM of GOB, so that the GOB metabolism changed from first to zero-order kinetics. Moreover, the model would fit with the assumption that N-hydroxyguanidines listed in table 4 are blockers of this enzymatic activity. If this model was true one would expect the proportion of high and low affinity GOB binding to increase with time. This is due to the fact that time would allow the concentration of the high affinity metabolite to increase even at zero-order kinetics. That this is indeed the case is shown in the experiment of fig. 1D. Thus, incubation of membranes with GOB for 30, 60, 120 and 180 min progressively increased the proportion of the high affinity com-

---

*For full chemical names of LW, LT and ATL compounds see Uhlén *et al.* 1993.

TABLE 4
Comparison of the minimal effective concentration (MEC) of some N-hydroxyguanidines for inducing reversal of the ability of 1.6 μM of GOB to compete for [^{3}H]RX821002 binding to spleen membranes (c.f. the curve marked with ▲ symbols of fig 1B) and the ability of these compounds to compete with [^{3}H]RX821002 in the very same membranes (c.f. the curve marked with ● in fig 1B). MEC is measured as the concentration of the compound giving a consistent increase of [^{3}H]RX821002 binding in the presence of 1.6 μM GOB, whereas the α_{2A}-AR K_d is estimated from the computer analysis of the competition curve of the compound in the absence of GOB using a one-site model.

Compound	*MEC-GOB-reversal (μM)*	*α_{2A}-AR K_d (μM)*	*Ratio*
LW04	1	28	28
LT11	1	20	20
LW03	1	15	15
LW11	3.2	77	24
ATL26	10	200	20
LW12	10	72	7.2
LT07	32	30	0.9
Hydroxyguanidine	320	660	2.1
LW01	1000	260	0.26
Hydroxyurea	3200	>3200	>1

ponent to 55, 64, 74 and 77%, respectively. (It may be noted that [^{3}H]RX821002 reaches steady state of binding within 8 min. Thus, the incubation time of 30 min is more than sufficient to afford equilibrium for the binding of the radioligand to the α_2-adrenoceptors). At present it is not clear what the nature is of the putative metabolic step for GOB conversion. The enzymatic reduction of hydroxyguanidine to guanidine has been described (Walker and Walker 1959) and GOB is known to be a quite labile compound that may be subjected to modifications such as cyclisation, reduction, dimerisation and chelation with divalent cations (Hoffelt and Bourdon 1980).

Induction of GOB high affinity may afford selectivity at the functional level

Although we have thus far not been able to directly show that GOB is subjected to metabolisation the presence of such a metabolism might be useful to afford selectivity also on a functional level. Thus, in fig. 2A are shown results for the electrically stimulated guinea pig ileum. Electrical stimulation of this tissue with single 1 ms shocks induces twitch contraction that are mediated by the activation of cholinergic neurones of the Auerbachs plexus. It is well established that such electrically induced contractions may be inhibited by α_2-AR agonists via the stimulation of α_{2A}-ARs located at the cholinergic neurones. By using the selective α_2-AR agonist BHT-920 the blocking abilities of yohimbine, BRL 44408 and ARC239 were assessed by using the Schild plot (for details of results see legend to fig. 2). The results clearly indicate that the α_2-ARs being activated by BHT-920 are of the α_{2A}-subtype since the α_{2A}-adrenoceptor blocker BRL 44408 was much more potent than the α_{2B}-adrenoceptor blocker ARC239 (c.f. Alberts 1993). When GOB was tested in the same system it first induced a small reduction of the twitch contraction. However, this effect was not sustained and therefore the ability of GOB to eventually block the effect of BHT-920 could be evaluated. As can be seen from fig. 2A GOB proved to be a very potent blocker of the effect of BHT-920 and its apparent K_b was estimated to be 21 nM. In separate experiments we also showed that [^{3}H]MK-912 labelled a single population of α_{2A}-ARs in the guinea pig ileum. Examples of these experiments are shown in fig. 2B where it can be seen that the

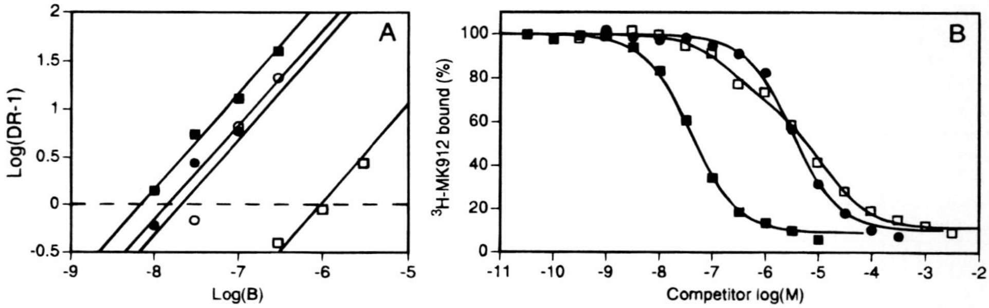

FIGURE 2A Schild plots for (from left to right) yohimbine (■), BRL44408 (●), guanoxabenz (○) and ARC239 (□) obtained on the electrically stimulated guinea pig ileum and using BHT-920 as the agonist. Electrical stimulation was afforded by single pulses at 0.1 Hz as described (Wikberg 1978). The K_b values obtained were for yohimbine 6.88 ± 0.52 nM, BRL 44408 14.5 ± 2.0 nM, guanoxabenz 20.9 ± 9.8 nM and for ARC239 971 ± 117 nM. The time of preincubation for yohimbine, BRL44408, guanoxabenz and ARC239 before the addition of BHT-920 was 30 minutes. **B**. Competition curves of guanfacine (■), prazosin (●) and guanoxabenz (□) using ~ 1nM [^{3}H]MK-912 on membranes prepared from the smooth muscle layers of the guinea pig ileum. (Data taken from manuscript in preparation by Tiger *et al.*).

α_{2A}-selective compound guanfacine is much more selective than the α_{2B}-selective compound prazosin. As can be also seen from fig. 2B the competition curve of GOB is clearly biphasic in membranes from the ileum; the two apparent K_ds being 140 and 7900 nM. Thus, from these experiments it seems possible to conclude that in the functional assays the high affinity component of GOB is being in action. Thus, to summarise the data of the present study show that the activation of GOB to its putative high affinity metabolite does not take place in the central nervous system and the neonatal lung, whereas it is present in the spleen, kidney and ileum. Thus, such activation of GOB might find use to afford a local selective action of GOB (or a similarly activatable compound) on α_2-ARs in these organs.

ACKNOWLEDGMENTS

Supported by the Swedish MRC (04X-05975). Dr. Tiger was supported by the Swedish Society of Medicine.

References

Alberts, P. (1993) Subtype classification of presynaptic α_2-adrenoceptors. *Gen. Pharmac.*, **24,** 1–8.

Lanier, S.M., Downing, S., Duzic, E. and Homcy, C.J. (1991) Isolation of rat genomic clones encoding subtypes of the α_2-adrenergic receptor. Identification of a unique receptor subtype. *J. Biol. Chem.*, **266,** 10470–10478.

Hoffelt, J. and Bourdon, R.R. (1980) Dosage du guanoxabenz dans les liquides biologiques. *Ann. Biol. Clin.*, **38,** 351–354.

Uhlén, S. and Wikberg, J.E.S. (1991) Delineation of rat kidney α_{2A}- and α_{2B}-adrenoceptors with [^{3}H]RX821002 radioligand binding: computer modelling reveals that guanfacine is an α_{2A}-selective compound. *Eur. J. Pharmacol.*, **202,** 235–243.

Uhlén, S. and Wikberg, J.E.S. (1991) Delineation of three pharmacological subtypes of α_2-adrenoceptors in the kidney. *Br. J. Pharmacol.*, **104,** 657–664.

Uhlén, S., Xia, Y., Chhajlani, V., Felder, C. C. and Wikberg, J.E.S. (1992) [^{3}H]-Mk912 binding delineates two α_2-adrenoceptors subtypes in rat CNS one of which is identical with the cloned pA2d α_2-adrenoceptor. *Br. J. Pharmacol.* **106,** 986–995.

Uhlén, S. and Wikberg, J.E.S. (1993) [^{3}H]MK-912: A novel α_{2C}-adrenoceptor selective ligand detects both α_{2A}- and α_{2C}-adrenoceptors in the cerebellum. *Biotech. Update*, **7,** 16–18.

Uhlén, S., Xia, Y., Chhajlani, V., Lien, E.J. and Wikberg, J.E.S. (1993) Evidence for the existence of two forms of α_{2A}-adrenoceptors in the rat. *Naunyn-Schmiedebergs Arch. Pharmacol.*, **347,** 280–288.

Voigt, M.M., McCune, S.K., Kanterman, R.Y. and Felder, C.C. (1991) The rat α_2-C4 adrenergic receptor gene encodes a novel pharmacological subtype. *FEBS Lett.*, **278,** 45–50.

Walker, J.H. and Walker, M.S. (1959) The enzymatic reduction of hydroxyguanidine. *J. Biol. Chem.*, **234,** 1481–1484.

Wikberg, J.E.S. (1978) Pharmacological classification of adrenergic α-receptors in the guinea pig. *Nature*, **273,** 164–166.

Xia, Y., Uhlén, S., Chhajlani, V., Lien, E.J. and Wikberg, J.E.S. (1993) Further evidence for the existence of two forms of α_{2B}-adrenoceptors in the rat. *Pharmacol. Toxicol.*, **72,** 40–49.

Zeng, D.W., Harrison, J.K., D'Angelo, D.D., Barber, C.M., Tucker, A.L., Lu, Z. and Lynch, K.R. (1990) Molecular characterization of a rat α_{2B}-adrenergic receptor. *Proc. Natl. Acad. Sci. USA*, **87,** 3102–3106.

ADRENERGIC RECEPTOR STRUCTURE AND ACTIVATION

BRIAN K. KOBILKA

Howard Hughes Medical Institute, 157 Beckman Center, Stanford University Medical Center, Stanford, CA 94305

INTRODUCTION

Adrenergic receptors are representative of the class of membrane bound receptors that are physically and functionally linked to guanine nucleotide regulatory proteins (G proteins) during signal transduction. Other members of this family of G protein-coupled receptors includes the muscarinic receptors, dopamine receptors, serotonin receptors as well as receptors for peptide and protein hormones such as angiotensin, endothelin, LH and FSH. These receptors are functionally and structurally closely related to the visual pigments. Agonist binding to the receptor from the extracellular surface promotes receptor mediated changes in a cytoplasmic G protein: the exchange of GTP for GDP on the G protein alpha subunit, and the dissociation of the heterotrimeric G protein. Over the past several years much has been learned about the structure and regulation of adrenergic receptors. This article will summarize studies which have led to a better understanding of receptor structure and in some cases provided clues to the structural changes that occur during signal transduction.

RECEPTOR STRUCTURE

The beta 2 adrenergic receptor was the first G protein coupled hormone or neurotransmitter receptor to be cloned (Dixon *et al.*, 1986; Kobilka *et al.*, 1987) and, along with the visual pigment rhodopsin, has been one of the most extensively studied members of this family of receptors. Analysis of the primary amino acid sequence reveals seven hydrophobic segments and a striking similarity in secondary structure to bacteriorhodopsin. There is experimental evidence that, like bacteriorhodopsin, the beta 2 adrenergic receptor has 7 membrane spanning segments (Fig. 1) (Wang *et al.*, 1989; Dholman *et al.*, 1987). Domains involved in ligand binding and G protein coupling have been identified through mutagenesis (Strader *et al.*, 1989; Strader *et al.*, 1991; O'Dowd *et al.*, 1988) and chimeric receptor studies (Kobilka *et al.*, 1988; Suryanarayana *et al.*, 1991; Guan *et al.*, 1992). In some cases it has been possible to identify single amino acids involved in binding specific classes of ligands (Strader, *et al.*, 1989, 1991; Suryanarayana). These studies demonstrate that the

---

Correspondence to: Brian K. Kobilka, Howard Hughes Medical Institute, Stanford University Medical Center.

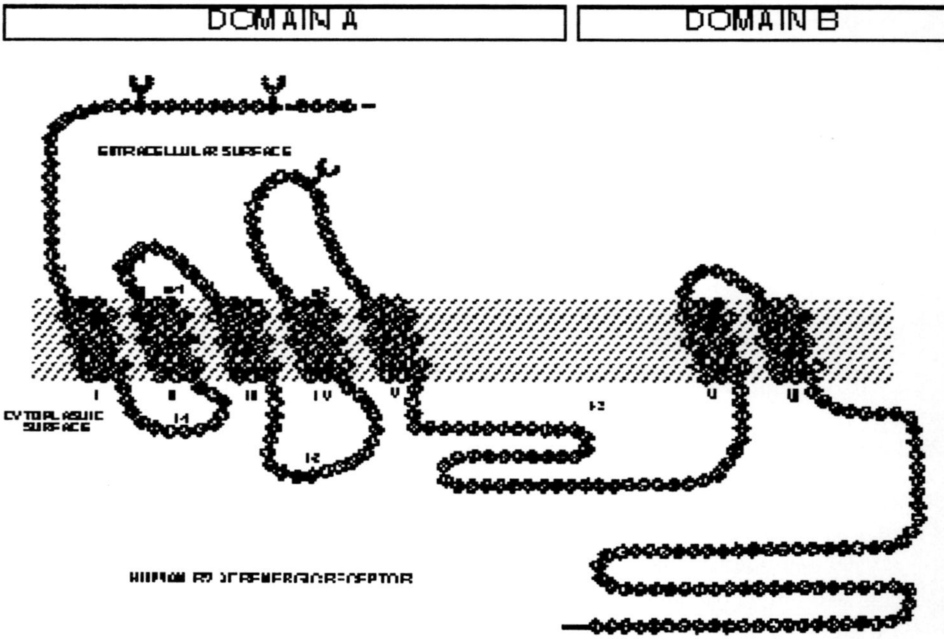

FIGURE 1 Diagram of the predicted secondary structure and membrane topology of the human beta 2 adrenergic receptor.

ligand binding domain is made up of several hydrophobic domains, while G protein coupling domains include the second and third intracellular domains and the carboxyl terminus.

Studies have also begun to shed light on intramolecular relationships that exist in the three dimensional structure of the receptor. The identification of different sites of interaction between receptor and ligands has placed constraints on the distance between hydrophobic domains. For example, Strader and colleagues provided evidence for specific interactions between the secondary amine of both agonists and antagonists and Asp 113 in the third hydrophobic domain (Strader, 1991). They also provided evidence for interactions between the hydroxyls on the aromatic ring of isoproterenol and serines 204 and 207 in the fifth hydrophobic segment, (Strader, 1989) and interactions between the catechol ring and aromatic amino acids in the sixth hydrophobic segment. These results provide information about the proximity of hydrophobic domains 3 and 5 in the agonist bound receptor. We have obtained evidence supporting a specific interaction between the ether oxygen on the antagonist alprenolol and Asn 312 in the seventh hydrophobic domain (Suryanarayana, 1991; 1992). This information therefore suggests a specific arrangement between hydrophobic domains 3 and 7 in the antagonist bound receptor.

Mutagenesis studies have also provided evidence for specific intramolecular interactions. We found that changing Asn 312 in the beta 2 receptor to Phe (the amino acid found in the homologous position of the alpha 2 receptor) led to a loss of function and retention of the mutated receptor in the endoplasmic reticulum. This mutation could be complemented by exchanging the amino terminus and the first two hydrophobic domains with alpha 2 receptor sequence (Suryanarayana and Kobilka, 1993). This provides evidence for an interaction between this phenylalanine residue and hydrophobic segments one and two in the alpha 2 receptor. This finding has recently been supported by similar studies on the muscarinic receptor (Pittel, Wess 1994), and the GNRH receptor (Zhou *et al.,* 1994). The identification of other intramolecular interactions and sites of receptor-ligand interactions will assist in the formulation of more accurate three dimensional models of G protein coupled receptors, and perhaps a better understanding of the structural changes that accompany agonist binding.

RECEPTOR ACTIVATION

The mechanism by which agonist binding activates receptors is not known. It is likely that the agonist induces a conformational change in the receptor that can be detected by the G protein. Many receptors transmit signals across the lipid bilayer by dimerization; however, the oligomeric state of G protein coupled receptors during signal transduction has not been well characterized. There is reason to believe that dimerization is possible. The structurally similar protein, bacteriorhodopsin, forms densely packed two dimensional crystals. Recently, Wess and colleagues demonstrated chimeric receptors derived from alpha 2 and muscarinic receptors can form dimers (Maggio *et al.,* 1993). When expressed separately, the chimeric receptors (alpha2 transmembrane 1-5)-(muscarinic transmembrane 6-7) and (muscarinic transmembrane 1-5)-(alpha2 transmembrane 6-7) are not functional; however, when these receptors are co-expressed, both alpha 2 and muscarinic specific binding can be detected. This suggests that receptor dimers are forming in which domains 1-5 from one chimer are associating with homologous receptor domains 6 and 7 from the other chimer. While this does not prove that dimerization occurs with the natural receptor, it indicates that functional receptor dimers could form.

Over the past 4 years several labs have reported mutations which lead to constitutive activity of G protein coupled receptors (i.e. receptors that stimulate G proteins in the absence of agonists) (Cotecchia *et al.,* 1990; Shenker *et al.,* 1993; Robbins *et al.,* 1993; Robinson *et al.,* 1992; Parma *et al.,* 1993). These receptors are active in the absence of agonists. These activating mutations have been observed in several different domains including the first (Suryanarayana *et al.*) and third (Suryanarayana *et al.*; Strader *et al.*) intracellular loops, and the second (Robbins *et al.,* 1993), sixth (Shenker *et al.,* 1993), and seventh (Robinson *et al.,* 1992) hydrophobic domains. In most of these mutations, an elevated basal level of receptor mediated G protein activation is observed in the absence of agonist. In some cases the addition of agonist causes further activation and antagonist can suppress the basal activity. However, a mutation in the second hydrophobic domain of the MSH receptor results in elevated basal activity, but no additional activation is seen with the addition of MSH (Robbins *et al.,* 1993). It has recently been shown that even the wild type beta 2 adrenergic receptor has some constitutive activity and can activate Gs in the absence of agonists, particularly when the receptor is expressed at high levels (Samama *et al.,* 1994; Chidiac *et al.,* 1994). This constitutive activity can be inhibited by some antagonists, but not by others (Samama *et al.,* 1994; Chidiac *et al.,* 1994). This observation suggests that both agonists and antagonists can induce conformational changes.

Biochemical studies also suggest that both agonists and antagonists induce changes in receptor conformation that can be distinguished from the conformation of the non-liganded receptor. Susceptibility to proteases can be used to monitor conformational changes in the beta 2 adrenergic receptor (Kobilka 1990). In the absence of ligand, cytoplasmic domains of the receptor are susceptible to proteolysis. Both agonist and antagonist reduce the susceptibility to proteolysis. Therefore, the non-liganded state and the antagonist state are not the same. An antagonist does not merely block agonist access, but is capable of inducing a detectable conformational change.

The third intracellular loop of the receptor appears to be the most flexible domain in the beta 2 adrenergic receptor, and it is likely that most of the ligand induced conformational changes involve this structural domain. We have previously demonstrated that the integrity of the third intracellular loop is not essential for receptor function, and that functional receptors can be made as heterodimers where the amino terminus through the middle of the third intracellular loop is expresses as one protein (domain A, Fig. 1) along with the remainder of the receptor expressed as a second protein (domain B, Fig. 1) (Kobilka *et al.*, 1988). The flexibility of the third loop and the important functional role of the amino and carboxyl terminal ends of the third loop in G protein coupling suggests that agonist activation may involve movement of domain A relative to domain B.

WORKING HYPOTHESIS

Based on these studies we have formulated a working hypothesis (Figure 2) to explain the consequences of agonist and antagonist binding on receptor structure. As discussed above, it has been shown experimentally (Kobilka *et al.*, 1988) that the beta 2 adrenergic receptor can be thought of as a heterodimer having two major domains (Figure 1): domain A (from

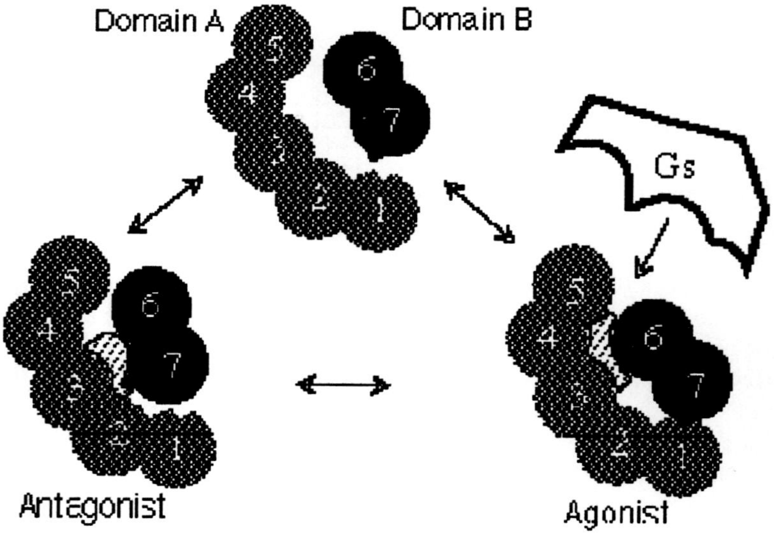

FIGURE 2 Model depicting a possible mechanism by which agonists and antagonists modify the structure of the beta 2 adrenergic receptor. The receptor is viewed from the cytoplasmic side of the membrane.

the amino terminus through the center of the first intracellular loop) and domain B (from the middle of the third intracellular loop through the carboxyl terminus). In Figure 2 these domains are viewed from the cytoplasmic side of the membrane. In the absence of agonist or antagonist, these two domains are relatively mobile with respect to each other, and the receptor may fluctuate between the active and the inactive conformation, with the inactive conformation being more stable. Agonists and antagonists stabilize the association of domains A and B. As discussed above, the non-ligand bound state is susceptible to proteolysis while both agonist and antagonist binding reduce the susceptibility of the receptor to protease. Ligands stabilize a particular conformation by interacting with specific amino acids in both domains. Agonists stabilize a specific arrangement of these domains that can be recognized by the G protein, while antagonists stabilize an arrangement that cannot be recognized by the G protein. Mutations which lead to constitutive activity are likely to either decrease the stability of the inactive state or increase the stability of the active state.

CONCLUSIONS

Over the past several years G protein-coupled receptors have been the subject of intense investigation. The results of these studies have begun to shed light on the structure of this family of receptors, and on the structural changes that mediate transmembrane signalling. The application of biochemical and biophysical techniques to the study of these receptors will be needed to more precisely determine the three dimensional structure and to identify the conformational changes that occur during signal transduction.

References

Chidiac, P., Hebert, T.E., Valiquette, M., Dennis, M., Bouvier, M. (1994) Inverse agonist activity of beta-adrenergic antagonists. *Mol. Pharm.*, **45,** 490–9.

Cotecchia, S., Exum, S., Caron, M.G. and Lefkowitz, R.J. (1990) Regions of the alpha 1-adrenergic receptor involved in coupling to phosphatidylinositol hydrolysis and enhanced sensitivity of biological function. *Proc.Natl. Acad. Sci. USA*, **87,** 2896–2900.

Dixon, R.A.F., Kobilka, B.K., Strader, D.J., Benovic, J.L., Dohlman, H.G., Frielle, T., Bolanowski, M.A., Bennett, C.D., Rands, E., Diehl, R.E., Mumford, R.A., Slater, E.E., Sigal, I.S., Caron, M.G., Lefkowitz, R.J. and Strader, C.D. (1986) Cloning of the gene and cDNA for mammalian beta-2-adrenergic receptor and homology with rhodopsin. *Nature*, **321,** 75–79.

Dholman, H.G., Bouvier, M., Benovic, J.L., Caron, M.G., Lefkowitz, R.J. (1987) The multiple membrane spanning topography of the beta 2 adrenergic receptor. *J. Biol. Chem.*, **262,** 16655–16662.

Guan, X-M., Peroutka, S.J., Kobilka, B.K. (1992) Identification of a single amino acid residue responsible for binding of a class of beta receptor antagonists to 5HT1A receptors. *Molecular Pharmacology*, **41,** 695–698.

Kobilka, B.K., Dixon, R.A.F., Frielle, T., Dohlman, H.G., Bolanowski, M.A., Strader, C.D., Benovic, J.L., Sigal, I.S., Yang-Feng, T.L., Francke, U., Caron, M.G. and Lefkowitz, R.J. (1987) cDNA for the human beta-2 adrenergic receptor: A protein with multiple membrane-spanning domains and encoded by a gene whose chromosomal location is shared with that of the receptor for platelet-derived growth factor. *Proc. Natl. Acad. Sci. USA*, **84,** 46–50.

Kobilka, B.K., Kobilka, T.S., Regan, J.W., Caron, M.G. and Lefkowitz, R.J. (1988). Chimeric alpha-2-beta-2-adrenergic receptors: Delineation of domains involved in effector coupling and ligand binding specificity. *Science*, **240,** 1310–1316.

Kobilka, B.K. (1990) The role of cytosolic and membrane factors in processing of the human beta-2 adrenergic receptor following translocation and glycosylation in a cell-free system. *J. Biol. Chem.*, **265,** 7610–7618.

Maggio, R., Vogel, Z., Wess, J. (1993) Coexpression studies with mutant muscarinic/adrenergic receptors provide evidence for intermolecular "cross-talk" between G-protein-linked receptors. *Proc. Natl. Acad. Sci. USA*, **90,** 3103–7.

O'Dowd, B.F., Hnatowich, M., Regan, J., Leader, W.M., Caron, M.G., Lefkowitz, R.J. (1988) Site-directed mutagenesis of the cytoplasmic domains of the human beta 2 adrenergic receptor. *J. Biol. Chem.*, **263,** 15985–92.

Parma, J., Duprez, L., Van Sande, J., Cochaux, P., Gervy, C., Mockel, J., Dumont, J., Vassart, G. (1993) Somatic mutations in the thyrotropin receptor gene cause hyperfunctioning thyroid adenomas. *Nature*, **365,** 649–651.

Pittel, Z., Wess, J. (1994) Intramolecular interactions in muscarinic acetylcholine receptors studied with chimeric m2/m5 receptors. *Mol. Pharm.*, **45,** 61–64.

Robbins, L.S., Nadeau, J.H., Johnson, K.R., Kelly, M.A., Roselli-Rehfuss, L., Baack, E., Mountjoy, K.G., Cone, RD. (1993) Pigmentation phenotypes of variant extension locus alleles result from point mutations that alter MSH receptor function. *Cell*, **72,** 827–834.

Robinson, P.R., Cohen, G.B., Zhukovsky, E. and Oprian, D.D. (1992) Constitutively active mutants of rhodopsin. *Neuron*, **9,** 719–725.

Samama, P., Pei, G., Costa, T., Cotecchia, S., Lefkowitz, R.J. (1994) Negative antagonists promote an inactive conformation of the beta 2-adrenergic receptor. *Mol. Pharm.*, **45,** 390–4.

Shenker, A., Laue, L., Kosugi, S., Merendino, J.J., Minegishi, T., Cutler, G.B. (1993) A constitutively activating mutation of the lutenizing hormone receptor in familial male precocious puberty. *Nature*, **365,** 652–654.

Suryanarayana, S., Daunt, D.A., Von Zastrow, M. and Kobilka, B.K. (1991) A point mutation in the seventh hydrophobic domain of the alpha 2 adrenergic receptor increases its affinity for a family of beta receptor antagonists. *J. Biol. Chem.*, **266,** 15488–15492.

Suryanarayana, S. and Kobilka, B.K. (1993) Amino Acid Substitutions at Position 312 in the Seventh Hydrophobic Segment of the b_2 Adrenergic Receptor Modify Ligand Binding Specificity. *Mol. Pharm.*, **44,** 111–114.

Suryanarayana, Von Zastrow, M. and Kobilka, B.K. 1992. Identification of intramolecular interactions in adrenergic receptors. *J. Biol. Chem.*, **267,** 21991–21994.

Strader, C.D., Candelore, M.R., Hill, W.S., Sigal, I.S., Dixon, R.A.F. (1989) Identification of two serine residues involved in agonist activation of the beta adrenergic receptor. *J. Biol. Chem.*, **264,** 13572–78.

Strader, C.D., Gaffney, T., Sugg, E.E., Candelore, M.R., Keys, R., Patchett, A.A., Dixon, R.A.F. (1991) Allele-specific activation of genetically engineered receptors. *J. Biol. Chem.*, **266,** 5–8.

Strader, C.D., Sigal, I.S., Dixon, R.A. (1989) Structural basis of beta adrenergic receptor function. *FASEB J.*, **3,** 1825–1832.

Wang, H., Lipfert, L., Malbon, C.C., Bahouth, S. (1989) Site-directed anti-peptide antibodies define the topography of the beta adrenergic receptor. *J. Biol. Chem.*, **264,** 14424–14431.

Zhou, W., Flanagan, C., Ballesteros, J.A., Konvicka, K., Davidson, J.S., Weinstein, H., Millar, R.P., Sealfon, S.C. (1994) A reciprocal mutation supports helix 2 and Helix 7 proximity in the gonadotropin-releasing hormone receptor. *Mol. Pharm.*, **45,** 165–170.

THE β-HYDROXYL GROUP OF CATECHOLAMINES MAY INTERACT WITH SER[90] OF THE SECOND TRANSMEMBRANE HELIX OF THE α_{2A}-ADRENOCEPTOR

YING-OU LI, J. PAUL HIEBLE, DERK J. BERGSMA, ANN M. SWIFT, SUBINAY GANGULY and ROBERT R. RUFFOLO, JR.

Departments of Pharmacology, Molecular Genetics and Gene Expression Sciences, SmithKline Beecham Pharmaceuticals, 709 Swedeland Road, King of Prussia, PA, 19406, USA

Molecular modeling, in some cases combined with site-directed mutagenesis, has predicted several potential sites for interaction of the β-hydroxyl group of the catecholamine neurotransmitters with the β_2-adrenoceptor. These include Ser[165], Asn[292] and Ser[319] in transmembrane helices 4, 6 and 7, respectively. Although certain conserved residues have been shown to be required for interaction of catecholamines with all adrenoceptors (*e.g.*, Asp[113] in transmembrane helix 3, and one or more serines in transmembrane helix 5), agonist-receptor interactions have not been studied extensively in the case of the α-adrenoceptors. Ser[90], located in trans-membrane helix 2, is conserved in all α_1-and α_2-adrenoceptor subtypes, but not in β-adrenoceptors or dopamine receptors. Mutation of this residue in the α_{2A}-adrenoceptor to alanine results in only a 1–4 fold reduction in affinity for a series of α_2-adrenoceptor agonists, representing a variety of structural classes. However, a greater reduction in affinity was observed for the active (-)-enantiomers of epinephrine and norepinephrine (13–15 fold). These results suggest that the β-hydroxyl group of the catecholamines may interact with the α_{2A}-adrenoceptor via a hydrogen bonding interaction with Ser[90] in transmembrane helix 2. Because this residue is not present in the β-adrenoceptor, it is possible that the catecholamine neurotransmitters may interact in a different manner with α- and β-adrenoceptors.

KEY WORDS: α_2-adrenoceptor, site-directed mutagenesis, Easson-Stedman Hypothesis, enantiomers of norepinephrine

INTRODUCTION

The catecholamine neurotransmitters, norepinephrine and epinephrine, interact with both α- and β-adrenoceptors. Although the catecholamines show some selectivity between the β-adrenoceptor subtypes, they act as full agonists in all α- and β-adrenoceptor model systems thus far examined. The molecular structure of the adrenoceptors, as well as that of the other G-protein coupled receptors, is assumed to be best represented by seven-transmembrane helices, arranged in a roughly circular array about a central channel. Nearly all of the studies directed toward elicidating the mode of interaction between the catecholamines and adrenoceptor proteins have been done using the β-adrenoceptor, most commonly the β_2-adrenoceptor subtype, as the model. Most models of this receptor assume an orientation of seven transmembrane helices similar to that demonstrated for bacteriorhodopsin, and as-

Correspondence to: Robert R. Ruffolo, Jr., Ph.D., Division of Pharmacological Sciences, UW 2523, SmithKline Beecham Pharmaceuticals, 709 Swedeland Road, King of Prussia, PA, 19406, USA.

sumed for mammalian rhodopsin. This is based on several analogies between these molecules, including the proposed ligand-receptor interaction, where the ligand can bind to several of the transmembrane domains of the receptor protein in the pocket created by the circular array of transmembrane helices.

Site-directed mutagenesis has demonstrated that an aspartic acid residue, Asp^{113}, located in the third transmembrane helix, is required for agonist binding to the β_2-adrenoceptor. The free carboxyl group of aspartic acid is postulated to interact with the protonated amino group of the adrenergic ligands. This is consistent with the observation that aspartic acid is found at this position in all of the adrenoceptors, and that any mutation removing the carboxyl moiety completely eliminates agonist affinity for the receptor. Although the potential sites of agonist-receptor interaction have been studied much less extensively for the α-adrenoceptors, an aspartic acid in transmembrane helix 3 is also required for the interaction of catecholamines with α_2-adrenoceptors, inasmuch as mutation of Asp^{113} to asparagine eliminated specific binding of $[^3H]$- yohimbine to the α_{2A}-adrenoceptor (Wang *et al.*, 1991).

The *meta*- and *para*- hydroxyl groups of the catecholamine agonists are postulated to form hydrogen bonds to Ser^{204} and Ser^{207}, respectively, located in the fifth transmembrane helix of the β-adrenoceptors (Strader *et al.*, 1988). Binding to both Ser^{204} and Ser^{207} appears to be required for full agonist activity at the β-adrenoceptor, and interference with these interactions, either by mutation-induced removal of a serine, or removal of a catechol hydroxyl group, results in reduced affinity of agonists for the β-adrenoceptor (Strader *et al.*, 1989).

The interaction of a catecholamine agonist with transmembrane helix 5 may be slightly different between the α_2- and β_2-adrenoceptors. Thus, mutation studies on the α_{2A}-adrenoceptor suggest an interaction of Ser^{204} (which corresponds to Ser^{207} of the β_2-adrenoceptor) with the *para*-hydroxyl group but not between Ser^{200} (corresponding to Ser^{204} on the β_2-adrenoceptor) and the *meta*-hydroxyl (Wang *et al.*, 1991). It has been postulated that the *meta*-hydroxyl group of the catecholamines interacts with Cys^{201} of the α_2-adrenoceptor.

Less is known concerning the potential binding site for the β-hydroxyl group of the catecholamine neurotransmitters. The interaction of this group with the receptor is responsible for the stereoselective activation of the adrenoceptors by the catecholamines. Experimental data supports the prediction of the Easson-Stedman hypothesis (Easson and Stedman, 1933) that the (-)-, but not the (+)- or β-desoxy catecholamines, can participate in a three-point interaction with the adrenoceptor (Ruffolo *et al.*, 1994).

MaloneyHuss and Lybrand (1992) have proposed a β_2-adrenoceptor model differing from that commonly proposed, with the transmembrane helices in a "mirror-image" orientation to that found in bacteriorhodopsin. This model will accommodate the binding of (-)-epinephrine to Asp^{113}, Ser^{204} and Ser^{207} as proposed in the other models, and shows a potential interation between the β-hydroxyl group with Ser^{319} in the seventh transmembrane helix. Mutation of Ser^{319} to alanine attenuates the ability of isoproterenol to activate adenylate cyclase (Dixon *et al.*, 1988), although the specificity of this effect was not examined. Alternatively, using the bacteriorhodopsin model, the β-hydroxyl group of norepinephrine has been proposed to hydrogen bond to Ser^{165} (Strosberg, 1993) in transmembrane helix 4.. The effects of mutation of Ser^{165} on agonist-receptor interaction have not yet been reported. Recently, mutation of Asn^{292} in transmembrane helix 6 to leucine was shown to produce a 5-fold reduction in the stereoselectivity of isoproterenol as an inhibitor of radioligand binding to the β_2-adrenoceptor (Wieland *et al.*, 1994). Conversion of Asn^{292} to aspartic acid had no effect, indicating that the β-hydroxyl group may interact in an equivalent fashion with a carboxyl or carboxamide group.

We were interested in evaluating potential molecular interactions of agonists with the α_{2A}-adrenoceptor, with the specific goal of determining a possible site of interaction for the β-hydroxyl group. As noted above, changes in its interaction with the receptor should be reflected in the stereoselectivity of the agonist-receptor interaction. As an initial guide to chosing sites for mutation, we compared the amino acid sequences of α-adrenoceptors and dopamine receptors, seeking potential hydrogen bonding sites conserved in all of the α-adrenoceptors but not in the dopamine receptors, whose physiological neurotransmitter, dopamine, lacks the β-hydroxyl group. Only one such site was found, $Ser,^{90}$ which is conserved in all six α-adrenoceptors but not in the three β-adrenoceptors or the five dopamine receptor subtypes. Ser^{90} was converted to alanine, to remove the potential hydrogen bonding interaction with the serine hydroxyl.

METHODS

Cloning of human α_{2A}-adrenoceptor

α_{2A}-adrenoceptor A 4.5 kb cDNA clone encoding the human α_{2A}-adrenoceptor was isolated from a human hippocampus cDNA library using as hydrodization probes a series of degenerate symthetic oligonucleotides homologous to the seven transmembrane domains of serotonin receptors. A 1.4 kb fragment encompassing the entire coding region of the cDNA (450 amino acids) was enzymatically excised from the original 4.5 kb clone and subcloned into Eco RI/Bam HI restriction sites of the pALTER plasmid vector for site-directed mutagenesis.

Preparation of mutant receptor

Site-directed mutagenseis was performed according to manufacturers directions (Altered Sites Mutagenesis System, Promega). The mutant gene was then inserted into the expression vector pCDN (Aiyar *et al.*, 1994) using the EcoRI and BamHI sites.

Transient expression in COS cells

COS cells were maintained in DMEM, 10% fetal bovine serum (GIBCO) at 37 °C in 5% CO_2. Cells were plated the day before transfection in multiple T150 flasks and transfected using the DEAE-dextran method (McCutchan and Pagano., 1968) with 40 μg DNA per flask. After 72 hours, cells were washed with BPS, pelleted, then frozen in a dry ice/ethanol bath, and stored at −70 °C.

Membrane preparation

Previously frozen cell pellets were defrosted and homogenized in a Brinkman Polytron with cold TME buffer (50 mM Tris, 12.5 mM $MgCl_2$, 5 mM EDTA, pH 7.4). The homogenate was centrifuged at 100,000 × g for 30 minutes at 4 °C. The membrane pellet was resuspended in TME buffer, pH 7.4.

Inhibition studies

Membranes were incubated with test antagonists at concentrations ranging from 0.1 nM to 100 μM. Assays were initiated by the addition of membrane protein, and incubated at 25 °C for 45 minutes. 50 mM Tris-EDTA buffer, pH 7.4, was used for [^{3}H] rauwolscine and [^{3}H] clonidine. Radioligands were present at concentrations near their K_d values. Non-specific binding was defined using 10 μM phentolamine. K_i values were calculated using the London software program for nonlinear regression.

RESULTS

When [^{3}H]-rauwolscine was used to label the α_2-adrenoceptors in COS cells transiently expressing the α_{2A}-adrenoceptor, relatively high (-)-norepinephrine concentrations were required to inhibit binding (K_i = 0.7 μM). Furthermore, dopamine was nearly equipotent with (-)-norepinephrine (K_i = 0.9 μM), indicating that this assay was not suitable for detecting agonist-receptor interactions dependent on the β-hydroxyl group of catecholamines.

Using the agonist radioligand, [^{3}H] -clonidine, the catecholamines were found to inhibit binding to the expressed α_{2A}-adrenoceptor with potencies consistent with their functional activity in α_2-adrenoceptor assays. When the mutant and wild-type α_{2A}-adrenoceptors were expressed in COS cells, the mutant receptors showed 13-fold lower affinity for (-)-norepinephrine.

The decrease in affinity induced by mutation was selective for (-) norepinephrine, inasmuch as the affinities of (+)-norepinephrine, dopamine and clonidine for Mutant 1 were nearly equal to those for the wild type receptor (Table 1). The affinity of several additional α-adrenoceptor agonists and antagonists for wild type and mutant α_{2A}-adrenoceptors were determined (Table 1). Agents tested included examples of several structural classes, including phenethylamines, aminotetralins and imidazolines. With the exception of the active enantiomers of norepinephrine and epinephrine, which showed potency ratios of 13 and 15, respectively, between the wild type and mutant receptors, all other agents were found to have potency ratios of 1.1 and 4.4 between the wild type and mutant receptor.

TABLE 1
Comparison of the Ability to Inhibit [^{3}H] Clonidine Binding to Wild Type and Mutated α_{2A}-Adrenoceptors

Compound	*K_i(WT)[a] (nM)*	*K_i (Mutant)[a] (nM)*	*Ratio*
(-)NE	31 ± 6	409 ± 47	13.2
(+)-NE	3549 ± 800	4512 ± 493	1.3
(-)-Epi	77 ± 9	1159 ± 143	15.1
Dopamine	329 ± 69	668 ± 99	2.0
Epinine	360 ± 139	1127 ± 380	3.1
(-) Ephedrine	2459 ± 606	5312 ± 2272	2.2
TL-99[b]	14 ± 5	35 ± 6	2.5
6,7-ADTN	155 ± 65	487 ± 160	3.1
M-7[c]	87 ± 25	97 ± 25	1.1
Tolazoline	167 ± 32	302 ± 68	1.8
(-) α-OH Tolazoline	565 ± 128	2468 ± 475	4.4
(+) α-OH Tolazoline	1264 ± 243	4903 ± 1351	3.8
Clonidine	4 ± 1	7 ± 3	1.8

[a] Inhibition of [^{3}H] clonidine binding to membrane homogenates of COS cells transiently expressing receptor.
[b] N,N-dimethyl- 6,7-ADTN.
[c] N,N-dimethyl- 5,6-ADTN.

DISCUSSION

Mutation of Ser,[90] located in transmembrane helix 2 of the α_{2A}-adrenoceptor, to alanine selectively reduces the affinity of the receptor for the active enantiomers of norepinephrine and epinephrine. While this does not conclusively prove that this amino acid is directly involved with the interaction between the receptor and β-hydroxyl group, some interaction involving the hydroxyl group of this serine appears to be required to configure the α_{2A}-adrenoceptor properly for interaction with these neurotransmitters.

The potential interaction between agonists and Ser[90] requires additional structural elements in addition to the β-hydroxyl group. The active (-)-enantiomer of ephedrine, which bears a β-hydroxyl, but no ring hydroxyl groups, shows only a 2.2-fold difference in potency between wild type and mutant α_{2A}-adrenoceptors, no different from that observed with agonists lacking an aliphatic hydroxyl group (Table 1). The (+) enantiomer of ephedrine shows virtually no affinity (K_i > 10 μM) for either mutant or wild type α_{2A}-adrenoceptor. It is possible that the ring hydroxyl groups are required to "lock" the molecule into a conformation bringing the β-hydroxyl group in proximity to Ser.[90]

While the introduction of a β-hydroxyl group increases the wild-type/mutant potency ratio of the imidazoline α-adrenoceptor antagonist, tolazoline, about twofold (Table 1), the effect is not stereoselective, with similar ratios being observed for the (-)- and (+)- enantiomers of the hydroxylated derivative. Hence this hydroxyl group is apparently not in the proper position to be influenced by Ser.[90]

Because Ser[90] is not present in the β-adrenoceptor subtypes, which also are activated by the catecholamines, a stereoselective effect of mutation at this position on catecholamine affinity for the α_{2A}-adrenoceptor implies a difference in the mode of interaction of these neurotransmitters with α- and β-adrenoceptors. Members of another structural class of adrenoceptor agonist, the catecholimidazolines, have been shown to interact differently with α-and β-adrenoceptors, since the relative potency of optically active analogs to activate β-, but not α-adrenoceptors adheres to the Easson-Stedman hypothesis (Ruffolo *et al.*, 1983).

It is possible that the catecholamine β-hydroxyl group is directed in opposite directions when interacting with α- (toward transmembrane helix 2) and β-(toward transmembrane helix 4) adrenoceptors (Figure 1). It is possible that the ring hydroxyl groups bind in a different orientation between α-and β-adrenoceptors (Wang *et al.*, 1991). Alternatively, even if the binding orientation of the ring hydroxyl groups is similar between the adrenoceptor classes, it is possible that catecholamine binding to α- and β-adrenoceptors occurs with the alkyl side-chain and aromatic ring in different rotameric conformations.

This premise is supported by functional data showing that the adrenoceptor selectivity of the 2- and 6-fluoro analogs of norepinephrine differ dramatically. 2-Fluoronorepinephrine is equipotent with norepinephrine at β-adrenoceptors, but 10–30 fold weaker at the α-adrenoceptor. In contrast, 6-fluoronorepinephrine is equipotent with norepinephrine at the α-adrenoceptor, but 10–100 fold weaker at the β-adrenoceptor (Nimit *et al.*, 1980). Because 2- or 6-fluoro substitution of dopamine, or 5-fluoro substitution of norepinephrine, had little effect on its activity at either α- or β-adrenoceptors (Nimit *et al.*, 1980), fluoro-substitution of norepinephrine must be influencing adrenoceptor selectivity via stabilization of a particular rotameric conformation through interaction between halogen and β-hydroxyl groups.

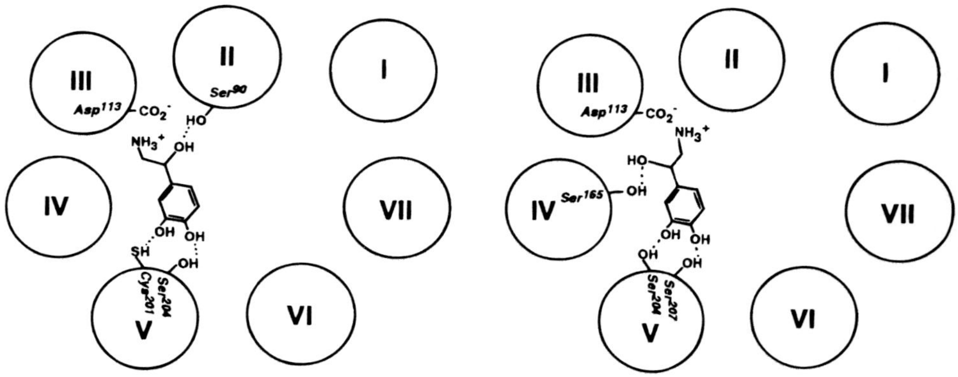

FIGURE 1 Schematic drawing showing the potential interaction of norepinephrine with the α_{2A}-adrenoceptor (A) and β_2-adrenoceptor (B).

A difference in rotameric preference between α- and β-adrenoceptors is also observed with the hydroxylated 2-aminotetrahydronaphthalenes. Although the nature of the amino substitutent will influence the absolute potency at both for α- and β-adrenoceptors, 6,7-dihydroxy substitution is consistently preferred for activity at the α-adrenoceptor (Kohli *et al.*, 1978), while 5,6-dihydroxysubstitution favors agonist activity at the β-adrenoceptor (Volkman *et al.*, 1977).

Hence, if the catecholamines do indeed interact with α- and β-adrenoceptors in a different agonist-receptor orientation, Ser,[90] in transmembrane helix 2 represents a likely site of interaction for the β-hydroxyl group of catecholamines with the α-adrenoceptors.

References

Aiyar, N., Baker, E., Wu H.-L., Nambi, P., Edwards, R.M., Trill, J.J., Ellis, C. and Bergsma, D.J. (1994) Human AT1 receptor is a single copy gene: Characterization in a stable cell line. *Mol. and Cell. Biochem.*, **131,** 75–86.

Dixon, R.A.F., Sigal, I.S. and Strader, C.D. (1988) Structure-function analysis of the β-adrenergic receptor. *Cold Spring Harbor Symposium on Quantitative Biology* **53,** 487–497.

Easson, L.H. and Stedman, E. (1933) Studies on the relationship between chemical constitution and physiological action. V. Molecular dissymmetry and physiological activity. *Biochem. J.*, **27,** 1257–1266.

Kohli, J.D., Volkman, P.H., Goldberg, L.I. and Cannon, J.G. (1978) Conformational requirements for vascular α-adrenergic activity. *Proc. Soc. Exp. Biol. Med.*, **158,** 28–31.

MaloneyHuss, K. and Lybrand, T.P. (1992) Three-dimensional structure for the β_2-adrenergic receptor protein based on computer modeling studies. *J. Mol. Bio.*, **225,** 859–871.

McCutchan, J.H. and Pagano, J.S. (1968) Enhancement of the infectivity of simian virus 40 deoxyribonucleic acid with diethylaminoethyl-dextran. *J. Natl. Cancer. Inst.*, **41,** 351–357.

Nimit, Y., Cantacuzene, D., Kirk, K.L., Creveling, C.R. and Daly, J.W. (1980) The binding of fluorocatecholamines to adrenergic and dopaminergic receptors in rat brain membranes. *Life Sci.*, **27,** 1577–1585.

Ruffolo, R.R., Jr., Patil, P.N. and Miller, D.D. (1983) Adrenoceptor-mediated effects of optically active catecholimidazolines. *Naunyn-Schmiedeberg's Arch. Pharmacol.*, **323,** 221–227.

Ruffolo, R.R., Jr., Stadel, J.M. and Hieble, J.P. (1994) α-Adrenoceptors: Recent developments. *Med. Res. Rev.*, **14,** 229–270.

Strader, C.D., Candelore, M.R., Hill, W.S., Sigal, I.S. and Dixon, R.A.F. (1989) Identification of two serine residues involved in agonist activation and the β-adrenergic receptor. *J. Biol. Chem.*, **264,** 13572–13578.

Strader, C.D., Sigal, I.S., Candelore, M.R., Rands, E., Hill, W.S. and Dixon, R.A.F. (1988) Conserved aspartic acid residues 79 and 113 of the β-adrenergic receptor have different roles in receptor function. *J. Biol. Chem.*, **263,** 10267–10271.

Strosberg, A.D. (1993) Structure, function and regularion of adrenergic receptors, *Protein Science*, **2,** 1198–1209.

Volkman, P.H., Kohli, J.D., Goldberg, L.I., Cannon, J.G. and Lee, T. (1977) Conformational requirements for dopamine-induced vasodilation. *Proc. Natl. Acad. Sci. USA*, **74,** 3602–3606.

Wang, C.-D., Buck, M.A. and Fraser, C.M. (1991) Site-directed mutagenesis of α_{2A}-adrenergic receptors: Identification of amino acids involved in ligand binding and receptor activation by agonists. *Mol. Pharmacol.*, **40,** 168–179.

Wieland, K., Ijzerman, A.P. and Lohse, M.J. (1994) Involvement of asparagine-292 in the human β_2-adrenergic receptor in stereospecific agonist binding., *Naunyn-Schmiedeberg's Arch. Pharmacol.*, **Suppl. 349,** R10.

REGULATION OF ALPHA$_2$-ADRENERGIC RECEPTOR SIGNALLING

STEPHEN M. LANIER

Department of Pharmacology, Medical University of South Carolina, Charleston, South Carolina

Alpha$_2$-adrenergic receptors (α_2-AR) mediate a number of cardiovascular and metabolic effects of catecholamines and are members of a large family of receptors which transduce their cell signals through heterotrimeric guanine nucleotide binding proteins. Utilizing the α_2-adrenergic receptor system as a representative subfamily of G-protein coupled receptors, we are attempting to define determinants of signalling specificity. One objective is to define the microenvironment of α_2-adrenergic receptors in different cell phenotypes and determine how differences in these microenvironments and cell architecture contribute to cell type specific signaling by the receptor.

The strategy entails the use of α_2-AR subtypes stably expressed in NIH-3T3 fibroblasts and the pheochromocytoma cell line PC-12, two cell types with distinct infrastructure and in which the receptors couple to different signalling pathways. Although R, G and E are the major entities involved in signal propagation, other associated proteins may be critical for efficient and specific signal transduction. These proteins may be cytoskeletal elements participating in compartmentalization and "localized signal transduction" or molecules that regulate events at the R-G or G-E interface. A more complete understanding of cellular plasticity with respect to signal transduction may lead to a new level of therapeutic specificity and the development of therapeutic approaches that perhaps target the R-G or G-E interface as opposed to the receptor's hormone binding site.

KEY WORDS: α_2-adrenoceptors, NIH-3T3 cells, PC-12 cells, receptor coupling, G-proteins, signal transduction

INTRODUCTION

Cell to cell communication by a variety of neurotransmitters and hormones involves three major entities: receptor (R), G-protein (G) and effector (E). Plasticity in this system is conferred by: 1) the existence of each member of the transduction pathway as isoforms/subtypes that are expressed in a tissue-specific and developmentally-regulated manner; 2) the recognition of multiple G-proteins by a single receptor; 3) the coupling of one G to multiple effectors and 4) the interaction of multiple G-proteins with a single effector. Factors that determine signal specificity for this system in the cell are poorly understood. Such factors include R-G or G-E affinity or stoichiometry and possibly the colocalization of R-G-E in cellular microdomains or accessory proteins that regulate R-G or G-E interaction. Utilizing the α_2-adrenergic receptor system as a representative subfamily of G-protein coupled receptors, our goal is to address the role of each of these factors within the natural environment of the cell.

Several experimental approaches have combined to provide our present understanding of cell signaling via G-protein coupled receptors, each with various advantages and dis-

Correspondence to: Stephen M. Lanier, Ph.D., Department of Pharmacology, Medical University of South Carolina, 171 Ashley Avenue, Charleston, South Carolina, 29425.

advantages. They are in fact complementary. A large portion of our basic understanding of the signalling process has come from reconstitution of the individual transducing entities in a defined environment. These studies have told us what is possible but they do not tell us what happens in the cell itself, and in many cases do not indicate the signaling specificity that is observed in the intact cell. Due in part to limited sources of different homogeneous cells/tissue expressing the receptor subtypes, it is difficult to determine factors that actually engineer this specificity. The issue of limited sources has been overcome in part by the use of cells stably transfected with the receptor gene/cDNA.

RESULTS

Utilizing the α_2-AR subtype genes/cDNA, we developed a defined system in which aspects of signal transduction specificity can be addressed within the natural environment of the cell (Duzic *et al.*, 1991; Duzic *et al.*, 1992; Coupry *et al.*, 1992). The cell types selected for transfection (NIH-3T3 fibroblasts, PC-12 pheochromocytoma cells) are widely used for analysis of different receptor-effector coupling systems and do not express endogenous α_2-AR as determined by radioligand binding and RNA blot analysis with probes derived from the conserved regions of the three rat subtype genes. Each cell type was stably transfected by calcium phosphate coprecipitation or electroporation. Isolated clonal cell lines were screened by binding with the α_2-AR radioligand ^{3}H-rauwolscine and subsequently by RNA blots as one indicator of appropriate gene insertion. To allow detailed analysis of the receptor coupling mechanisms and to confirm observations in more than one transfectant we generated a number of clonal cell lines expressing each receptor subtype over a range of receptor densities.

Transfectants were evaluated by several criteria to establish that the receptor gene is not modified during gene insertion. In addition to screening the transfectants by expected ligand recognition properties the clonal cell lines were analyzed by RNA blot hybridization. Only transfectants expressing appropriate size of receptor mRNA are utilized for further studies. The expected size of receptor mRNA can be calculated from the expression vector construct and this analysis serves as a strong indicator of appropriate gene insertion (i.e. that the promotor elements, the coding region and the 3′untranslated sequence through the SV40 polyadenylation site in the vector remain intact). To provide additional evidence for proper gene insertion and appropriate processing of the receptor, the expressed protein was identified by photoaffinity labeling with ^{125}I-AZPC. The size of the identified receptor by SDS-PAGE is similar to that observed previously in tissue homogenates.

Our work in this system (Lanier *et al.*, 1991; Duzic *et al.*, 1992; Coupry *et al.*, 1992; Duzic and Lanier, 1992; Tian *et al.*, 1994) indicates that 1) there are differences in R-G coupling among the subtypes when expressed in the same environment and 2) the receptor subtype adapts to its environment in terms of signalling pathways. The question is how is this specificity generated?

Receptor-effector coupling in NIH-3T3 cells expressing the three receptor subtypes. Based on differences in R-G interaction and effector coupling, the α_{2C}-AR differs from the α_{2B} and $\alpha_{2A/D}$-AR subtypes in its ability to recognize pertussis toxin (PT)-sensitive G-proteins expressed in NIH-3T3 fibroblasts (Gi2, Gi3) (Duzic *et al.*, 1992). To determine if the α_{2C}-AR subtype can interact with PT sensitive G-proteins not expressed in these cells, we reanalyzed R-G coupling after co-transfecting NIH-3T3 cells (which express G_{i2} and G_{i3})

with the α_{2C}-AR gene and $G_{o\alpha1}$ cDNA. These experiments indicate that the α_{2C}-AR preferentially couples to $G_{o\alpha1}$ (Coupry *et al.*, 1992). These data suggest that R-G affinity is one of the first determinants of signalling specificity.

Cell specific coupling of the receptor subtypes This issue was initially tested by determining receptor coupling to adenylylcyclase following stable expression of the rat α_{2B}- and $\alpha_{2A/D}$-AR subtypes in three different cell phenotypes: NIH-3T3 fibroblasts, DDT_1 MF-2 smooth muscle cells, and pheochromocytoma PC-12 cells (Duzic and Lanier, 1992). These data indicate that the cellular response mediated by an α_2-AR subtype is cell specific and thus depends on its environment. These observations can be summarized as follows:

1) In NIH-3T3 cells the α_{2B} and α_{2D}-AR subtypes mediate inhibition of forskolin induced elevation of cellular cAMP with similar efficiency. This inhibition is PT sensitive and is primarily meditated through Giα2.
2) However, in PC-12 cells the α_{2B}-AR only potentiates the effect of forskolin on cellular cAMP, whereas both inhibition and potentiation are observed with the α_{2D}-AR transfectants depending on agonist concentration. The potentiating effect is PT insensitive, does not involve altered extracellular cAMP extrusion and is enhanced following PT treatment suggesting that both "stimulatory PT-insensitive" and "inhibitory PT-sensitive" G-proteins are competing for coupling to the receptor. The levels of Giα2 and G11/q per ug membrane protein are similar in PC-12 and NIH-3T3 transfectants. Receptor activation does not augment the increase in cAMP elicited by PGE_2 acting through Gs suggesting that the effect of α_2-AR activation does not involve "Gsα-primed" enzyme and G$\beta\gamma$.
3) In PC-12 transfectants, both receptor subtypes mediate inhibition of membrane adenylylcyclase activity—the potentiating effect requires the intact cell.
4) In PC-12 cells, but not NIH-3T3, both receptor subtypes elicit measurable increases in intracellular calcium.
5) Chelation of intracellular calcium prevents the potentiating effect of receptor activation in PC-12 cells and reveals the more "classical" inhibition of the forskolin induced increases in cellular cAMP.
6) The potentiating effect of receptor activation on forskolin-elevated cAMP in α_{2B}-AR but not $\alpha_{2A/D}$-AR transfectants is blocked by reduction of extracellular calcium by preincubation with EGTA.
7) PC-12 cells but not NIH-3T3 cells express the Ca^{2+}/calmodulin regulated adenylylcyclase Type III which may account for the ability of receptor activation to augment adenylylcyclase activity. However, an additional determinant of specificity operates reflected in cell-specific increases in intracellular calcium.

DISCUSSION

Although R, G and E are the major entities in the signalling process, other proteins may be involved in facilitating this process either indirectly by structural support or compartmentalization, crosstalk or directly by regulating events at the R-G or G-E interface. Cell-specific structure may operate as a determinant of signalling specificity either by spatial or temporal considerations or through functional organization of cellular microdomains. The interaction of membrane bound receptors, G-proteins, and effectors with the

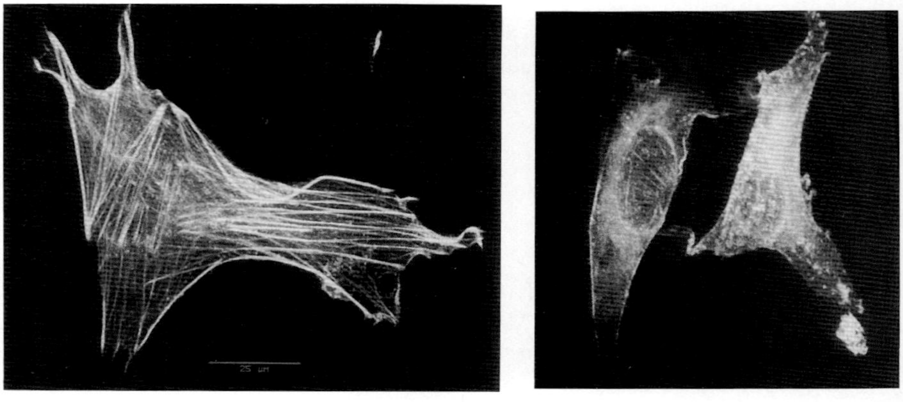

(A)

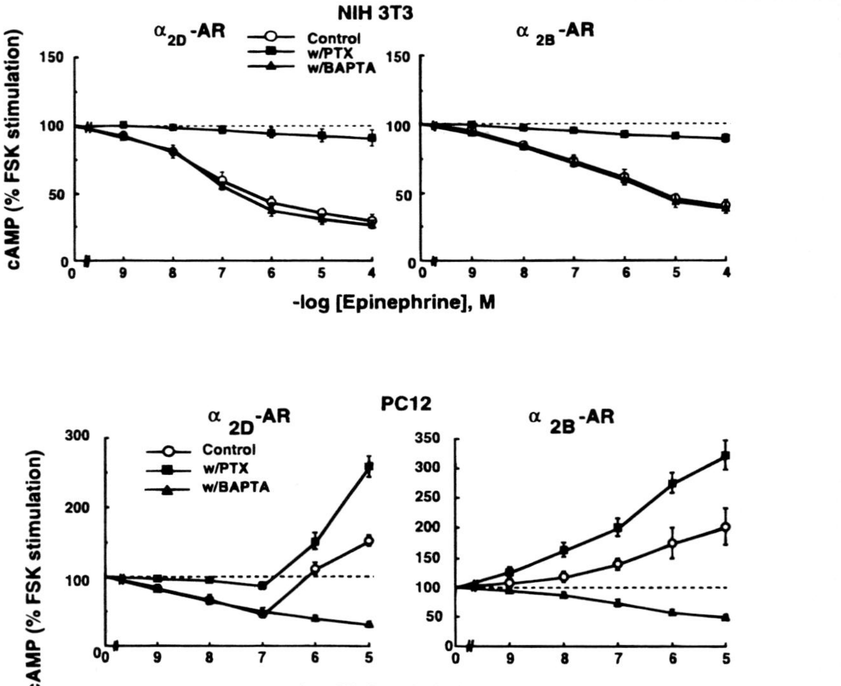

(B)

FIGURE 1 Cell architecture and effects of receptor activation on cellular cAMP in PC-12 and NIH-3T3 transfectants. (A) Actin microfilaments in NIH-3T3 fibroblasts (left) and PC-12 cells (right). Cells were grown on as monolayer on plastic removable slides and fixed with paraformaldehyde-glutaraldehyde by standard procedures. Actin was stained with rhodamine phalloidin and visualized by confocal microscopy. (B) Confluent plates of transfected NIH-3T3 and PC-12 cells were incubated with 10 μM forskolin plus increasing concentrations of epinephrine for 10 min at 37°C. cAMP was extracted and assayed as described (2). The receptor densities in NIH-3T3 and PC-12 transfectants range from 1100–2000 fmol of receptor/mg of protein. The levels of cAMP (nmol/mg/protein) in the absence and presence of forskolin in the NIH-3T3 and PC-12 transfectants were 0.77 ± 0.09, 0.12 ± 0.04 and 7.93 ± 1.14, 1.22 ± 0.23. Data are expressed as percent inhibition or augmentation of forskolin-induced elevation of cAMP and represent the mean ± S.E. of four separate experiments performed in duplicate. Panel B is modified from (Duzic *et al.*, 1992).

cytoskeleton is not clearly understood. Is there a signal transduction network whose assembly is facilitated by agonist or does agonist activate a preformed network (i.e. precoupled receptor)? Is the signal transduction network for this system analogous to that observed for receptors with a single transmembrane span (i.e. PDGF receptor) wherein binding of agonist stimulates a series of protein-protein interactions dependent on protein phosphorylation? These issues can be addressed by defining the receptor's microenvironment in various cell types.

The molecular cloning of the various members of the signal transduction pathway associated with G-protein coupled receptors has provided a wealth of information regarding how these molecules actually elicit a cellular response. However, the rapid progress achieved over the last six years has provided surprising and unexpected information, particularly the existence of each member of the transduction pathway (R-G-E) as isoforms/subtypes that are expressed in a tissue specific manner. These observations present a major challenge to investigators in this field. What are the factors involved in regulating the apparent plasticity of signal transduction? Can this information be "exploited" to provide a new level of therapeutic specificity? A more complete understanding of cellular plasticity with respect to signal transduction by each receptor subtype may lead to the development of therapeutic approaches that perhaps target the R-G or G-E interface as opposed to the receptor's hormone binding site. Development of the latter will require precise knowledge of the events that occur at the R-G and G-E interface, the identity of important accessory proteins in the "signal transduction network", and the development of special mechanisms to place such agents at their site of action.

ACKNOWLEDGMENTS

The work summarized represents the efforts of Sean Downing (Georgetown University) and Drs. Emir Duzic (Lederle Laboratories) and Isabelle Coupry (I.S.E.R.M. #0308 Toulouse, France). The assistance of Bradley Miller and Dr. Steven Rozensweig (Department of Pharmacology, Medical University of South Carolina) in preparing the confocal micrograph is appreciated. Due to space consideration, the citation list is limited; a complete reference list is found in the manuscripts cited. This work was supported by National Institute of Health Grant ROI-NS24821 and The Council for Tobacco Research—2235.

References

Duzic, E. and Lanier, S.M. (1992) Factors determining the specificity of signal transduction by G-protein coupled receptors. III. Coupling of α_2-adrenergic receptor subtypes in a cell type-specific manner. *J. Biol. Chem.*, **267,** 24045–24052.

Duzic, E., Coupry, I., Downing, S. and Lanier, S.M. (1992) Factors determining the specificity of signal transduction by G-protein coupled receptors. I. Coupling of α_2-adrenergic receptor subtypes to distinct G-proteins. *J. Biol. Chem.*, **267,** 9844–9851.

Coupry, I., Duzic, E., Lanier, S.M. (1992) Factors determining the specificity of signal transduction by G-protein coupled receptors. II. Preferential coupling of the alpha$_{2C}$-adrenergic receptor to the guanine nucleotide binding protein, G_o. *J. Biol. Chem.*, **267,** 9852–9857.

Lanier, S.M., Downing, S., Duzic, E. and Homcy, C.J. (1991) Isolation of rat genomic clones encoding subtypes of the α_2-adrenergic receptor: Identification of a unique receptor subtype. *J. Biol. Chem.*, **266,** 10470–10478.

Tian, W., Duzic, E., Lanier, S.M. and Deth, R.C. (1994) Determinants of α_2-adrenergic receptor activation of G proteins: Evidence for a precoupled receptor/G protein state. *Mol. Pharmacol.*, **45,** 524–531.

SPECIFICITY IN α_2-ADRENOCEPTOR SIGNAL TRANSDUCTION: RECEPTOR SUBTYPES, COUPLING TO DISTINCT SIGNAL TRANSDUCTION PATHWAYS AND LOCALIZATION TO DISCRETE SUBDOMAINS IN TARGET CELLS

LEE E. LIMBIRD, LEIGH B. MACMILLAN and JEFFREY R. KEEFER

Vanderbilt University Medical Center, Department of Pharmacology, Nashville, Tennessee 37240, USA

KEY WORDS: α_2-adrenoceptors, signal transduction, chimeric receptors, receptor coupling, receptor localization

Specificity in signal transduction can be conferred by at least three mechanisms. The existence of multiple subtypes for receptors, particularly when they couple to distinct signal transduction pathways, offers one mechanism for specificity in signal transduction via endogenous agonists. However, the three α_2-adrenergic receptor subtypes, α_{2A}/α_{2D}, α_{2B} and α_{2C}, all have been shown to couple to the same signal transduction pathways in a variety of target cells (MacKinnon *et al.*, 1994), and it is premature to conclude whether subtle subtype-selective differences in coupling to various GTP binding proteins (Duzic *et al.*, 1992) or in α_2-receptor turnover (Eason and Liggett, 1992; Pleus *et al.*, 1993) noted to date provide important mechanisms by which epinephrine and norepinephrine would encode for functional specificity in various target cells. However, two additional mechanisms for conferring specificity may represent reasonable strategies by which catecholamines coordinate cellular function in a highly specific way, i.e. receptor coupling to discrete signal transduction pathways in one physiological setting versus another and receptor localization to discrete microdomains on the cell surface. This chapter will emphasize the latter two mechanisms for conferring specificity of cellular regulation by epinephrine and norepinephrine on different target cells.

COUPLING OF α_{2A}-ADRENERGIC RECEPTORS TO MULTIPLE SIGNAL TRANSDUCTION PATHWAYS

The α_{2A}-adrenergic receptor (α_{2A}AR) was first categorized as a receptor population capable of coupling to inhibition of adenylyl cyclase and attenuating cyclic AMP production in target cells. However, it soon became clear that although attenuation of cyclic AMP

Correspondence to: L. E. Limbird, Vanderbilt University Medical Center, Department of Pharmacology, Nashville, Tennessee.

accumulation could account for α_{2A}AR-mediated control of metabolic events such as decreased glycogenolysis or lipolysis or transepithelial modulation of Na^+/H^+ exchange in the kidney, this biochemical event was not sufficient to account for α_{2A}AR-elicited effects on excitation-secretion and excitation-contraction coupling. In many target cells, particularly electrically-excitable cells, the α_{2A}AR also has been demonstrated to couple to activation of receptor-operated K^+ currents as well as to inhibition of voltage-gated Ca^{++} currents, both representing electrical events that could effect attenuation of secretion from neuroendocrine and neuronal cells. Additional signaling pathways also have been demonstrated for the α_2-adrenergic receptor in native cells or target tissues, including activation of phospholipase A_2 (Sweatt *et al.*, 1986) and, via independent mechanisms, activation of phospholipase D (MacNulty *et al.*, 1992). In all cases, α_2-adrenergic receptor activation of these pathways is pertussis-toxin sensitive, suggesting that GTP-binding proteins in the Gi/Go subfamily are responsible for eliciting these effects.

We have been particularly interested in identifying the precise signal transduction pathways utilized by α_{2A}-adrenergic receptors in various target cells as a first step toward identifying novel targets for therapeutic intervention. In early studies in our laboratory, we established that activation of human platelet secretion via α_{2A}-adrenergic receptors resulted from activation of a pathway involving Na^+/H^+ exchange. These studies, via a circuitous experimental route, led us to the discovery that binding to homogenous preparations of α_{2A}AR is allosterically modulated by monovalent cations ($Na^+ \geq Li^+ > K^+$) and amiloride analogs (Nunnari *et al.*, 1987), although there is no evidence that the α_2-receptor itself evokes Na^+ or H^+ counter-transport (Keefer *et al.*, 1994a; Wilson *et al.*, 1992).

In studies intended to eliminate allosteric modulation of the α_{2A}AR by cations, we mutated an aspartate residue predicted to be embedded in the second transmembrane domain of the receptor to a corresponding asparagine residue (designated the D79N mutant α_{2A}AR) and demonstrated that cation modulation of ligand binding was eliminated, although adrenergic ligand binding and allosteric modulation by amiloride analogs were comparable to the wild-type receptor (Horstman *et al.*, 1990). Introduction of the D79N mutant α_{2A}AR into AtT20 anterior pituitary cells in culture demonstrated that the wildtype α_{2A}AR was able to couple to inhibition of adenylyl cyclase, activation of K^+ currents, and inhibition of ψ-gated Ca^{++} currents. In contrast, the D79N α_{2A}AR selectively lost its ability to activate receptor-operated K^+ currents (Surprenant *et al.*, 1992).

The selective loss of receptor coupling to K^+ currents by the D79N α_{2A}AR may indicate that the receptor coupling to K^+ currents requires a higher stoichiometry of receptor-G protein encounters than coupling to other effectors, perhaps due to the role of $\beta\gamma$ subunits of GTP-binding proteins in mediating receptor-operated K^+ currents (Clapham and Neer, 1993). Whatever the molecular mechanism, however, this selective elimination of receptor coupling to one but not another electrical consequence of α_{2A}AR activation suggests that introduction of this mutation into the mouse genome might provide a powerful tool for determining which electrical consequence of the α_{2A}AR activation, namely suppression of voltage-gated Ca^{++} currents or activation of receptor-operated K^+ currents, is involved in α_{2A}AR-mediated control of multicellular processes, such as suppression of epileptogenesis, modification of pain perception, central control of blood pressure, and influence of mood. It is for this reason that we have undertaken homologous recombination strategies to substitute a mutated gene coding for a D79N α_{2A}AR into the mouse genome at the native α_{2A}AR locus in such a way that this mutant receptor will be expressed at the appropriate time and place during development, at the density characteristic of WT α_{2A}AR expression in various target cells, and in the cell surface microdomain characteristic of the WT α_{2A}AR. The targeting vector that we have designed for this purpose is shown in Figure 1, and corresponds to the "hit and run" strategy developed for creating targeted subtle mutations in the mouse genome (Ramirez-Solis *et al.*, 1993).

Electroporation-mediated introduction of this targeting vector into totipotent embryonic stem cells has, as confirmed by Southern analysis, resulted in targeted insertion of this vector into the appropriate α_{2A}AR locus in 15 neomycin-resistant embryonic stem cell lines selected for further analysis. Of these, two targeted cell lines were expanded for treatment with FIAU to select for those colonies where reversion and excision of non-homologous plasmid DNA resulted in a substitution of the mutant D79N α_{2A}AR for the wild type α_{2A}AR at its locus in the mouse genome. Six such cell lines were developed,

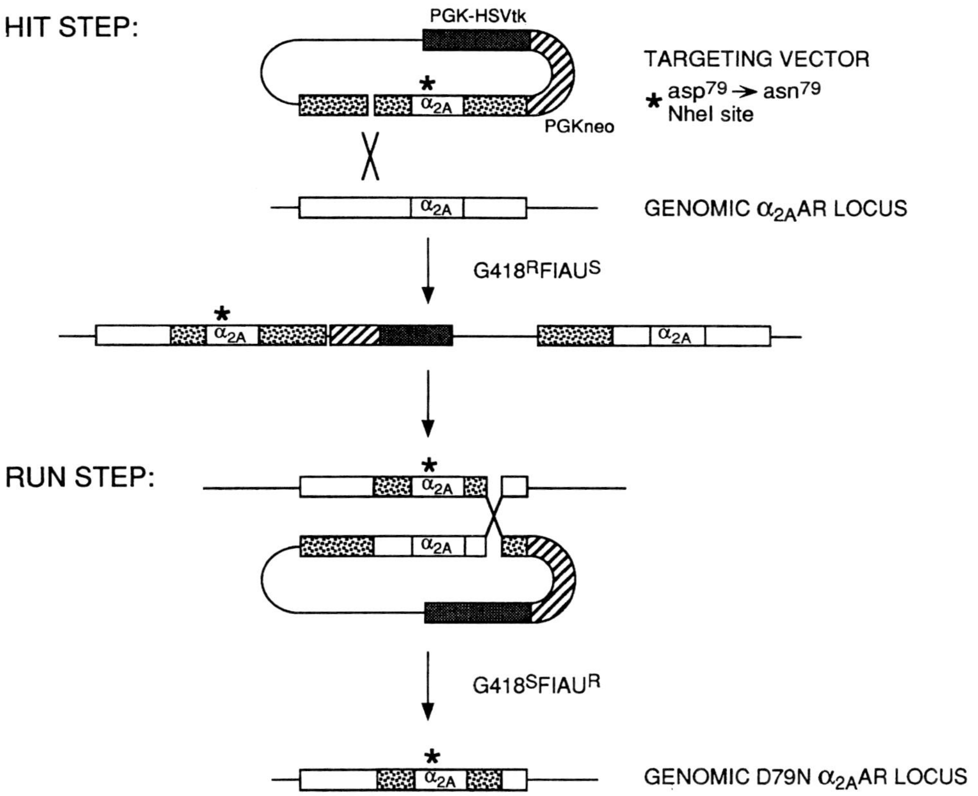

FIGURE 1 Hit and run gene targeting is a two step process for creating subtle mutations in the mouse genome. In the "hit" step, single reciprocal recombination results in the insertion of the entire targeting vector into the α_{2A}AR locus. The duplicated homologous regions can then realign, leading to spontaneous excision ("run" step) of one homologous region along with the non-homologous vector sequences. In selected cells, the D79N α_{2A}AR mutation will replace the wild-type α_{2A}AR gene sequence.

and introduction of three of these cell lines into early stage embryos and transfer to a pseudo-pregnant female mouse have given rise to significant chimerism in the offspring of these mice. If the mutant embryonic stem cells contribute to the germ line, then mating of these chimeric mice will give rise to heterozygous and, in subsequent matings, homozygous offspring.

The advantages of introducing this subtle substitution of D79N α_{2A}AR for WT α_{2A}AR via homologous recombination instead of introduction of mutant D79N α_{2A}AR by overexpression using standard transgenic technologies are many. The α_2AR, like all G protein-coupled receptors, is regulated in a number of ways, including the relative density of receptors and cognate GTP-binding proteins utilized for coupling to diverse effectors. Thus, overexpression of the receptors would confound interpretation of ultimate functional responses. Furthermore, homologous recombination allows the D79N α_{2A}AR to be expressed in a spatially and temporally faithful way because expression will be regulated by the appropriate endogenous regulatory sequences. Thus, after confirmation that selective loss of coupling to K^+ currents has occurred for any population of target cells of interest in homozygous offspring, these animals will provide novel experimental tools to reveal which electrical consequence of α_2 receptor activation is responsible for coordination of various multicellular responses elicited by α_{2A}AR *in vivo*.

RECEPTOR SPECIFICITY CONFERRED BY LOCALIZATION OF RECEPTORS IN SPECIFIC MICRODOMAINS ON THE CELL SURFACE

Although specificity in coordinating cellular functions may arise, at least in part, by selective activation of various signal transduction pathways via coupling to distinct populations of G proteins in target cells, data in the literature suggest that specificity of receptor-G-protein-effector coupling cannot be dictated solely by cell-specific expression of discrete G-protein populations. For example, the findings of Kleuss *et al.* (1991, 1992 and 1993) utilizing antisense strategies suggest that distinct $\alpha\beta\gamma$ subunit complexes of G proteins couple muscarinic versus somatostatin receptors to the suppression of the same population of voltage-gated Ca^{++} currents in GH_3 pituitary cells in culture. Morphological evidence indicates that G-proteins themselves are non-homogeneously expressed on the cell surface (Muntz *et al.*, 1992). Similarly, α_2-adrenergic, opiate and muscarinic receptors in NG108 cells show a functional compartmentation in terms of G-protein interactions, based on studies of guanine nucleotide regulation of receptor-agonist interactions (Ott *et al.*, 1989; Graeser and Neubig, 1993). These various lines of evidence are reconciled by suggesting that in addition to the intrinsic affinity of varying protein-protein interactions in defining the specificity of receptor-effector interactions in various target cells, the trafficking of signal transducing proteins to particular subdomains of cells may confer a higher order of specificity in receptor-mediated signal transduction events.

We have begun to investigate the mechanisms responsible for α_{2A}AR targeting to subdomains of polarized cells. As an initial model system, we have introduced the α_{2A}AR as well as an epitope-tagged version of the α_{2A}AR into Madin-Darby canine kidney (MDCK) epithelial cells, which polarize functionally and morphologically when grown in Transwell culture. Combined biochemical and immunochemical strategies have demonstrated that the α_{2A}AR is more than 90% localized to the basolateral domain of these cells, which parallel its localization *in vivo* in renal and intestinal epithelial cells (Keefer and Limbird, 1993). Immunolocalization has revealed that the α_{2A}AR is enriched in the lateral subdomain of MDCK cells and pulse-labeling studies demonstrate that the receptor is directly delivered to this surface, rather than being randomly inserted into all membrane surfaces of MDCK cells and achieving basolateral localization by a preferential retention at the lateral membrane. Finally, mutagenesis strategies have demonstrated that considerable structural modification of the α_{2A}AR can occur without perturbing α_{2A}AR targeting to the basolateral domain, including elimination of glycosylation of the NH_2 terminus or palmitoylation of the C-terminus, truncation of the C-terminus and removal of the large third cytoplasmic loop of the α_{2A}AR. These findings suggest that sequences in or near the bilayer might be more important for directing the α_{2A}AR to the lateral subdomain of MDCK cells than sequences predicted to be accessible to extracellular or to cytoplasmic compartments (see Figure 2). Interestingly, mutation of tyrosine and phenylalanine sequences characteristic of endocytosis motifs and predicted to reside in the 7th transmembrane span of the α_{2A}AR also does not disrupt direct receptor delivery to the basolateral surface of cultured MDCK cells. Thus, novel structural motifs must be important for α_{2A}AR targeting and hopefully will be revealed in ongoing studies of α_{2A}AR chimeras with other G protein-coupled receptors that achieve apical localization (Keefer *et al.*, 1994b).

The studies on α_{2A}AR trafficking revealed an unexpected finding. Although removal of the predicted third cytoplasmic loop of the α_{2A}AR is not critical for α_{2A}AR delivery to the basolateral surface, α_{2A}AR mutated to delete this region demonstrate a much shorter half-life on the basolateral surface (see Figure 2), suggesting that sequences in this loop might be important for receptor interaction with cytoskeletal or other proteins that might tether the α_{2A}AR to this lateral microdomain of renal epithelial cells and thus potentially serve the role of coordinating α_{2A}AR-mediated signal transduction in these cells (Keefer *et al.*, 1994b).

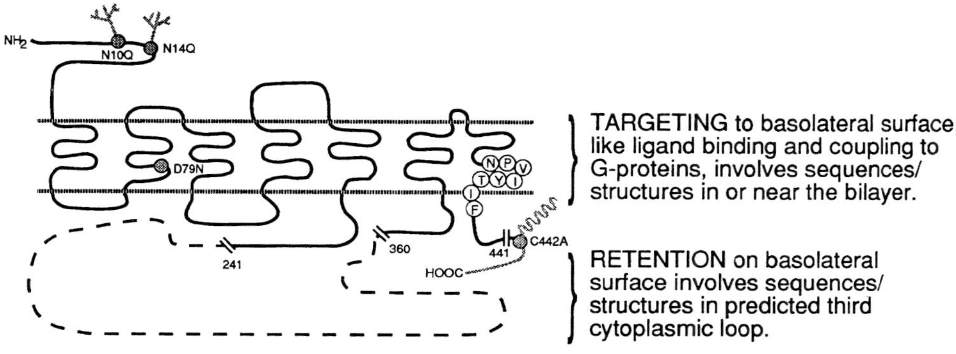

FIGURE 2 Mutagenesis strategies have revealed that mutations to eliminate glycosylation (N10Q/N14Q) or acylation (C442A), to perturb G-protein coupling (D79N), to truncate the C-terminus (L441 stop), to delete the third cytoplasmic loop (Δ241-360), or to substitute alanine for aromatic residues in the seventh transmembrane span implicated in β-receptor sequestration (NPVIYTIF) do not perturb delivery of the α_{2A}AR to the basolateral surface of MDCK cells. In contrast, deletion of the third cytoplasmic loop results in accelerated turnover of the α_{2A}AR when compared with the wild-type α_{2A}AR and any of the mutant α_{2A}AR structures studied.

Current studies are examining possible differences in delivery and steady-state localization of α_2-adrenergic receptor subtypes both in polarized renal epithelial cells as well as in neuronal preparations. The focus on neuronal preparations derives from hypotheses, such as those forwarded years ago by Schoffelmeer and Mulder (1984), that α_2AR localized to the somatotendritic membrane may play an important role in suppression of neuronal signaling by activation of receptor-operated K^+ currents and limiting propagation of excitatory signals whereas α_2AR localized to the synaptic terminals might preferentially couple to inhibition of voltage-gated Ca^{++} currents in an effort to limit intracellular Ca^{++} critical for fusion of secretory vesicles. Testing of this hypothesis first requires evaluation of whether or not α_2AR subtypes are differentially localized and the mechanisms that might effect that localization.

SUMMARY

Studies in a number of laboratories including our own have demonstrated the α_{2A}AR couples to a variety of signal transducing pathways in different physiological settings. More recently, we have attempted to address what molecular features of the α_{2A}AR are involved in receptor coupling to distinct chemical and electrical signaling pathways and, as well, the structural features of the receptor that guide its localization to particular microdomains on the cell surface and determine the lifetime of the α_{2A}AR in those settings. Taken together, the overall goal of our studies is to understand in as precise detail as possible the multiple mechanisms that contribute to the specificity of α_{2A}AR-elicited signal transduction in various target cells with the ultimate goal of identifying novel targets for therapeutic intervention in pathological situations where these pathways have gone awry.

References

Clapham, D.E. and Neer, E.J. (1993) New roles for G-protein $\beta\gamma$-dimers in transmembrane signalling. *Nature*, **365,** 403–406.

Duzic, E., Coupry, I., Downing, S. and Lanier, S.M. (1992) Factors determining the specificity of signal transduction by guanine nucleotide-binding protein-coupled receptors. *The Journal of Biological Chemistry*, **267,** 9844–9851.

Eason, M.G. and Liggett, S.B. (1992) Subtype-selective desensitization of α_2-adrenergic receptors. *The Journal of Biological Chemistry*, **267,** 25473–25479.

Graeser, D. and Neubig, R.R. (1993) Compartmentation of receptors and guanine nucleotide-binding proteins in NG108-15 cells: lack of cross-talk in agonist binding among the α_2-adrenergic, muscarinic, and opiate receptors. *Molecular Pharmacology*, **43,** 434–443.

Horstman, D.A., Brandon, S., Wilson, A.L., Guyer, C.A., Cragoe, E.J., Jr. and Limbird, L.E. (1990) An aspartate conserved among G-protein receptors confers allosteric regulation of α_2-adrenergic receptors by sodium. *The Journal of Biological Chemistry*, **265,** 21590–21595.

Keefer, J.R. and Limbird, L.E. (1993) The α_{2A}-adrenergic receptor is targeted directly to the basolateral membrane domain of Madin-Darby canine kidney cells independent of coupling to pertussis toxin-sensitive GTP-binding proteins. *The Journal of Biological Chemistry*, **268,** 11340–11347.

Keefer, J.R., Nunnari, J., Pang, I.H. and Limbird, L.E. (1994a) Introduction of purified α_{2A}-adrenergic receptors into uniformly oriented, unilamellar phospholipid vesicles: productive coupling to GTP-binding proteins but lack of receptor-dependent ion transport. *Molecular Pharmacology*, in press.

Keefer, J.R., Kennedy, M.E. and Limbird, L.E. (1994b) Unique structural features important for stabilization *versus* polarization of the α_{2A}-adrenergic receptor on the basolateral membrane of Madin-Darby canine kidney cells. *The Journal of Biological Chemistry*, **269,** 16425–16433.

Kleuss, C., Hescheler, J., Enel, C., Rosenthal, W., Schultz, G., and Wittig, B. (1991) Assignment of G-protein subtypes to specific receptors inducing inhibition of calcium currents. Vahere **353,** 43–48.

Kleuss, C., Scherübl, H., Herscheler, J., Schultz, G. and Wittig, B. (1993) Selectivity in Signal Transduction determined by β subunits of Heterotriminic G Proteins. *Science* **259,** 832–834.

Kleuss, C., Scherübl, H., Hescheler, J., Schultz, G. and Wittig, B. (1992) Different β-subunits determine G-protein interaction with transmembrane receptors. *Nature*, **358,** 424–426.

MacKinnon, A.C., Spedding, M. and Brown, C.M. (1994) α_2-Adrenoceptors: more subtypes but fewer functional differences. *Trends in Pharmacological Sciences*, **15,** 119–123.

MacNulty, E.E., McClue, S.J., Carr, I.C., Jess, T., Wakelam, M.J.O. and Milligan, G. (1992) α_2-C10 Adrenergic receptors expressed in Rat 1 fibroblasts can regulate both adenylylcyclase and phospholipase D-mediated hydrolysis of phosphatidylcholine by interacting with pertussis toxin-sensitive guanine nucleotide-binding proteins. *The Journal of Biological Chemistry*, **267,** 2149–2156.

Muntz, K.H., Gilman, A.G. and Mumby, S.M. (1992) Distribution of endogenous G-protein subunits in the plasma membrane of cultured cells. *Circulation*, **86,** 1–774.

Nunnari, J.M., Repaske, M.G., Brandon, S., Cragoe, E.J., Jr. and Limbird, L.E. (1987) Regulation of porcine brain α_2-adrenergic receptors by Na^+, H^+ and inhibitors of Na^+/H^+ exchange. *The Journal of Biological Chemistry*, **262,** 12387–12392.

Ott, S., Costa, T. and Herz, A. (1989) Opioid receptors of neuroblastoma cells are in two domains of the plasma membrane that differ in content of G proteins. *Journal of Neurochemistry*, **52,** 619–626.

Pleus, R.C., Shreve, P.E., Toews, M.L. and Bylund, D.B. (1993) Down-regulation of α_2-adrenoceptor subtypes. *European Journal of Pharmacology*, **244,** 181–185.

Ramirez-Solis, R., Davis, A.C. and Bradley, A. (1993) Gene targeting in embryonic stem cells. *Methods in Enzymology*, **225,** 855–878.

Schoffelmeer, A.N. and Mulder, A.H. (1984) Presynaptic opiate receptor- and α_2-adrenoceptor-mediated inhibition of noradrenaline release in the rat brain: role of hyperpolarization?. *European Journal of Pharmacology*, **105,** 129–135.

Surprenant, A., Horstman, D.A., Akbarali, H. and Limbird, L.E. (1992) A point mutation of the α_2-adrenoceptor that blocks coupling to potassium but not calcium currents. *Science*, **257,** 977–980.

Sweatt, J.D., Connolly, T.M., Cragoe, E.J., Jr. and Limbird, L.E. (1986) Evidence that Na^+/H^+ exchange regulates receptor-mediated phospholipase A2 activation in human platelets. *The Journal of Biological Chemistry*, **261,** 8667–8673.

Wilson, A.L., Womble, S.W., Prakash, C., Cragoe, E.J., Jr., Blair, I.A. and Limbird, L.E. (1992) Novel amiloride analog allosterically modulates the α_2-adrenergic receptor but does not inhibit Na^+/H^+ exchange. *Molecular Pharmacology*, **42,** 175–179.

PREPARATION AND CHARACTERIZATION OF ANTIBODIES DIRECTED AGAINST G-PROTEIN-COUPLED RECEPTORS

KEVIN R. LYNCH

Departments of Pharmacology & Biochemistry, University of Virginia Health Sciences Center, 1300 Jefferson Park Avenue, Charlottesville, Virginia 22908 USA

It has long been recognized that antibodies are useful tools in studying protein chemistry and in localizing of proteins in the context of tissue histology. More recently it has become apparent that receptors coupled to guanyl nucleotide binding proteins (G-proteins) mediate the signalling of a large number of peptide hormones, neurotransmitters and autacoids. Despite an obvious need for antibody probes selective for these receptor proteins, such reagents have proven rather difficult to generate. In this paper, I discuss approaches that have been successful (or unsuccessful) in generating useful antibodies directed against G-protein-coupled receptors and strategies for testing the efficacy of receptor antibodies. An emphasis will be on antibodies directed against catecholamine receptors, particularly α_2-adrenoceptors. There is no attempt to provide precise protocols for the variety of techniques mentioned, rather this is meant to discuss the rationale that lie behind the approaches taken to develop antisera.

KEY WORDS: Guanyl nucleotide binding protein (G-protein)-coupled receptor, antisera, antibodies, immunogen, immunohistochemistry, fusion protein

INTRODUCTION

G-protein-coupled receptors (GCRs) are low abundance, integral membrane proteins. Although some of these receptors (e.g. those responding to catecholamines) have been studied intensely for over a decade by radioligand binding techniques, and more recently by measurements of their underlying mRNAs using recombinant DNA probes, there is no known reagent that can duplicate the sensitivity and subtype selectivity of a good antisera. Because GCR purification is such an arduous task, the obvious approach of purifying milligram amounts of protein and using this material as an immunogen (and for screening antisera or antibodies) is generally not available. However, the molecular cloning of over one hundred unique GCRs in the past decade has made available their underlying DNAs. Nevertheless, purified GCR protein in any amount remains difficult to obtain, largely because of difficulties in expressing recombinant GCRs to high levels in cultured cells and the inherent hydrophobic nature of these proteins, which makes purification difficult. Thus investigators have turned to indirect methods for raising antibodies against GCRs. Among most often used strategies are epitope tagging, immunizing with synthetic peptides and immunizing with fusion proteins expressed in bacteria. Although an optimum GCR antisera,

Correspondence to: Kevin R. Lynch, Ph.D., Departments of Pharmacology & Biochemistry, University of Virginia Health Sciences Center, 1300 Jefferson Park Avenue, Charlottesville, Virginia 22908 USA.

i.e. one that *specifically* i) stains tissues, ii) immunoprecipitates the receptor and iii) detects the receptor on Western blots, has not, to our knowledge, been reported, there have been developed a number of useful GCR antibodies. [Throughout the following discussion, "antisera" will refer to whole serum, while "antibodies" will refer to material immuno-affinity purified from antiserum, or, when noted, monoclonal antibodies. It should be noted that most anti-GCR sera, whether raised in response to peptides or fusion proteins, are affinity purified and thus it is antibodies that are used in most applications.]

CHARACTERIZING GCR ANTISERA (ANTIBODIES)

Immunoprecipitation

There are generally two approaches useful for radiolabelling GCRs prior to immunoprecipitation; one is to synthesize the receptor protein, or a fragment thereof *in vitro* by cell-free translation in the presence of [^{35}S]methionine (and/or cysteine), the other is to overexpress the receptor in cultured cells, label the receptor with a specific, photoactivated, radionuclide-containing (usually iodine-125) ligand, followed by solubilization of the receptor with detergent. In either case, the receptor is recovered by reacting with antibodies (controls consisting of pre-immune sera or pre-absorbing antibodies with immunogen) and the complex is recovered with protein A or protein G coated beads. Obviously, the second method is preferred in that it is the more stringent test (and it tests for recognition of native receptor), but it depends the availability of the cross linking ligand. A further test of antibody quality would be to test for selectivity by biosynthetically labelling cellular protein by incubation of cells with radionuclide-containing amino acids and determining the number and size of bands on SDS-PAGE autoradiograms following solubilization and immunoprecipitation.

Immunohistochemistry

This can be performed with cultured cells or tissues. The former often involves transfected cells that express large amounts of receptor as compared to what is observed in tissues. Nevertheless, the transfected cells, when they lack endogenous receptor, provide a method for testing selectivity of antibodies regarding other subtypes of the same receptor. The controls for immunohistochemistry are problematic because of difficulty in mimicking the target. That is, the cells are exposed to some sort of fixative (e.g. 4% paraformaldehyde) to preserve histology and thus the receptor is neither native (as detected in immunoprecipitation protocols) nor fully denatured (as in Western blotting protocols). The common control for immunohistochemistry is to block reactivity by pre-absorption of immune sera with immunogen, however, it should be noted that this does not guarantee that signal observed without pre-absorption is due to the presence of the GCR. Confidence in immunohistochemical detection rises when two different antisera (or antibodies) are applied and coincident signals are observed. Another control strategy will become available as GCR null mice become available using transgenic gene "knock-out" technologies.

Western blotting

This is perhaps the least demanding test for anti-GCR sera in that the protein is denatured to some degree (often completely) prior to analysis. The experiment most often performed is to express GCRs individually (and to high levels) in COS or HEK293 cells, prepare a crude membrane fraction and fractionate the proteins therein by SDS-PAGE. However, COS cells in particular are notorious for high levels of glycosylation (GCRs are almost invariably glycoproteins), thus relative migration observed in COS cell extracts and tissue extracts might differ substantially. The migration of the protein core can be revealed by extensive deglycosylation of the receptor, or expression of a receptor mutated to remove N-linked glycosylation consensus sequences.

COS cells thus provide a cogent system whereby a GCR can be expressed via transfection of its cDNA and receptor expression level monitored by binding. These cultures can used to obtain membranes for Western blot analysis and cross-linking to radio-iodinated ligands for immunoprecipitation. Furthermore, transfected cell cultures can be fixed and stained with antibodies; since not all the cells in the culture express the exogenous DNA, the non-expressing cells provide a convenient measure of background staining. My group has used COS cell cultures transfected with rat α_{2A}-, α_{2B}- and α_{2C}-adrenoceptors to test antibodies (raised against bacterial fusion proteins) in all three assay systems (Rosin *et al.*, unpublished), which illustrates a further advantage to this approach, i.e. antisera can be tested regarding subtype selectivity. I emphasize that reactivity of antibodies with antigen in one format does not predict reactivity in other formats. For example, our antibodies directed against α_{2A}-adrenoceptor immunoprecipitate labelled receptor and stain transfected cells, but detect receptor protein only weakly in on Western blots. The latter finding illustrates a general problem, i.e. some antibodies, although detecting high levels of receptor expressed in transfected cells such as COS cells, are not sensitive enough to detect receptor protein in tissues or tissue extracts.

RAISING GCR ANTISERA

It is usually desirable to raise polyclonal antisera and immuno-affinity purify the antibodies, rather than to generate monoclonal antibodies. The primary reason is that immunohistochemical application are usually more successful with polyclonal antibodies.

Epitope-tagged GCRs

The simplest route to antibodies directed against GCRs is to construct a recombinant receptor wherein a specific amino acid sequence is added for which an antibody is available. Since experience with a variety of GCRs indicates that their amino or carboxyl terminal regions can be modified without deleterious effects, this strategy can be applied readily to this receptor family. A particularly useful octapeptide tag is Asp-Tyr-Lys-$(Asp)_4$-Lys (DYKDDDDK) (Prickett *et al.*, 1989), which is recognized by a commercially available monoclonal antibody. This sequence is often inserted immediately after the initiation methionine residue, and can be cleaved from the expressed protein by the specific serine protease, enterokinase. This strategy has been used, for example, to immuno-stain COS-7 cells expressing human platelet activating factor receptor (Kunz *et al.*, 1992) and to immuno-stain Madin-Darby Canine Kidney cells expressing α_{2A}-adrenergic receptors (Keefer and Limbird, 1993). The obvious disadvantage of epitope-tagging is that it is limited to recombinant receptors. An exciting possible application of this approach is to use the epitope tag as a handle to purify recombinant receptor. It is conceivable that, given a cell line expressing high levels of receptor (i.e. 25–50 pmol/mg membrane protein), amounts of purified receptor (i.e. tens of micrograms) sufficient to raise monoclonal antibodies could be realized. The usefulness of this strategy could be increased by including a second tag sequence such as the hexahistidine motif that binds proteins to nickel chelating resins (Hochuli *et al.*, 1987).

Peptide immunogens

In theory, antisera directed against any protein wherefor amino acid sequence information is available can be raised by using synthetic peptides as immunogens. One wishes to use hydrophilic peptides (to avoid water solubility problems), as well as peptides that are exposed on the protein surface. Since structural information is lacking, the choice of peptide can by made by computer algorithms that are designed to identify, from primary structure, stretches of amino acids that are likely to be on the surface of the protein. [Proteins for which there is structural information derived from physical measurements are available in substantial amounts at a high degree of purity, and thus had long since been used as immunogens.] In the case of GCRs, the commonly accepted model of seven membrane spanning regions dictates that peptide immunogens be chosen from one of the termini, or from one of the loop regions, since these contain the amino acids with hydrophilic side chains. A further advantage to the non-transmembrane regions is that these are the most divergent regions among GCR subtypes, and thus afford the greatest chance of obtaining subtype selective antibodies. If one desires an antisera that will stain whole cells or tissues without prior permeabilization, then the peptide is chosen from either the amino terminus or one of three exofacial loops. The length of the synthetic peptide is dictated by the capability of the peptide synthesizing machine as well as cost, and is usually 10–20 residues. The peptides are linked covalently (often through a carboxyl terminal cysteine residue) to a carrier protein such keyhole limpet hemocyanin, and introduced into rabbits using a standard immunization/boost protocol. The antisera can be enriched readily by immunoaffinity chromatography on a column support containing the peptide, and specific reactions are those competed by peptide.

It is somewhat difficult to judge the success rate of anti-peptide antibodies directed against GCRs since the failures are not reported in print. However, anecdotes abound regarding money and time spent to little discernable effect, i.e. the antisera detects the peptide, but not the receptor even when it is expressed to high levels (e.g. in COS cells). Furthermore, the peptide immunogen strategy is obvious and within the ken of any investigator (with sufficient finances, since the entire operation can be performed commercially), but relatively few anti-GCR peptide sera have been described. Nevertheless, there have been reported antisera raised with peptide immunogens that have proven useful in some applications. For example, Sibley and colleagues (McVittie *et al.*, 1991) used seven different peptides to the rat D_2 dopamine receptor (two amino terminal, four third intracellular loop, one carboxyl terminal) and tested these in an ELISA assay; all but the carboxyl terminal peptide antisera were active. Three of the peptide antibodies were shown to immunoprecipitate D_2 dopamine receptors from transfected CHO cells and to stain these cells. Furthermore, several of the antibodies stained medium spiny neurons in striatal slices from rat brain, as would be expected of antibodies recognizing D_2 dopamine receptors. The use of multiple peptide immunogens as used in this study is perhaps the best assurance of success in obtaining useful antibodies for a given application.

Fusion proteins

The alternative to using receptor peptides as immunogens is to express a portion of the receptor in bacteria, purify this protein fragment, and use this as an immunogen. To facilitate purification, a DNA molecule is constructed that encodes the receptor fragment fused at the carboxyl terminus of another protein that is stable and soluble in *Escherichia coli*, and is readily purified. Among the most popular is vectors are the pGEX series (Smith and Johnson, 1988), which encode a glutathione S-transferase (GST) from *Schistosoma japonicum*. GST is a 26 kDa protein that is purified readily purified by affinity chromatography on glutathione-agarose beads. It is necessary to avoid using transmembrane regions to construct GST fusion proteins since the resulting proteins are very often insoluble and lack of solubility negates the advantage of affinity purification. Antisera raised to GST/receptor fusion proteins are tested initially against the fusion protein, and if reactivity is found, antibodies can be affinity purified on immobilized fusion protein and tested as outlined previously. Soluble fusion protein and GST are used to determine specificity of the immune reaction. Although fusion proteins are more laborious to generate than peptides, the former has the advantage that many more epitopes are presented by the fusion protein, the receptor portion of which is usually at least five times the length of a peptide immunogen. The pGEX vectors are available commercially, as other several similar expression vector systems.

GST/receptor fusion proteins using third intracellular loop regions of muscarinic acetylcholine (Levey *et al.*, 1992; Dorje *et al.*, 1991), α_2-adrenergic (Rosin *et al.*, 1993) and dopamine (Levey *et al.*, 1993) receptors have been used successfully to raise antisera. In each of these cases, the antibodies immunoprecipitated labelled recombinant receptor and stained discrete cell groups in rat brain sections. Our experience with both the anti-A and anti-C reagents is that they are very sensitive with regard to receptor denaturation. Thus overfixation of tissue sections or cells leads to loss of reactivity, and neither antibody performs well in a Western blot format.

It is worth noting that all of these receptors have large (i.e. > 100 amino acids) third intracellular loops that are both highly divergent (among receptor subtypes) and hydrophilic. A diagram of the rat α_{2A}- and α_{2C}-adrenoceptors showing the region of the receptors that we used to construct GST fusion protein is shown in the Figure 1. Most GCRs have considerably smaller third intracellular loops, and the remaining loop regions (both exofacial and intracellular) are small (< 30 amino acids) also. Another obvious target for fusion protein construction is the carboxyl terminus region, which is often at 50 amino acids long. However, to our knowledge there are no reports of GCR antibodies made to fusion proteins using this region. My group attempted to raise and characterize antibodies using GST/COOH terminal fusion proteins using amino acid sequence from both the AT_1 angiotensin receptor and an orphan receptor (D'Angelo, Harrison and Lynch, unpublished), and despite our experience and success with anti-α_2-adrenoceptor sera, we met with abject failure regarding the COOH terminal immunogens. We do not understand the reason for this failure.

SUMMARY

Although antisera directed against GCRs are clearly highly informative reagents in probing receptor biochemistry, these reagents have been difficult to generate. Difficulties in purifying GCRs in adequate amounts have precluded using whole receptors as immunogens,

α_{2A} – Adrenergic Receptor

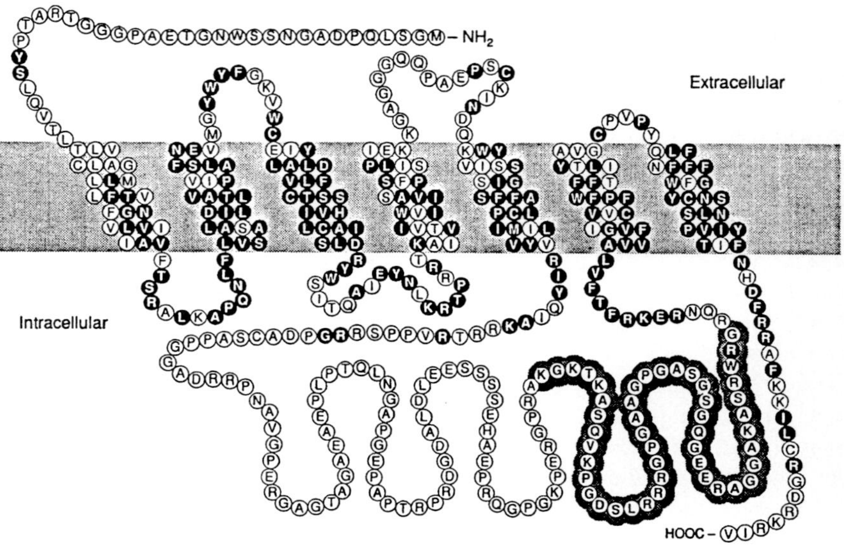

α_{2C} – Adrenergic Receptor

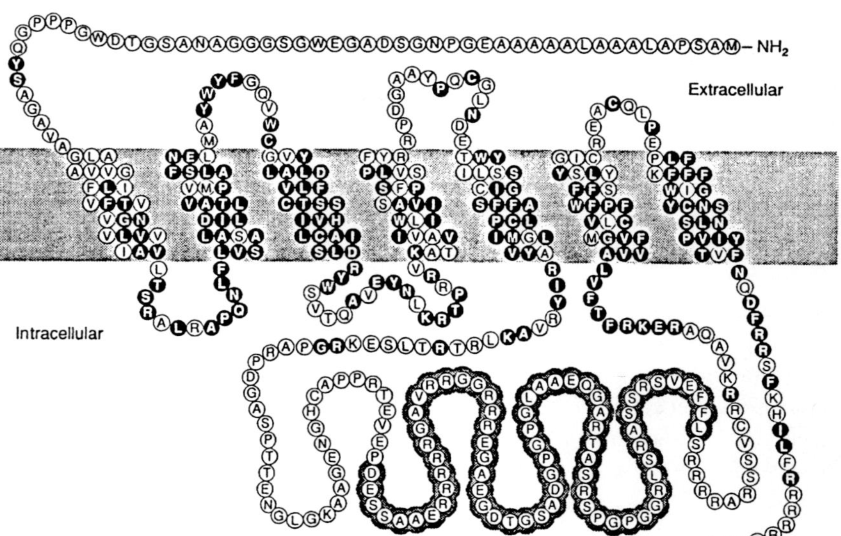

FIGURE 1 This cartoon shows the commonly accepted model of a GCR passing the cell membrane seven times, with the amino terminus exofacial. The amino acids (single letter code) shown in reverse video are identical in all three α_2-adrenoceptor subtypes. The shaded amino acids are those that were coupled (via the underlying DNAs) to GST. Both receptor sequences are from the rat.

forcing investigators to use either receptor peptides or receptor fusion proteins as substitutes. Peptides antibodies are simpler to generate, but there have been numerous failures to realize antibodies that are adequate for demanding applications such as staining low levels of receptor in tissue sections. Although impressive antibodies directed against dopamine, muscarinic and α_2-adrenergic receptors have been generated using GST/third intracellular loop fusion proteins, this success apparently has not been duplicated when other regions of GCRs are used to construct GST fusion proteins.

References

Dorje, F., Levey, A.I. and Brann, M.R. (1991) Immunological detection of muscarinic receptor subtype proteins (m1–m5) in rabbit peripheral tissues. *Mol. Pharmacol.*, **40,** 459–462.

Hochuli, E., Dobeli, H. and Schacher, A. (1987) New metal chelate adsorbent selective for proteins and peptides containing neighboring histidine residues. *J. Chromatog.*, **411,** 177–184.

Keefer, J.R. and Limbird, L.E. (1993) The α_{2A}-adrenergic receptor is targeted directly to the basolateral membrane domain of Madin-Darby Canine Kidney cells independent of coupling to pertussis toxin-sensitive GTP-binding proteins. *J. Biol. Chem.*, **268,** 11340–11347.

Kunz, D., Gerard, N.P. and Gerard, C. (1992) The human leukocyte platelet-activating factor receptor. cDNA cloning, cell surface expression, and construction of a novel epitope-bearing analog. *J. Biol. Chem.*, **267,** 9101–9106.

Levey, A.I., Hersch, S.M., Rye, D.B., Sunahara, R.K., Niznik, H.B., Kitt, C.A., Price, D.L., Maggio, R., Brann, M.R. and Ciliax, B.J. (1993) Localization of D_1 and D_2 dopamine receptors in brain with subtype-specific antibodies. *Proc. Natl. Acad. Sci. USA*, **90,** 8861–8865.

Levey, A.I., Kitt, C.A., Simonds, W.F., Price, D.L. and Brann, M.R. (1992) Identification and localization of muscarinic acetylcholine receptor proteins in brain with subtype-specific antibodies. *J. Neurosci.* **11,** 3218–3226.

McVittie, L.D., Ariano, M.A. and Sibley, D.R. (1991) Characterization of anti-peptide antibodies for the localization of D_2 dopamine receptors in rat striatum. *Proc. Natl. Acad. Sci. USA*, **88,** 1441–1445.

Prickett, K.S., Amberg, D.C. and Hopp, T.P. (1989) A calcium-dependent antibody for identification and purification of recombinant proteins. *Biotechniques*, **7,** 580–589.

Rosin, D.L., Zeng, D., Stornetta, R.L., Riley, T., Okusa, M.D., Guyenet, P.G. and Lynch, K.R. (1993) Immunohistochemical localization of α_{2A}-adrenergic receptors in catecholaminergic and other brainstem neurons in the rat. *Neuroscience*, **56,** 139–155.

Smith, D.B. and Johnson, K.S. (1988) Single-step purification of polypeptides expressed in *Escherichia coli* as fusions with glutathione S-transferase. *Gene*, **67,** 31–40.

MEMBRANE ORGANIZATION AND MOBILITY OF α_2 ADRENERGIC RECEPTORS IN MDCK CELLS

STAFFAN UHLÉN[1], DANIEL AXELROD[2], JEFFREY R. KEEFER[3], LEE E. LIMBIRD[3] and RICHARD R. NEUBIG[1,4]

[1]Department of Pharmacology, [2]Department of Physics, [2]Biophysics Research Division, [4]Department of Internal Medicine (Hypertension), The University of Michigan, Ann Arbor, MI 48109-0626, [3]Department of Pharmacology, Vanderbilt University, School of Medicine, Nashville, TN

A porcine α_{2A} adrenergic receptor with a viral epitope engineered onto its amino-terminus was expressed in Madin-Darby canine kidney (MDCK) cells. The receptor was visualized by staining with a tetramethylrhodamine (TMR) labelled monoclonal antibody directed against the extracellular epitope. In fixed cells, the receptor was localized to the sides and top surfaces of the cells in a non-homogenous distribution. Bright spots were seen, especially on the top of the cell, indicating large clusters of receptors. Living cells showed a similar distribution of fluorescence for up to approximately 1 hour but at 2 hours there was accumulation of fluorescence in large intracellular aggregates. The rate of this internalization was significantly enhanced with an α_2 receptor mutant in which aa 241–360 from the third intracellular loop were deleted. Receptor mobility was measured at 25°C in low intensity regions at the bottom of the cells by fluorescence recovery after photobleaching (FRAP). Both the wild type and deletion mutant receptor were entirely mobile (80–91%) and had diffusion coefficients of ~2.2×10^{-10} cm^2 s^{-1}. These data indicate that the α_{2A} adrenergic receptor exists in both clustered and diffuse mobile forms. The role of receptor organization is discussed.

ABBREVIATIONS: AR, adrenergic receptor; DT, Translational diffusion coefficient; FRAP, Fluorescence recovery after photobleaching; G protein, Guanine nucleotide binding protein; Gi, Inhibitory G protein; Gs, Stimulatory G protein; TMR, Tetramethylrhodamine

KEY WORDS: alpha$_2$ adrenergic receptor, G protein, fluorescence microscopy, diffusion, MDCK cells, receptor localization, receptor internalization, receptor mutants

INTRODUCTION

Guanine nucleotide binding proteins (G proteins) play a central role in adrenergic receptor (AR) mediated signal transduction beyond simple regulation of adenylyl cyclase (Limbird, 1988; Freissmuth *et al.*, 1989; Brown, 1991; Gupta *et al.*, 1992). The α_1 AR activates phospholipase C through a G_q/G_{11} type G protein (Wu *et al.*, 1992). The α_2 AR activates potassium channels via pertussis-toxin sensitive G proteins (Suprenant *et al.*, 1992) and may activate sodium-hydrogen exchange through pertussis toxin insensitive G proteins (Clark *et al.*, 1990). In addition to adenylyl cyclase, β AR activates calcium channels directly through G_s (Yatani and Brown, 1989).

A tremendous amount is known about the multitude of different proteins that are

Correspondence to: Richard Neubig, Dept. of Pharmacology, University of Michigan, M6322 Medical Science Bldg. 1, Ann Arbor, MI 48109-0626.

Dohlman *et al.*, 1987; Simon *et al.*, 1991; Rodbell, 1992). There are more than 200 receptor types and subtypes (including at least 9 AR subtypes), at least 21 G protein α subunits, 4 β subunits and 6 γ subunits (Hepler and Gilman, 1992). There are at least 6 different subtypes of mammalian adenylyl cyclase (Tang and Gilman, 1992), many phospholipases A and C (Rhee and Choi, 1992) and numerous potassium and calcium channels that are regulated by G proteins (Brown, 1991). How these diverse components are organized in the cell is an important area for future exploration.

"Classical" G protein model

The mechanism of receptor activation of G proteins has been reviewed extensively (Freissmuth *et al.*, 1989; Gilman, 1987; Chabre and Deterre, 1989). In the generally accepted view of this model, the inactive G protein contains bound GDP. Binding of agonist causes the receptor to have a higher affinity for the G protein, it diffuses to the G protein and forms a complex of ligand, receptor, G protein, and GDP. The affinity of GDP in this complex is low, so GDP dissociates allowing GTP to bind. After GTP binds, the affinity of G protein for receptor is lowered so the G protein dissociates from the receptor permitting the receptor to catalytically activate more GDP-bound G proteins. The α and βγ subunits of the G protein then dissociate from each other releasing the "active" α subunit (or perhaps active βγ) to diffuse either in the membrane or the cytosol to activate the effector enzyme or channel. Most details of this model are supported in detail in the case of the rhodopsin-transducin system where the degree of amplification is large (Chabre, 1985; Stryer, 1986).

There is much less information on the organization and mobility of components of the G protein-coupled receptor systems in more typical mammalian cells. Indeed, numerous pieces of evidence suggest that there is a more complex organization of signalling components that is essential for the profound specificity of signal transduction (Neubig, 1994).

Evidence for precoupling of adrenergic and other receptors

Rather than receptors and G proteins interacting by random diffusion after agonist binding, there is evidence from ligand binding studies for precoupling of receptors with G proteins (i.e. presence of RG complexes in the absence of agonist). This is suggested by data for dopamine D_2 (Wreggett and De Leon, 1984), β adrenergic (Nerme *et al.*, 1986), and adenosine A1 (Leung *et al.*, 1990) receptors and most compellingly from kinetic studies of α_2 adrenergic (Neubig *et al.*, 1988; Gantzos and Neubig, 1987) neutrophil formylpeptide (Sklar *et al.*, 1989; Fay *et al.*, 1991; Posner *et al.*, 1994) receptors. Also, increases in basal GTPase activity by receptors in the absence of agonist are seen in reconstitution systems with β AR (Pedersen and Ross, 1985; Cerione *et al.*, 1984) and in cell membranes with opiate receptors (Costa and Herz, 1989; Costa *et al.*, 1990), but see also (Costa *et al.*, 1992). More recently, this has been confirmed through the identification of "constituitive" activity of α_2 and β adrenergic receptors (Samama *et al.*, 1993; Ren *et al.*, 1993). A mutant with increased basal adenylyl cyclase activity further supports a role for RG complexes which form in the absence of ligand (i.e. precoupled RG) (Samama *et al.*, 1993). Two interesting implications of these results are: 1) a prediction that antagonists with "negative intrinsic activity" should exist because they shift the equilibrium away from precoupled RG to free R and G and 2) that constituitively active receptors could contribute to human disease states (Samama *et al.*, 1993).

Mechanism of receptor-G protein coupling

What is the mechanism of coupling of receptors and G proteins? Is it simply diffusion of the components (receptor, G protein, and effector) in a fluid plasma membrane? Based on recent models of tyrosine kinase receptor signalling (Schaller and Parsons, 1993), complex interactions of the receptor and G protein with cytoskeleton or other cellular machinery may play a bigger role than is usually discussed. Precoupling of receptors with G proteins could occur through simple diffusion but evidence for more complex organization of the signalling components in the cell membrane has been reviewed recently (Rodbell, 1992; Neubig, 1994). It is conceivable, perhaps even likely, that both diffusion and complex organizational mechanisms play a role in intact cells.

In this report, we describe the distribution and mobility of epitope-tagged α_2 AR expressed in Madin-Darby canine kidney (MDCK) cells (Keefer and Limbird, 1993) using both morphological methods and the technique of fluorescence recovery after photobleaching. We describe a clustered distribution of the receptor on the on the top and sides of the cells. In contrast, the receptor is present in low density on the bottom surface of cells and those receptors are highly mobile. Thus the receptor appears to have two distinct populations (a diffuse, mobile population and a clustered and probably immobile population) as has been found for nicotinic acetylcholine receptors (Axelrod *et al.*, 1976). The relation of these data to similar observations of a clustered distribution of G protein $\beta\gamma$ subunits (Kwon *et al.*, 1994) is discussed.

MATERIALS AND METHODS

Cell culture

Madin-Darby Canine Kidney cells (MDCK cells) were cultured in Dulbecco's Modified Eagle's Medium (Irvine Scientific) supplemented with 10% bovine fetal serum, 2 mM l-glutamine, 100 IU/ml penicillin, and 100 μg/ml streptomycin in an atmosphere of 95% air and 5% CO_2 at 37 °C in a humified incubator. Cells were subcultured by incubating with 0.05% trypsin. For the FRAP-, and CCD camera-experiments the cells were cultured on 25 mm, 1# circular coverslips (Bellco Glass Inc.), and for the confocal experiments on rectangular coverslips (Thomas Scientific), in 5 ml culture dishes. The cells were used 16–22 hours after seeding on the coverslips.

Immunochemical reagents

The monoclonal mouse anti-hemaglutinin antibody 12CA5 (Berkley Antibody Co.) was used. This antibody recognizes the viral epitope tagged to the porcine α_2-adrenoceptor used in our studies (Keefer and Limbird, 1993). Control MDCK cells with no receptor (WT) were from ATCC. Both the normal porcine α_2AR (Keefer and Limbird, 1993) and a mutant with the bulk of the third intracellular loop deleted (aa 241–360, notl) (Keefer *et al.*, 1994) were used. The 12CA5 antibody was conjugated with TMRSE (5-(and 6)-carboxytetramethylrhodamine succimidyl ester) (Molecular Probes) by the following procedure: 1ml of 1 mg/ml antibody, 80 μl of 1 mg/ml TMRSE in N,N-dimethyl-formamide (Fisher), and 180 μl of 1M NaHCO3 (pH 8.2) were mixed and incubated at room temperature for 1 hour, then 100 μl of 1.5 M hydroxylamine (pH 8.0) was added and the incubation continued for further 30 minutes. Labeled antibody was isolated from free dye by gelfiltration through a 6.5 ml column of Sephadex-G50 (Pharmacia). The early pink band was collected and bulk-washed for a minute with Bio-Beads SM-2 (Bio-Rad). The supernatant containing the rhodamine-labeled antibody was stored at 4 °C for up to 4 months until used.

Immunostaining of tagged α_2-adrenoceptors expressed in MDCK cells

Non-fixed cells used in CCD camera and FRAP experiments. The MDCK cells (on coverslips) were stained with a 1/32 dilution of the rhodamine-labeled anti-tag antibody for 8 min at room temperature with 5% goat serum as non-specific blocking agent. The cells were then washed 2 times for 7 min, and quickly mounted into a 650 μl cell culture chamber (Bellco Glass Inc.). Hank's balanced salt solution without phenol red was used in these procedures. During the short time of our staining and washing procedures the pH was fairly stable, i.e. pH 7.3–7.7.

Fixed cells used in CCD camera experiments. The MDCK cells were fixed for 25 min with 2% paraformaldehyde in PBS (pH 7.5), then incubated for 10 min in 50 mM NH_4Cl-PBS, washed for 5 min in PBS, and thereafter stained as described above for the non-fixed cells.

Fixed cells used in confocal microscopy experiments. The MDCK cells were fixed for 1 hour with 3% paraformaldehyde in PBS, then incubated for 20 min in 50 mM NH_4Cl-PBS, washed for 10 min, stained for 42 min with a 1/20 dilution of the rhodamine-labeled antibody in 5% goat serum and PBS, washed 2 times for 15 min, and then mounted in 90% glycerol (with PBS, pH 9.3).

Fluorescence microscopy using the CCD camera

The sample was illuminated by a krypton laser (Laser Ionics) at 520.8 nM. Using epifluorescence optics, the fluorescence was collected by a Nikon 63X oil-immersion objective (Leitz Ortholux II). The images were recorded digitally through the CCD camera system (Photometrics, Star I). The exposure time of the camera was 4 sec which was short enough to avoid excessive bleaching of the rhodamine labeled sample.

Fluorescence microscopy using the confocal microscope

Images were obtained on a Bio-Rad 600 confocal microscope equipped with an Argon laser and a 100X oil immersion objective set in a Leitz microscope stage. Digital images were processed with a gamma filter of 3.

Fluorescence recovery after photobleaching (FRAP) experiments

The fluorescence recovery after photobleaching (FRAP) measurements were performed using a 63 × 1.4 Zeiss objective and the apparatus described previously (Greenberg and Axelrod, 1993). A 100 ms pulse bleached the flurophores and the fluorescence recovery was probed with the same focussed beam attenuated by a factor of about 10,000. Probe-induced fluorescence was monitored by a Hitatchi photomultiplier (R 943-02). The bleach depth was maintained at approximately 40–60% in all experiments by use of appropriate neutral density filters. Only one set of data was collected from a single spot on a cell. The beam was moved to another spot on the same cell or to a different cell before initiating another run. Up to 10 FRAP experiments were performed on each coverslip, all within 50 minutes after addition of primary antibody. Data were collected in bins of 100 msec for 10 seconds pre-bleach, 0.1 sec bleach pulse, and 40 seconds post-bleach. The argon beam (514 nM) was focused through an epi-illumination fluorescence microscope to a spot radius of approximately 0.52 μM on the lower surface of the cell. The focusing was facilitated by first focusing on the edge of the cell, and then moving just inside the edge.

Recovery curves were fit using Prism (GraphPad Software, San Diego CA) to equation 1 which is an approximation of the theoretical recovery curve (Axelrod *et al.*, 1976)

$$F(t) = F_O + F_r\, t/(t + \tau) \tag{1}$$

where t is the time after bleach, $F(t)$ is the fluorescence as a function of t, F_O is the fluorescence immediately after bleach, F_r is the amount of fluorescence recovery, and τ is the time for half-maximal recovery. The prebleach fluorescence F_p was determined by fitting the prebleach data to a constant. The fractional recovery of fluorescence was calculated from the expression $F_r/(F_p - F_O)$. Diffusion coefficients were calculated to be $D = (\gamma/4)w^2/\tau$ where γ is related to the bleaching depth and is approximately 1.2 here (Axelrod *et al.*, 1976), and w is the e^{-2} spot radius.

RESULTS

Fixed cells vs living cells

In order to study mobility of the α_2 adrenergic receptors we wished to label receptors in living cells. Preliminary experiments with sandwich methods showed that use of unlabelled primary and fluorescently tagged secondary antibody resulted in rapid (< 20 minutes) internalization of most of the receptors on the MDCK cells (data not shown). This has also been described by Kobilka *et al.*, (von Zastrow *et al.*, 1993) for β receptors in HEK 293 cells. Attempts to stain the epitope-tagged receptor with fluorescently labelled Fab fragments of the 12CA5 antibody did not show any specific staining of fixed cells. In contrast, TMR-labelled 12CA5 primary antibody showed clear staining of fixed MDCK cells expressing the epitope-tagged α_2 adrenergic receptor (Figure 1, Alpha$_2$—Tag/Fixed). Control MDCK cells not expressing the epitope-tagged receptor did not show significant staining under these conditions (Figure 1, WT/Fixed).

Cells stained with primary antibody showed a clustered distribution of receptors in both fixed and living cells up to 40 minutes after staining. There appeared to be some internalization of the α_2 receptors at 2 hours with the normal epitope tagged receptor (Figure 1, Alpha$_2$—Tag/2 hours). A receptor mutant (Not1) which lacks a portion of the third intracellular loop and shows more rapid degradation (Keefer *et al.*, 1994) also showed more rapid internalization. The bulk of TMR-antibody-labelled receptor appeared in a tight "supercluster" by 2 hours after staining with the labelled primary antibody (Figure 1, Not1—Tag/2 hours).

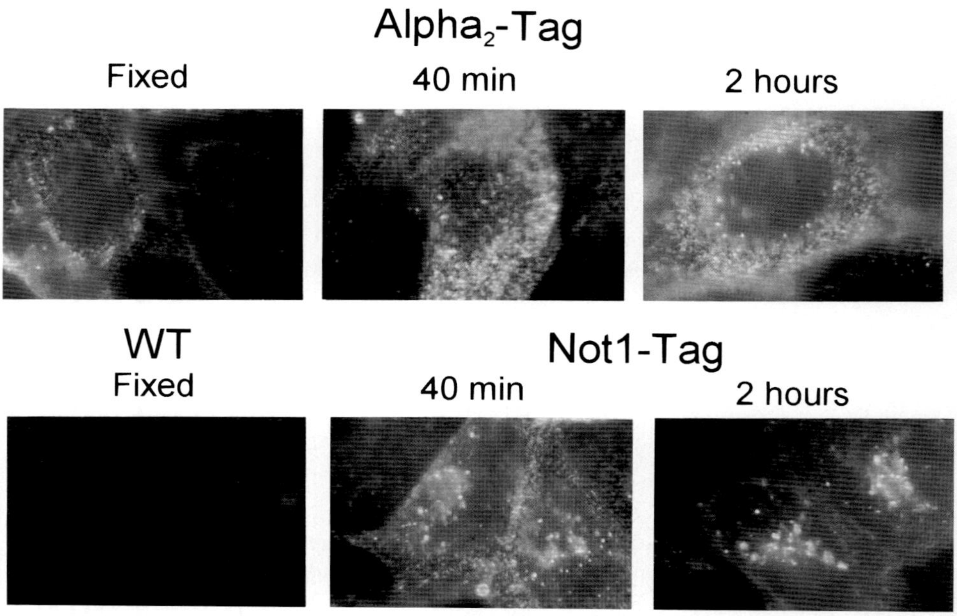

FIGURE 1 CCD camera images of α_2-adrenoceptors on fixed or living MDCK cells at different times after staining. α_2-Adrenoceptors stained with Rhodamine-anti-tag antibodies. Cells expressing the epitope-tagged porcine α_{2A}-adrenoceptor (Alpha$_2$-Tag), no receptor (WT), or the Not1 mutant of the tagged receptor (Not1-Tag). As indicated, cells were either fixed before staining, or living cells were observed 40 min or 2 hours after the initiation of staining.

Distribution of α_2 adrenergic receptors in cells

Using CCD camera imaging, the top and side surfaces of fixed cells showed an inhomogenous staining pattern (Figure 2) as has been previously seen for β adrenergic receptors in A431 cells (Wang *et al.*, 1989). MDCK cells grown on coverslips were labelled on the top and side surfaces of the cells (Figure 2) as previously reported for cells grown on Transwells (Keefer and Limbird, 1993). There was minimal fluorescence detectable on the bottom surface of the cells with either the CCD camera imaging or with confocal microscopy (Figure 3). In addition, confocal images of cells fixed with a more rigorous fixation procedure also showed the speckled staining pattern on the top and sides of the cell (Figure 3).

FRAP measurements of receptor mobility

Fluorescence recovery of the antibody-labelled receptor after photobleaching was measured as described in Materials and Methods. Data were collected from the lower surface of the cell. Attempts to collect data from the top surfaces were unsuccessful because dramatic oscillations of fluorescence (period ~10 seconds) precluded analysis of FRAP recovery curves (data not shown). Despite the low fluorescence observed on the bottom of the cells by microscopic examination, there was specific receptor-associated fluores-

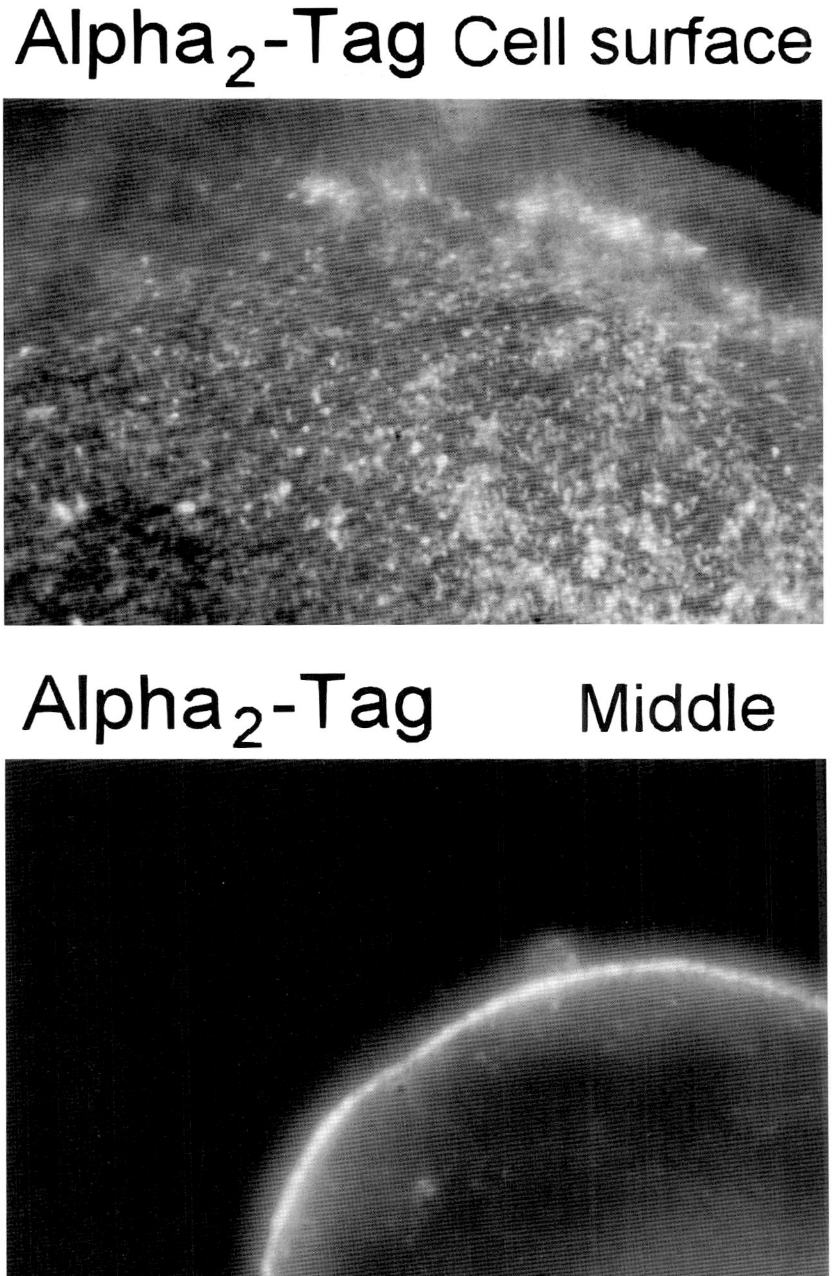

FIGURE 2 CCD camera images of α_2-adrenoceptors on MDCK cells at different focal planes. Shown are Alpha$_2$-Tag cells stained with Rhodamine-anti-tag antibodies as described in Methods. The upper image is focused on the upper surface of a fairly flat cell. The lower image is focused approximately the middle of a more rounded cell. These were living cells viewed about 40 minutes after staining.

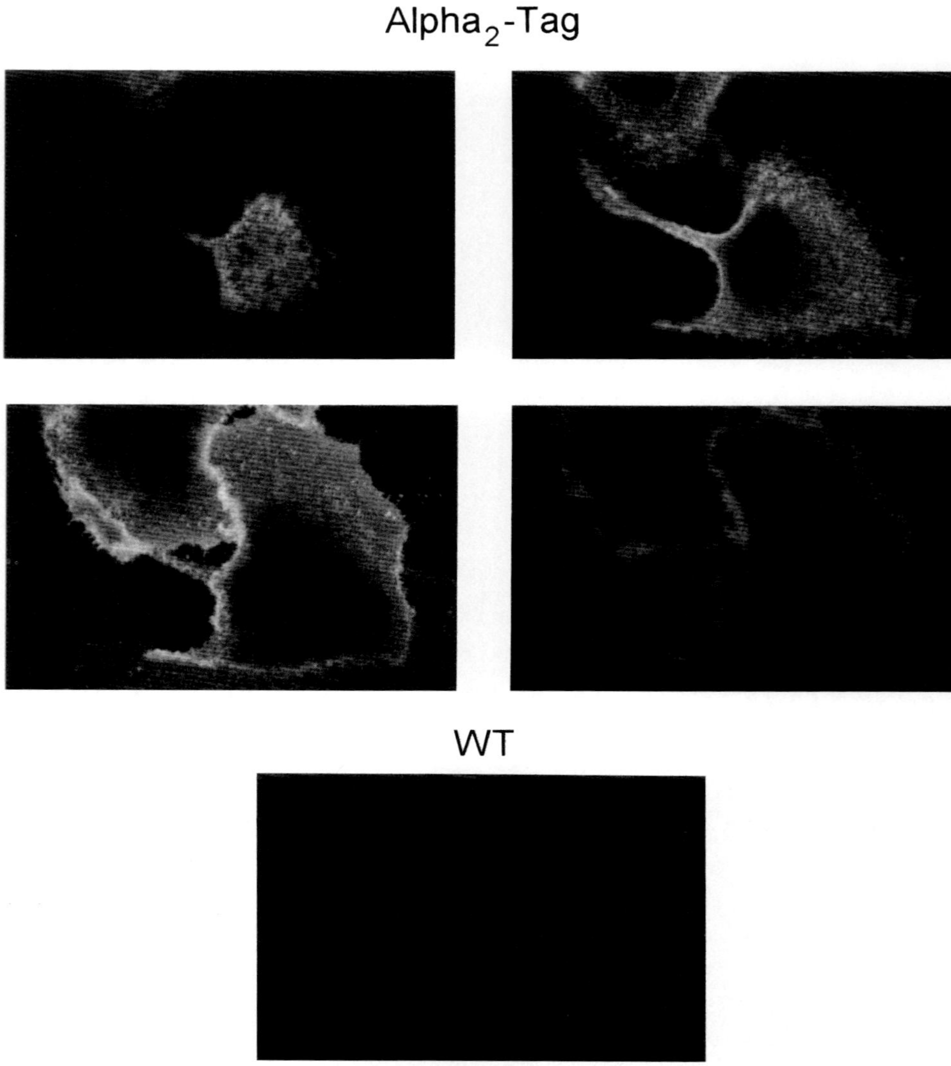

FIGURE 3 Confocal images of α_2-adrenoceptors on fixed MDCK cells. Fixed cells were stained with anti-Tag antibody and visualized on the confocal microscope as described in Methods. The focus was on the top of the cells (top left), 2 μm lower (top right), 2 μm further down (middle left), and additionally 2 μm down at the very bottom of the cell (middle right). The bottom image (WT) shows the minimal non-specific staining of non-transfected cells.

cence detectable by the photomultiplier tube in the FRAP apparatus. Cells without the epitope-tagged receptor (Figure 4C) show much less fluorescence than cells with either the normal receptor (Figure 4A) or the Not1 fragment deleted receptor (Figure 4B). Both receptor-containing cell types showed substantial recovery of fluorescence after bleaching. There was no significant difference between the percent of recovery ($80 \pm 6\%$, $n = 5$

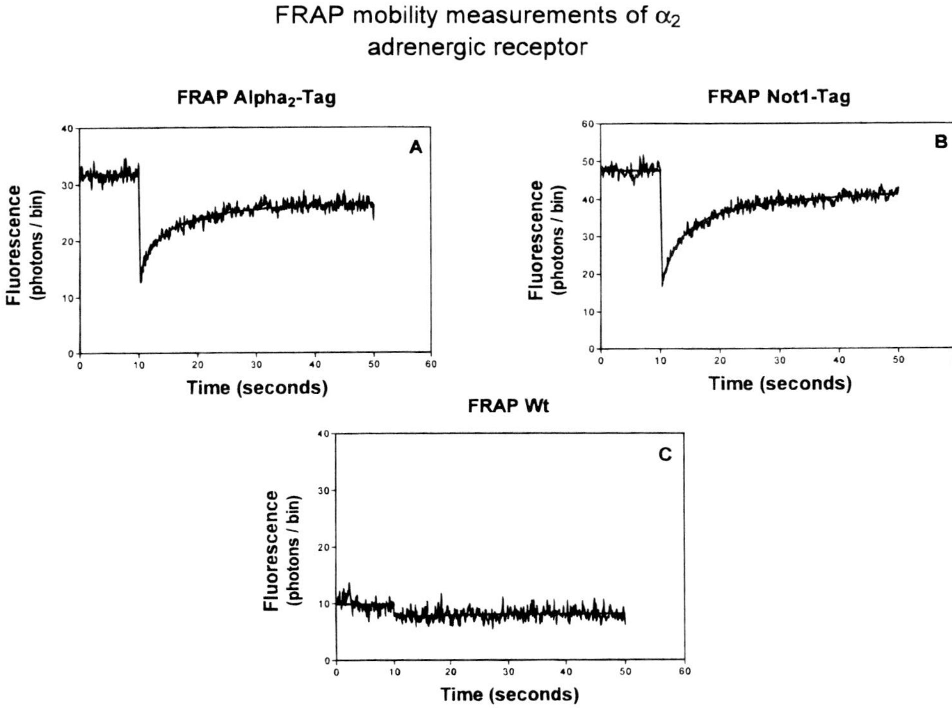

FIGURE 4 Fluorescence recovery after photobleaching (FRAP) of α_2-adrenoceptors on living MDCK cells. MDCK cells with (A) tagged α_2-adrenoceptors, (B) the Not1 mutant receptor, and (C) no receptor were labelled with anti-Tag antibody and studied at room temperature (Leung *et al.*, 1990; Bylund and U'Prichard, 1983; Jagadeesh *et al.*, 1990). Fluorescence was measured for 10 seconds prior to bleaching then for 40 seconds after bleaching. The two data points containing the bleaching artifact were deleted. The laser spot was focused near the edges on the bottom of the cells. Experiments were performed between 25 and 50 min after the initialization of staining. Data shown are averages of 20 to 50 individual traces.

for normal and 91 ± 5%, n = 4 for Not1 receptors) or recovery half-times (3.6 ± 0.8 seconds for normal and 3.7 ± 1.0 seconds for Not1 receptors) between the normal and Not1 receptors. This corresponds to a diffusion coefficient of approximately 2.2×10^{-10} cm^2 s^{-1}. Also, there was no difference in mobile fraction or diffusion coefficient after treatment of cells with epinephrine for 30–60 minutes (data not shown).

DISCUSSION

The distribution and mobility of epitope-tagged α_2 adrenergic receptors expressed in MDCK cells show some surprising features. First, there is a striking inhomogeneity to the pattern of α_2 adrenergic receptors on the cells. It has previously been reported that epitope-tagged α_2 adrenergic receptors are oriented on the basolateral surface of polarized MDCK cells (Keefer and Limbird, 1993). In addition to this "macro" scale targeting, we now report a distinctive clustered pattern of the fluorescence on the top and side surfaces of the cells grown on coverslips. The clustered distribution of receptors is not likely to be an artifact of the fixation of the cells. Clustered receptors were seen in unfixed cells as well as in lightly fixed (2% paraformaldehyde, 20 minutes) and heavily fixed (3% paraformaldehyde, 1 hour) cells. Also, clustering was observed with both the CCD camera and in the confocal microscope. A similar clustered distribution of β adrenergic receptors has been reported in A431 cells (Wang *et al.*, 1989) indicating that it is not just expressed receptors which exhibit this clustering phenomenon. It is possible that it occurs at high receptor density as A431 cells contain a large concentration of β receptors but the answer to this question will depend on studies of endogenous receptors in other cell types. It was recently reported that GPI-linked proteins can be induced by antibodies to cluster into cavolae even after fixation (Mayor *et al.*, 1994) however a more rigorous fixation which was shown to block clustering of GPI-linked receptors did not eliminate the clustering of the α_2 adrenergic receptors.

The functional significance of the clustering is unclear but G_i and G_o also exhibit a "clustered distribution" in plasma membranes and may be associated with microspikes enriched in actin and in neuronal growth cones (for review see Neubig, 1994). Immunofluorescence localization of G protein β subunits in A431 cells (Wang *et al.*, 1989) and in transfected COS cells (Muntz *et al.*, 1992) also showed a speckled pattern. We recently found that fluorescently labelled G protein α and βγ subunits (Kwon *et al.*, 1994; Kwon *et al.*, 1993) introduced into NG 108-15 neuroblastoma-glioma cells (Kwon *et al.*, 1994) by cell-vesicle fusion showed a strong degree of clustering. Also, the mobility of the tetramethylrhodamine-labelled βγ (TMR-βγ subunit of G_o in NG cells as determined by FRAP was very low. Only 16% of the fluorescence of TMR-βγ recovered after photobleaching (Kwon *et al.*, 1994). For α subunits this fraction was 34%. Thus the clustered distribution and restricted mobility of βγ subunit indicates that much of it is virtually immobilized. These observations make it necessary to reexamine current models of mobile receptors, G proteins, and effectors in cells.

The parallel occurrence of clusters of adrenergic receptors and G protein βγ subunits is consistent with a functional role for the clusters. Unfortunately attempts to stain permeabilized MDCK cells with G protein subunit antisera were unsuccessful due to high nonspecific staining so we could not determine whether the clusters of receptor and G protein coincide. This will be important in understanding the function of receptor clusters.

The FRAP data obtained with the α_2 adrenergic receptor show a relatively low diffusion coefficient of $2–3 \times 10^{-10}$ cm^2s^{-1} but nearly complete recovery of fluorescence. This is similar to the diffusion coefficient reported for the vasopressin V1 and V2 receptors (Jans *et al.*, 1990; Jans *et al.*, 1990). In the case of the adenylyl cyclase-activating V2 receptor, nearly all of the receptor was mobile. In contrast, the V1 receptor, which is coupled to phospholipase C was only about 40–50% mobile. Our FRAP data for the α_2 adrenergic receptor may overestimate the mobile fraction of that receptor. Data were collected from the lower surface of the receptor where densities are much lower and there is no evidence of a clustered distribution. If it were possible to collect FRAP data from the top surface of the cells, the mobile fraction might be lower.

Thus the α_2 AR appears to have both a clustered (and probably immobile) pool and a freely mobile pool (Figure 5). This is reminiscent of the subsynaptic clustering of the

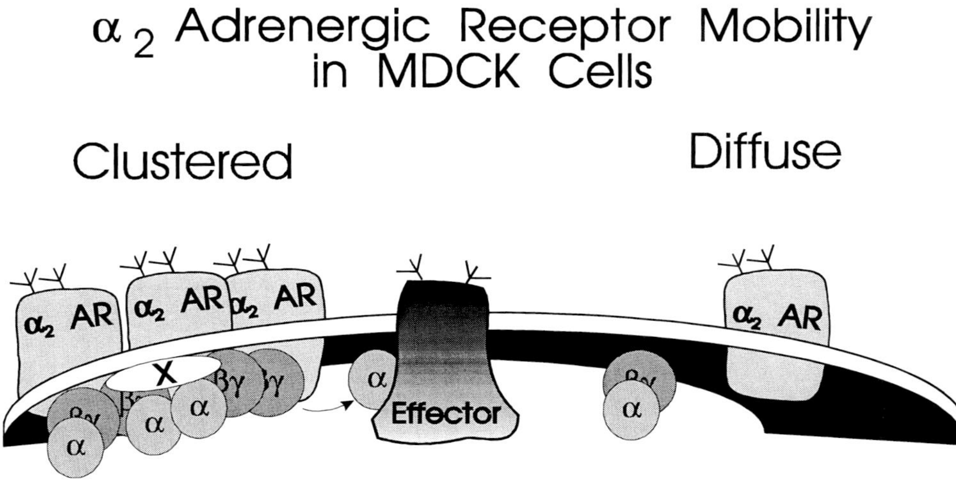

FIGURE 5 Schematic diagram of α_2-adrenoceptor organization in MDCK cells. Clusters of receptors were identified morphologically and the existence of a diffuse mobile pool observed by FRAP measurements. Previous studies showed clustering of G protein $\beta\gamma$ subunits associated with a very low mobility (Kwon *et al.*, 1994) however, co-clustering of the receptor with G protein subunits has not been demonstrated directly. For the G_s coupled vasopressin V2 receptor, the mobile pool has been shown to be functional however for G_i coupled receptors the function of clustered vs. diffuse receptors are not known (Neubig, 1994). It will be important to establish whether other proteins (designated as X) are involved in establishing or maintaining the clusters of receptors and G proteins.

nicotinic acetylcholine receptor in that a freely mobile pool of the nAcChR exists and on long time scales can move into the clusters (see (Axelrod, 1983) for review). It will be interesting to determine if G proteins co-cluster with the α_2 AR and if other proteins (X in Figure 5) are involved in the clustering process. Possible roles for caveolin or non-muscle spectrin have been proposed (Neubig, 1994).

It is intriguing to speculate that receptors that couple to G_s and stimulate adenylyl cyclase may follow the classical pattern of mobile receptors and significant signal amplification while G_i (and perhaps G_q) coupled receptors may be more selectively organized to limit interactions to defined G_i subtypes and effector systems. The exquisite specificity of coupling of muscarinic and somatostatin receptors to unique G_i heterotrimers (Kleuss *et al.*, 1993; Kleuss *et al.*, 1992; Kleus *et al.*, 1991) may be a reflection of cellular organization as well as structural information encoded in the receptor-G protein interaction sites.

ACKNOWLEDGMENTS

Support was from the National Institutes of Health (GM39561 to RRN, NS14565 to DA, and DK43879 to LEL) and National Science Foundation (DMB-8805296 to DA).

References

Axelrod, D. (1993) Lateral motion of membrane protein and biological function. *J. Membr. Biol.*, **75,** 1–10.

Axelrod, D., Koppel, D.E., Schlessinger, J., Elson, E.L. and Webb, W.W. (1976) Mobility measurements by analysis of fluorescence photobleaching recovery. *Biophys. J.*, **16,** 1055–1069.

Axelrod, D., Ravdin, P., Koppel, D.E., Schlessinger, J., Webb, W.W. and Elson, E.L. (1976) Lateral motion of fluorescently labeled acetylcholine receptors in membranes of developing muscle fibers. *Proc. Natl. Acad. Sci. USA* **73,** 4594–4598.

Brown, A.M. (1991) A cellular logic for G protein-coupled ion channel pathways. *FASEB J.*, **5,** 2175–2179.

Bylund, D.B. and U'Prichard, D.C. (1983) Characterization of alpha 1- and alpha 2-adrenergic receptors. *Int. Rev. Neurobiol.*, **24,** 343–431.

Cerione, R.A., Codina, J., Benovic, J.L., Lefkowitz, R.J., Birnbaumer, L. and Caron, M.G. (1984) The mammalian β_2-adrenergic receptor: Reconstitution of functional interactions between pure receptor and pure stimulatory nucleotide binding protein of the adenylate cyclase system. *Biochemistry*, **23,** 4519–4525.

Chabre, M. (1985) Trigger and amplification mechanisms in visual phototransduction. *Annu. Rev. Biophys. Biophys. Chem.*, **14,** 331–360.

Chabre, M. and Deterre, P. (1989) Molecular mechanism of visual transduction. *Eur. J. Biochem.*, **179,** 255–266.

Clark, J.D., Cragoe, E.J., Jr. and Limbird, L.E. (1990) α_2-Adrenergic receptors regulate Na^+-H^+ exchange via a cAMP-dependent mechanism. *Am. J. Physiol. Renal, Fluid Electrolyte Physiol.*, **259,** F977–F985.

Costa, T. and Herz, A. (1989) Antagonists with negative intrinsic activity at delta opioid receptors coupled to GTP-binding proteins. *Proc. Natl. Acad. Sci. U.S.A.*, **86,** 7321–7325.

Costa, T., Lang, J., Gless, C. and Herz, A. (1990) Spontaneous association between opioid receptors and GTP-binding regulatory proteins in native membranes: specific regulation by antagonists and sodium ions. *Mol. Pharmacol*, **37,** 383–394.

Costa, T., Ogino, Y., Munson, P.J., Onaran, H.O. and Rodbard, D. (1992) Drug efficacy at guanine nucleotide-binding regulatory protein-linked receptors: thermodynamic interpretation of negative antagonism and of receptor activity in the absence of ligand. *Mol. Pharmacol*, **41,** 549–560.

Dohlman, H.G., Caron, M.G. and Lefkowitz, R.J. (1987) A family of receptors coupled to guanine nucleotide regulatory proteins. *Biochemistry*, **26,** 2657–2664.

Fay, S.P., Posner, R.G., Swann, W.N. and Sklar, L.A. (1991) Real-time analysis of the assembly of ligand, receptor, and G protein by quantitative fluorescence flow cytometry. *Biochemistry*, **30,** 5066–5075.

Freissmuth, M., Casey, P.J. and Gilman, A.G. (1989) G proteins control diverse pathways of transmembrane signalling. *FASEB J.*, **3,** 2125–2131.

Gantzos, R.D. and Neubig, R.R. (1987) Low temperatures prevent α_2 agonist induced receptor-N_i coupling. *Fed. Proc.*, **46,** 702(Abstr.)

Gilman, A.G. (1987) G proteins: transducers of receptor-generated signals. *Annu. Rev. Biochem.*, **56,** 615–649.

Greenberg, M.L. and Axelrod, D. (1993) Anomalously slow mobility of fluorescent lipid probes in the plasma membrane of the yeast *Saccharomyces cervesiae*. *J. Membr. Biol.*, **131,** 115–127.

Gupta, S.K., Gallego, C. and Johnson, G.L. (1992) Mitogenic pathways regulated by G protein oncogenes. *Mol. Biol. Cell*, **3,** 123–128.

Hepler, J.R. and Gilman, A.G. (1992) G proteins. *Trends Biochem. Sci.*, **17,** 383–387.

Jagadeesh, G., Cragoe, E.J., Jr. and Deth, R.C. (1990) Modulation of bovine aortic alpha-2 receptors by Na+, 5′-guanylylimidodiphosphate, amiloride and ethylisopropylamiloride: evidence for receptor G-protein precoupling. *J. Pharmacol. Exp. Ther.*, **252,** 1184–1196.

Jans, D.A., Peters, R. and Fahrenholz, F. (1990) An inverse relationship between receptor internalization and the fraction of laterally mobile receptors for the vasopressin renal-type V2-receptor. An active role for receptor immobilization in down-regulation? *FEBS Lett.*, **274,** 223–226.

Jans, D.A., Peters, R. and Fahrenholz, F. (1990) Lateral mobility of the phospholipase C-activating vasopressin V1-type receptor in A7r5 smooth muscle cells a comparison with the adenylate cyclase-coupled V2-receptor. *EMBO J.*, **9,** 2693–2699.

Keefer, J.R., Kennedy, M.E. and Limbird, L.E. (1994) Unique structural features important for stabilization *versus* polarization of the α_{2A}-adrenergic receptor on the basolateral membrane of Madin-Darby canine kidney cells. *J. Biol. Chem.*, **269,** 16425–16433.

Keefer, J.R. and Limbird, L.E. (1993) The α_{2a}-adrenergic receptor is targeted directly to the basolateral membrain domain of Madin-Darby canine kidney cells independent of coupling to pertussis toxin-sensitive GTP-binding proteins. *J. Biol. Chem.*, **268,** 11340–11347.

Kleuss, C., Hescheler, J., Ewel, C., Rosenthal, W., Schultz, G. and Wittig, B. (1991) Assignment of G-protein subtypes to specific receptors inducing inhibition of calcium currents. *Nature*, **353,** 43–48.

Kleuss, C., Scherubl, H., Hescheler, J., Schultz, G. and Wittig, B. (1992) Different β-subunits determine G-protein interactions with transmembrane receptors. *Nature*, **358,** 424–426.

Kleuss, C., Scherubl, H., Hescheler, J., Schultz, G. and Wittig, B. (1993) Selectivity in signal transduction determined by gamma subunits of heterotrimeric G proteins. *Science*, **259,** 832–834.

Kwon, G., Axelrod, D. and Neubig, R.R. (1994) Lateral mobility of tetramethylrhodamine-labelled G protein α and βτ subunits in NG 108-15 cells. *Cell. Signal.* (In Press)

Kwon, G., Remmers, A.E., Datta, S. and Neubig, R.R. (1993) Synthesis and characterization of fluorescently labelled subunits of brain G protein. *Biochemistry*, **32,** 2401–2408.

Leung, E., Jacobson, K.A. and Green, R.D. (1990) Analysis of agonist-antagonist interactions at A_1 adenosine receptors. *Mol. Pharmacol.*, **38,** 72–83.

Limbird, L.E. (1988) Receptors linked to inhibition of adenylate cyclase: additional signaling mechansims. *FASEB J.*, **2,** 2686–2695.

Mayor, S., Rothberg, K.G. and Maxfield, F.R. (1994) Sequestration of GPI-anchored proteins in caveolae triggered by cross-linking. *Science*, **264,** 1948–1951.

Muntz, K.H., Sternweis, P.C., Gilman, A.G. and Mumby, S.M. (1992) Influence of gamma subunit prenylation on association of guanine nucleotide-binding regulatory proteins with membranes. *Mol. Biol. Cell*, **3,** 49–61.

Nerme, V., Severne, Y., Abrahamsson, T. and Vauquelin, G. (1986) Spontaneous coupling of the beta-adrenergic receptor to Ns in mammalian cardiac membranes. *Mol. Pharmacol.*, **30,** 1–5.

Neubig, R.R. (1994) Membrane organization in G protein mechanisms. *FASEB J*

Neubig, R.R., Gantzos, R.D. and Thomsen, W.J. (1988) Mechanism of agonist and antagonist binding to α_2 adrenergic receptors: evidence for a precoupled receptor-guanine nucleotide protein complex. *Biochemistry*, **27,** 2374–2384.

O'Dowd, B.F., Lefkowitz, R.J. and Caron, M.G. (1989) Structure of the adrenergic and related receptors. *Annu. Rev. Neurosci.*, **12,** 67–83.

Pedersen, S.E. and Ross, E.M. (1985) Functional activation of beta-adrenergic receptors by thiols in the presence or absence of agonists. *J. Biol. Chem.*, **260,** 14150–14157.

Posner, R.G., Fay, S.P., Domalewski, M.D. and Sklar, L.A. (1994) Continuous spectrofluorometric analysis of formyl peptide receptor ternary complex interactions. *Mol. Pharmacol.*, **45,** 65–73.

Ren, Q., Kurose, H., Lefkowitz, R.J. and Cotecchia, S. (1993) Constitutively active mutants of the alpha 2-adrenergic receptor [published erratum appears in J Biol Chem 1994 Jan 14;269(2):1566]. *J. Biol. Chem.*, **268,** 16483–16487.

Rhee, S.G. and Choi, K.D. (1992) Multiple forms of phospholipase C isozymes and their activation mechanisms. *Adv. Second Messenger Phosphoprotein Res.*, **26,** 35–61.

Rodbell, M. (1992) The role of GTP-binding proteins in signal transduction: from the sublimely simple to the conceptually complex. *Curr. Top. Cell. Reg.*, **32,** 1–47.

Samama, P., Cotecchia, S., Costa, T. and Lefkowitz, R.J. (1993) A mutation-induced activated state of the β_2-adrenergic receptor. Extending the ternary complex model. *J. Biol. Chem.*, **268,** 4625–4636.

Schaller, M.D. and Parsons, J.T. (1993) Focal adhesion kinase: an integrin-linked protein tyrosine kinase. *Trends Cell. Biol.*, **3,** 258–262.

Simon, M.I., Strathmann, M.P. and Gautam, N. (1991) Diversity of G proteins in signal transduction. *Science*, **252,** 802–808.

Sklar, L.A., Mueller, H., Omann, G. and Oades, Z. (1989) Three states for the formyl peptide receptor on intact cells. *J. Biol. Chem.*, **264,** 8483–8486.

Stryer, L. (1986) Cyclic GMP cascade of vision. *Annu. Rev. Neurosci.*, **9,** 87–119.

Surprenant, A., Horstman, D.A., Akbarali, H. and Limbird, L.E. (1992) A point mutation of the alpha 2-adrenoceptor that blocks coupling to potassium but not calcium currents. *Science*, **257,** 977–980.

Tang, W.-J. and Gilman, A.G. (1992) Adenylyl cyclases. *Cell*, **70,** 869–872.

von Zastrow, M., Link, R., Daunt, D., Barsh, G. and Kobilka, B. (1993) Subtype-specific differences in the intracellular sorting of G protein-coupled receptors. *J. Biol. Chem.*, **268,** 763–766.

Wang, H.Y., Berrios, M. and Malbon, C.C. (1989) Indirect immunofluorescence localization of beta-adrenergic receptors and G-proteins in human A431 cells. *Biochem. J.*, **263,** 519–532.

Wreggett, K.A. and De Lean, A. (1984) The ternary complex model. Its properties and application to ligand interactions with the D2-dopamine receptor of the anterior pituitary gland. *Mol. Pharmacol.*, **26,** 214–227.

Wu, D., Katz, A., Lee, C.H. and Simon, M.I. (1992) Activation of phospholipase C by alpha 1-adrenergic receptors is mediated by the alpha subunits of Gq family. *J. Biol. Chem.*, **267,** 25798–25802.

Yatani, A. and Brown, A.M. (1989) Rapid β-adrenergic modulation of cardiac calcium channel currents by a fast G protein pathway. *Science*, **245,** 71–74.

EXPRESSION AND REGULATION OF RENAL α_2-ADRENOCEPTORS

MARK L. JOHNSON, GUODONG GONG and WILLIAM A. PETTINGER

Midwest Hypertension Research Center, Creighton University School of Medicine, Omaha, NE 68131

Renal α_2-adrenoceptors (α_2-ARs) are overexpressed in many forms of genetic hypertension, but despite intensive study their role in the pathophysiology of high blood pressure remains a subject of debate. In this article we discuss recent studies which demonstrate that renal α_2-adrenoceptors and blood pressure are regulated in a parallel fashion by genes located on the sex chromosomes and autosomes and are similarly modulated by androgens. We propose that the overexpression of renal α_2-adrenoceptors (and more specifically, renal α_{2B}-AR mRNA) is a biochemical marker for multiple genes that cause high blood pressure.

KEY WORDS: α_2-adrenoceptors, testosterone, kidney, hypertension

Renal α_2-adrenoceptors (α_2-ARs) are overexpressed in a number of animal models of genetic hypertension, including the Spontanteously Hypertensive Rat (SHR) (Sanchez and Pettinger, 1981), Dahl salt-sensitive hypertensive (DS) (Gong *et al.*, 1992), and Sabra hypertensive (Parini *et al.*, 1983) rats when compared to their respective normotensive control strains. At present the mechanisms which regulate the normal expression of the α_2-AR genes and contribute to the altered pattern of regulation in hypertension are poorly understood. Studies regarding the altered pattern of regulation are made even more difficult by the complex nature of the relationships that exist between the genes that cause high blood pressure and the modulating influences of other factors that regulate these genes.

There is now accumulated evidence suggesting that the overexpression of renal α_2-adrenoceptors (α_2-AR) density is one component of the genetic hypertension pathway. Understanding how the α_2-AR genes are regulated will therefore provide valuable insight into the mechanisms which produce high blood pressure. Our first observation was that renal α_2-AR density is upregulated SHR (Sanchez and Pettinger, 1981). More recent studies have shown that male SHR renal α_2-AR density is significantly higher than the level in female SHR. We also observed a slightly higher density in the normotensive Wistar-Kyoto (WKY) male rat compared to the WKY female. Male SHR receptor density is higher than the level of WKY males. This indicates that overexpression of renal α_2-ARs is a sex-linked or a sex-influenced trait (Gong *et al.*, 1994). This pattern of male-female differences in renal α_2-AR density is identical to the pattern of blood pressure differences between sexes and strains.

Correspondence to: Dr. Mark L. Johnson, Creighton University School of Medicine, Midwest Hypertension Research Center, 601 North 30th Street, Suite 6730, Omaha, NE 68131.

Recently, it has been shown that in the SHR, about one-half of the hypertension is caused by genes located on the autosomes (also includes the X chromosome) and the other half is due to a gene on the Y chromosome (Ely and Turner, 1990). Turner, Ely and colleagues have developed two new substrains of SHR designated SHR/a and SHR/y (Turner *et al.*, 1991; Ely *et al.*, 1993). The SHR/y strain was developed by breeding F_1 males from the original WKY female X SHR male cross to a WKY female. The F_2 males were bred in a likewise fashion and the process repeated to the current point of the F_{15} generation. The SHR/y strain contains >99.99% WKY autosomes and SHR Y chromosome. The SHR/a strain was produced in an analogous fashion and now contains >99.99% SHR autosomes and the Y chromosome from WKY. The difference in blood pressure between SHR and WKY is an additive function of the two genetic compartments separated into these two substrains of SHR.

We examined the renal α_2-AR density in the parental strains and these two substrains using [^{3}H]-rauwolscine binding to whole kidney membranes (unpublished observations). In agreement with our previous observations using SHR purchased from the Charles River Laboratories, the Akron SHR male had a higher renal α_2-AR density than females and WKY males or females. The densities in the SHR/y and SHR/a strain were intermediate and significantly different from SHR and WKY. Furthermore, the two compartments influence receptor density in an additive manner. We conclude that receptor density in SHR is influenced by both the hypertensive Y chromosome and autosomal compartments.

These data and several other pieces of data suggest one of two possible explanations for the upregulation of renal α_2-AR in SHR: 1) the increase in renal α_2-AR density is part of the hypertension pathway, or 2) receptor density is somehow regulated by blood pressure directly. We investigated the second possibility by attempting to dissociate receptor density from blood pressure (Gong *et al.*, 1994). SHR males and females were treated with enalapril from 5 weeks to 14 weeks of age. Animals from each group had their blood pressure and receptor density measured at 5, 14 and 18 weeks of age. Although enalapril lowered blood pressure in both males and females, α_2-AR density did not change. The sexual dimorphism in blood pressure and receptor density was maintained in all treatment groups. These data indicate that the increase in blood pressure must be occurring at a step distal to the increase in receptor density in SHR males. If the reverse were true, we would have expected the receptor density to follow blood pressure in this experiment. Therefore, we conclude that the increase in renal α_2-AR density is not the result of high blood pressure *per se*, but appears to be one step in the pathway leading to the development of high blood pressure.

If blood pressure alone appears not to be driving receptor density, what factor(s) might be responsible for this increase? The testes, androgens and the androgen receptor have been shown to play a critical role in the development of SHR hypertension (Ely *et al.*, 1991). We investigated their role in the upregulation of renal α_2-AR density (Gong *et al.*, 1994). Animals were gonadectomized or sham operated at 4 weeks of age. Testosterone or placebo therapy was given in the respective, operated groups from 4 to 18 weeks of age. Sham males had a higher receptor density and blood pressure than sham females. These data again confirmed the sex-influenced component of receptor density and blood pressure in SHR. Castration reduced male receptor density and blood pressure. Interestingly, castrated males had a slightly higher receptor density than intact females, suggesting that androgens alone do not account for the total male-female difference. Testosterone increased blood pressure and receptor density in both gonadectomized males and females to that of sham operated males. Therefore, androgens are important positive regulators of renal a_2-AR density. Androgens have been shown to alter the level of a_2-AR gene transcription in adipocytes (Saulnier-Blache *et al.*, 1992). A similar mechanism in the kidney would be a highly probable explanation for our data. To partially test this possibility we developed a quantitative reverse transcriptase—polymerase chain reaction (RT-PCR) based assay to measure α_{2B}-AR mRNA levels in the kidneys from these various treatment groups (unpublished observations). Sham operated male α_{2B}-AR mRNA levels were significantly higher than sham operated female levels. Castration reduced the male levels to one-half of the sham level. Testosterone replacement restored male α_{2B}-AR mRNA levels to control values. Female α_{2B}-AR mRNA levels changed in a precisely parallel fashion to the previously observed changes in receptor density. Thus, all of the effects of gonadectomy and testosterone treatment on receptor density can be accounted for by changes in receptor mRNA concentration. The most likely explanation for the observed effects of testosterone, given what we know about how steroid hormones generally function, is a mechanism involving regulation of transcription, although we cannot totally exclude an affect on mRNA half-life based on these data.

Given what we now know about the factors which regulate the expression of the renal α_2-AR genes and the development of hypertension we propose the model shown in Figure 1.

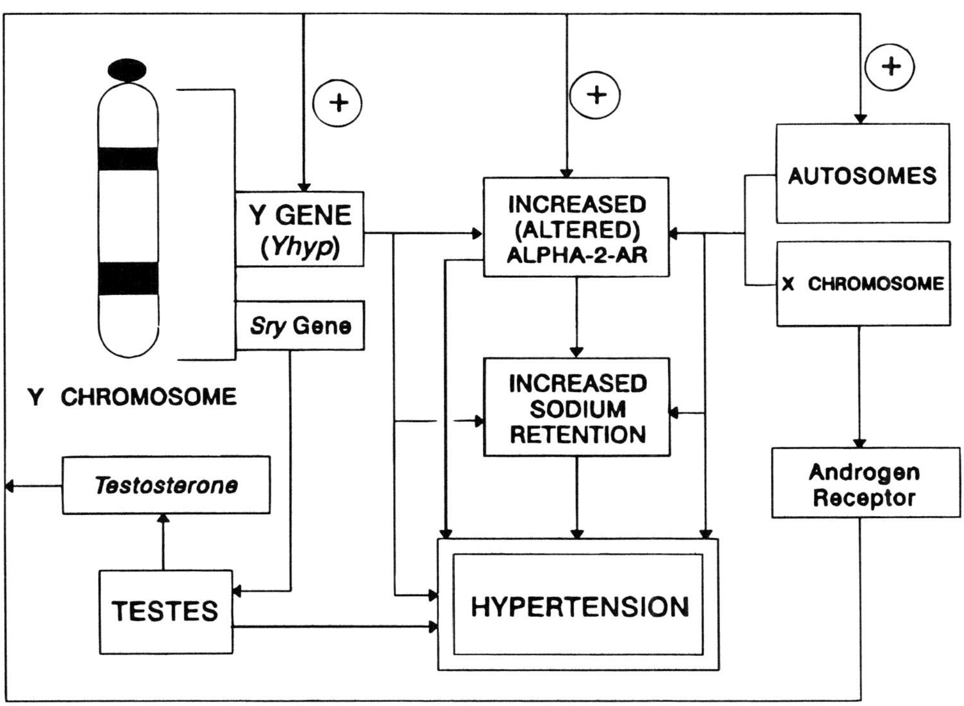

FIGURE 1 Proposed pathway illustrating the regulation of the renal α_2-adrenoceptor genes which leads to the development of hypertension in the Spontaneously Hypertensive Rat.

The Y chromosome contains a gene, *Sry*, which is responsible for the development of the testes (Page *et al.*, 1987). The testes play an important role in the development of high blood pressure, through the production of testosterone. Testosterone binds to the androgen receptor [encoded by a gene on the X chromosome] and acts as a positive regulator of the Y chromosome and autosomal components of hypertension (Ely and Turner, 1990; Turner *et al.*, 1991; Ely *et al.*, 1993) and the renal α_{2B}-AR gene. The SHR Y chromosome contains a gene, which we designate *Yhyp*, that contributes about 50% of the blood pressure increase in SHR males versus WKY males [Y linked, not a part of the sex-influenced component, (Ely and Turner, 1990; Turner *et al.*, 1991; Ely *et al.*, 1993)]. We hypothesize that the *Yhyp* gene produces a transcriptional regulator protein which may have multiple gene targets whose protein products affect blood pressure. Our data indicate that one of these targets is the α_{2B}-AR gene. Although not shown in this model, there is likely to be one or more intermediate steps involving other genes and their products that contribute to the increased sodium retention that leads to hypertension (see below, for example). Indeed, there may be parallel pathways that feed into the pathway we have proposed in this model.

The density of α_2-ARs is one determinant of receptor function. α_2-AR agonists attenuate forskolin-stimulated cAMP accumulation in a receptor density-dependent manner (Fraser *et al.*, 1989). Maximal stimulation of high affinity GTPase activity correlates with α_2-AR density (Duzic *et al.*, 1992). Also, it has been shown that stimulation by α_2-adrenoceptor agonists stimulates the Na^+/H^+ antiporter in renal proximal tubules (Nord *et al.*, 1987). This supports our hypothesis that the α_2-AR genes are part of the pathway that leads to the development of hypertension.

Understanding the regulation of the renal α_2-AR genes represents a major challenge for researchers in this field for the next several years. We have recently isolated a genomic clone for SHR α_{2B}-AR gene and have begun the molecular dissection of its regulatory elements. The results of these studies are expected to provide new insight into not only how this gene is regulated, but also may prove invaluable to our ongoing effort to identify the *Yhyp* gene and define the pathway(s) that lead to the development of hypertension.

References

Duzic, E., Coupry, I., Downing, S. and Lanier, S.M. (1992) Factors determining the specificity of signal transduction by guanine nucleotide-binding protein-coupled receptors. I. Coupling of α_2-adrenergic receptor subtypes to distinct G-proteins. *J. Biol. Chem.*, **267,** 9844–9851.

Ely, D.L. and Turner, M. (1990) Hypertension in the Spontaneously Hypertensive Rat is Linked to the Y Chromosome. *Hypertension*, **16,** 277–281.

Ely, D.L., Daneshvar, H., Turner, M.E., Johnson, M.L. and Salisbury, R.L. (1993) The Hypertensive Y Chromosome Still Elevates Blood Pressure in F11 Normotensive Rats. *Hypertension*, **21,** 1071–1075.

Ely, D., Salisbury, R., Hadi, D., Turner, M. and Johnson, M. (1991) Androgen Receptor and the Testes Influence Hypertension in a Hybrid Rat Model. *Hypertension*, **17,** 1104–1110.

Fraser, C.M., Arakawa, S., McCombie, W.R. and Venter, J.C. (1989) Cloning, sequence analysis, and permanent expression of a human alpha$_2$-adrenergic receptor in Chinese hamster ovary cells. Evidence for independent pathways of receptor coupling to adenylate cyclase attenuation and activation. *J. Biol. Chem.*, **264,** 11754–11761.

Gong, G., Pettinger, W.A., McArdle, S. and Jeffries, W.B. (1992) Renal α_2-adrenoceptor overexpression in Dahl salt sensitive rats: regulation by dietary NaCl and dissociation from salt-insensitive hypertension. *Hypertension Res.*, **15,** 85–92.

Gong, G., Dobin, A., McArdle, S., Sun, L., Johnson, M.L. and Pettinger, W.A. (1994) Sex Influence on Renal α_2-Adrenergic Receptor Density in the Spontanteously Hypertensive Rat. *Hypertension*, **23,** 607–612.

Nord, E.P., Howard, M.J., Hafezi, A., Moradeshagi, P., Vaystub, S. and Insel, P.A. (1987) α_2-adrenergic agonists stimulate Na^+-H^+ antiport activity in the rabbit renal proximal tubule. *J. Clin. Invest.*, **80,** 1755–1762.

Page, D.C., Mosher, R., Simpson, E.M., Fisher, E.M.C., Mardon, G., Pollack, J., McGillivray, de la Chapelle, A. and Brown, L.G. (1987) The Sex-Determining Region of the Human Y Chromosome Encodes a Finger Protein. *Cell*, **51,** 1091–1104.

Parini, A., Diop, L., Dausse, J., Meyer, P. and Ben-Ishay, D. (1983) α-adrenoceptors in Sabra hypertensive (SBH) and normotensive (SBN) rats: Effect of sodium. *J. Hypertens.*, **1,** 204–206.

Sanchez, A. and Pettinger, W.A. (1981) Dietary sodium regulation of blood pressure and renal α_1- and α_2-receptors in WKY and SH rats. *Life Sci.*, **29,** 2795–2802.

Saulnier-Blache, J.S., Bouloumie, A., Valet, P., Devedjian, J.C. and Lafontan, M. (1992) Androgenic regulation of adipocyte α_2-adrenoceptor expression in male and female Syrian hamsters: proposed transcriptional mechanism. *Endocrinology*, **130,** 316–327.

Turner, M.E., Johnson, M.L., and Ely, D.L. (1991) Separate sex-influenced and genetic components in spontaneously hypertensive rat hypertension. *Hypertension*, **17,** 1097–1103.

ANESTHETIC AND ANALGESIC ACTIONS OF α_2 ADRENOCEPTOR AGONISTS

MERVYN MAZE§, LAWRENCE POREEΩ and BRAD C. RABIN¶

Department of Anesthesia, Stanford University, Stanford, CA., & Anesthesiology Service, Veterans Affairs Medical Center, Palo Alto, CA; § Associate Professor of Anesthesiology; Ω Research Associate; ¶Research Assistant

The α_2 adrenergic agonists have proven clinical utility in the perioperative period. Sedation, anxiolysis, analgesia, reduction in anesthetic requirements, and hemodynamic stability, are among the desirable properties that this class of compound exhibit in the surgical patient. Additionally, the α_2 adrenergic agonists may possess ischemia-protective effects for both the brain and the heart which could greatly enhance its clinical benefits perioperatively. The elucidation of the molecular mechanism for the anesthetic action of the α_2 agonists provides the basis for understanding regulation of responsiveness as well as a target for rational drug discovery.

KEY WORDS: Adrenergic receptor, α2, agonists, clonidine, dexmedetomidine, anesthesia, analgesia

INTRODUCTION

Anesthesiologists utilize the sedative, anxiolytic, analgesic, and hemodynamic-stabilizing properties of α_2 adrenergic agonists (such as clonidine, dexmedetomidine and xylazine) to facilitate the anesthetic management of the surgical patient. The α_2 adrenergic agonists also reduce the need for, and the complications associated with, other anesthetic agents. Recently-acquired knowledge of the site and mechanism of action of the α_2 adrenergic agonists along with models of the 3-dimensional structure of the adrenergic receptor provide an opportunity for structure-based drug design to achieve the specificity which is often lacking in anesthetic drugs. This review will address the clinical utility of α_2 agonists and the molecular mechanism for its anesthetic and analgesic actions.

CLINICAL UTILITY OF α_2 AGONISTS IN THE SURGICAL PATIENT

1. Pre-anesthetic administration

Sedation is one of the most consistent behavioral effects mediated by central α_2 adrenoceptors. While sedation is undesirable in some settings (e.g. treatment of ambulatory patients with hypertension), this property has been used to clinical advantage as a premedication for the surgical patient (Segal *et al.*, 1991). A second requirement of premedicant

---

Correspondence to: Mervyn Maze, M.B., Anesthesiology Service 112A, PAVAMC, 3801 Miranda Ave., Palo Alto, CA 94304.

agents is to ameliorate the anxiety which is often evident in the surgical patient. α_2 agonists exert a prominent anxiolytic action in several animal paradigms (Salonen *et al.*, 1992*a*) and clinical syndromes in which anxiety is present. However, higher doses of α_2 agonists may produce an anxiogenic response probably through non-selective activation of α_1 receptors. The sedative effects of α_2 agonists is significantly potentiated by co-administration with a benzodiazepine (Salonen *et al.*, 1992*b*) (Figure 1).

Another benefit of α_2 agonists as a premedication is its ability to potentiate the anesthetic action of other agents and to reduce anesthetic requirements during surgery. While there is a ceiling to the anesthetic-sparing properties of clonidine (≤ 50%), more selective α_2 agonists are able to reduce the volatile anesthetic requirements by more than 95% in animals (Vickery *et al.*, 1988). Thus, providing the future hope that more selective and efficacious α_2 agonists may be independently capable of producing anesthesia. The reduction in anesthetic requirement produced by α_2 agonists can also be demonstrated in humans, regardless of the type of anesthetic technique (intravenous, volatile or regional blockade).

2. Intraoperative administration

While European investigators have utilized continuous infusion techniques to maintain plasma clonidine levels in the therapeutic range during the intraoperative and immediate postoperative period, the lack of a parenteral preparation in this country precludes such an

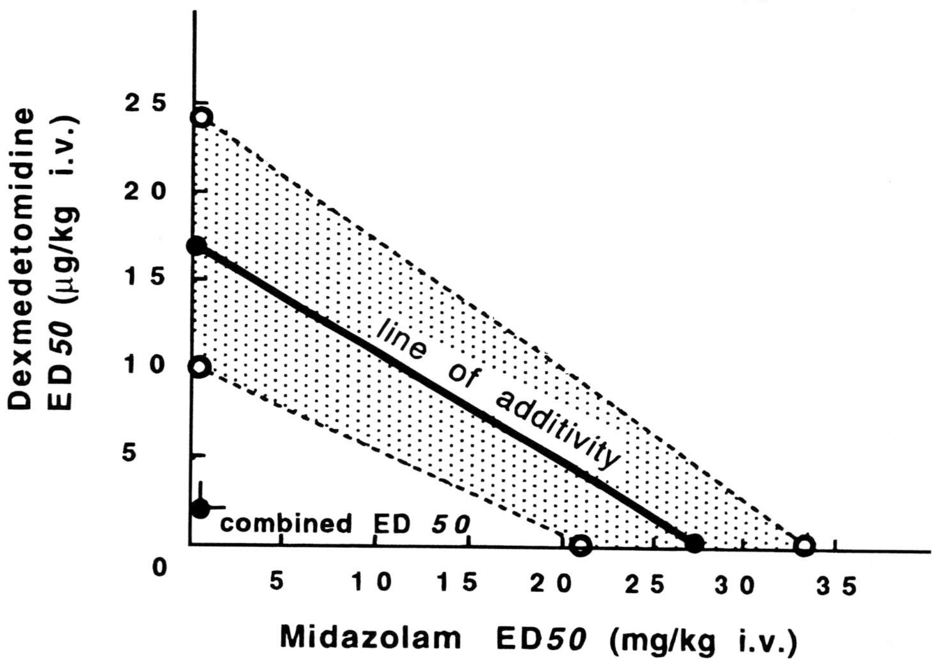

FIGURE 1 Isobolographic analysis of the hypnotic dose response curves of midazolam and dexmedetomidine alone and in combination. The ED_{50} values for dexmedetomidine and midazolam alone were obtained by probit analysis of the individual drug's dose response curve and are plotted on the ordinate and abscissa respectively. A line of additivity (Σ-Σ) connects these points and is banded by 95% confidence limits (shaded area) obtained from the variance of the individual dose-response curves. The ED_{50} value of the combination (Σ) and its 95% confidence limits are also plotted.

approach; thus we have employed a combination of oral and transdermal clonidine. Patients on this regimen required a lower concentration of anesthetic agent, exhibited greater hemodynamic stability, and recovered more rapidly from anesthesia (Segal *et al.*, 1991).

3. Postoperative administration

Anesthesiologists have advocated the use of α_2 agonists for control of postsurgical pain and to reduce postanesthesia emergence phenomena.

Analgesia. α_2 agonists provide improved pain control in postoperative patients whether administered directly into the neuraxis or systemically. The efficacy of postoperative clonidine depends on the dose of the agonist and the severity of postoperative pain.

Neuraxial: Clonidine produced analgesia in a dose-dependent manner when administered epidurally to patients undergoing total knee arthroplasty or abdominal surgery. Higher doses were associated with some disadvantages, including hypotension, bradycardia and transient sedation (Eisenach *et al.*, 1989). Several reports describe the effectiveness of combining epidural administration of clonidine with local anesthetics or opioids. Adding 150 μg of clonidine to epidural fentanyl, morphine and bupivacaine resulted in longer duration of postoperative analgesia (Rostaing *et al.*, 1991). Furthermore, these combinations can reduce the effective dose of clonidine, resulting in less side effects. Recently, intrathecal clonidine was shown to be an effective analgesic agent after cesarean section without remarkable side effects (Filos *et al.*,).

Systemic: The onset and duration of analgesia after intramuscular clonidine is comparable to that seen with epidural clonidine. In addition, although the peak plasma concentration of clonidine was higher in the intramuscular group, side effects were similar. Clonidine has also been used successfully *via* the intravenous and transdermal routes for postoperative analgesia.

Hypotension, bradycardia, and drowsiness are common side effects with this form of pain control in the postoperative period; unlike the opiate narcotics, the α_2 agonists do not produce clinically significant respiratory depression (Jarvis *et al.*, 1992). While the combination of opiate narcotics and α_2 agonists exhibit a marked synergistic interaction for analgesia, this does not extend to respiratory depression (Jarvis *et al.*, 1992). Therefore, the combination of these two classes of analgesics, at lower doses than when administered alone, appears to be a most prudent approach to postoperative pain control.

Emergence phenomena Shivering, with the attendant increase in oxygen utilization, is quite common following general anesthesia with volatile anesthetic agents. The α_2 agonists attenuate postoperative shivering probably by a central action on thermal regulatory mechanisms (Joris *et al.*, 1993).

α_2 agonists have been administered to provide sympatholysis in the postoperative period. This strategy limits sympathetically-mediated hypertension and tachycardia thereby decreasing the ischemic burden to the heart. In a recently concluded Phase II study clonidine, a highly-selective α_2 agonist, showed promising ability to decrease ischemic episodes in vulnerable post-surgery patients (Dorman *et al.*, 1993).

4. Special clinical settings

Hypertension Because of its hypotensive properties and its compatibility with other blood pressure-lowering medications, α_2 agonists can be used to limit elevations in blood pressure in hypertensive patients undergoing surgery.

Glaucoma α_2 agonists are effective at reducing the elevated intra-ocular pressure which is seen in patients with glaucoma. Anesthetic and surgical stimuli provoke further elevations which can be counteracted by pretreatment with α_2 agonists.

Cerebrovascular Disease We recently demonstrated that dexmedetomidine, administered at the time that focal cerebral ischemia is induced, can decrease neuronal death in anesthetized rabbits (Maier *et al.*, 1993). Others have shown that α_2 agonists are also effective in models of global ischemia. These animal studies have yet to be corroborated by studies involving humans.

Chronic Pain States The most intractable chronic pain states appear to be those comprising the "sympathetically mediated pain" (SMP) states. Examples of SMP include causalgia and trigeminal neuralgia which are refractory to most analgesic agents including opiate narcotics. The α_2 agonists appear to be quite effective at relieving pain in these patients (Rauck *et al.*, 1993). In patients with cancer pain that have become tolerant to the opiate narcotics, the α_2 agonists are useful substitute or adjunct to therapy with opiate narcotics.

MECHANISM OF ANESTHETIC ACTION

Because the locus coeruleus (LC) contains pathways involved in the maintenance of vigilance as well as a high prevalence of α_2 adrenoceptor, we investigated the role of this brainstem nucleus in the hypnotic response to dexmedetomidine (Correa-Sales *et al.*, 1992*a*). Dexmedetomidine, administered into the LC *via* a previously-sited cannula, increased the hypnotic response (as defined by the loss of the righting reflex), in a dose-dependent fashion. When dexmedetomidine was injected 2 mm lateral to the LC it did not cause the animals to lose their righting response. LC-administered atipamezole, a selective α_2 adrenergic antagonist, dose-dependently blocked the hypnotic response to either LC or ip dexmedetomidine (Figure 2). These data suggest that α_2 adrenergic receptors in the LC are a major site for the hypnotic action of dexmedetomidine. The localization of the hypnotic response in the LC subsequently allowed us to perturb various molecular components, in a site-specific manner, to define the steps in the signal transduction pathway of anesthetic action. We examined the role of pertussis toxin-sensitive (PTX) G proteins in the LC by pretreating stereotactically-cannulated rats with PTX, 0.34 μg and testing their hypnotic response 5 days later. *In vivo* treatment with PTX into the LC blocked the hypnotic response to LC-administered dexmedetomidine. The *in vivo* PTX treatment effectively ribosylated the G proteins. These data established a pivotal role for PTX-sensitive G proteins in the locus coeruleus in the hypnotic response to α_2 agonists in the rat (Correa-Sales *et al.*, 1992*b*).

Candidates for the effector mechanism of the α_2-mediated hypnotic response include inhibition of adenylate cyclase which has been shown to be pivotal to the cellular response of α_2 agonists in some, but not in all, cases. Both dibutyryl cAMP and rolipram, drugs which functionally sustain intracellular cAMP content, reversed the hypnotic response to dexmedetomidine. To test for the specificity of the hypnotic reversing effects of dibutyryl cAMP and rolipram, animals were pretreated with Rp-adenosine-3′,5′-cyclic phosphorothioate (Rp-cAMPS), a cAMP-dependent protein kinase inhibitor, and the experiments were repeated. Reversal of the hypnotic response to an α_2 agonist by either dB cAMP or rolipram could be prevented by blocking cAMP-dependent protein kinase activity with Rp-cAMPS (Figure 3). These data establish a pivotal role for inhibition of adenylate cyclase

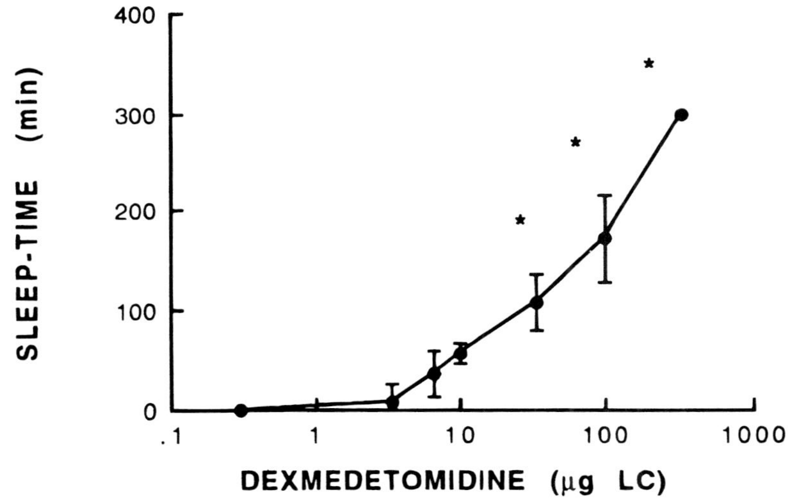

FIGURE 2 Dose-response curve for the hypnotic action of dexmedetomidine. Dexmedetomidine, was delivered into the left LC of chronically-cannulated rats and the duration of loss of righting reflex was measured. Each value represents the mean ± SD. n = 6–15/group. At the highest dose, the observation was terminated at 300 min. * denotes signficiantly different from the antecedent dose.

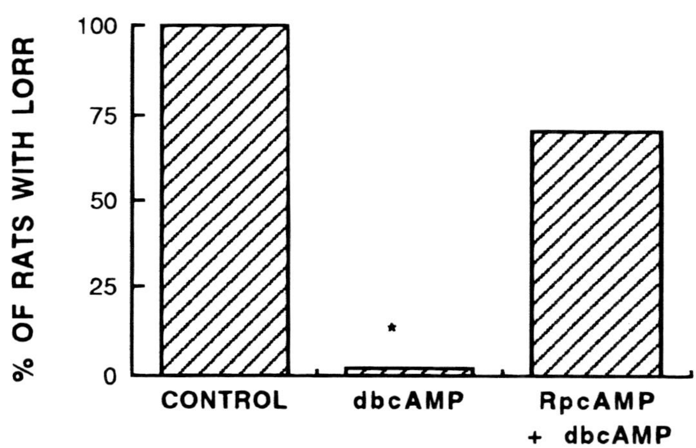

FIGURE 3 Effect of Rp-adenosine cyclic 3′, 5′-cyclic phosphorothioate (Rp-cAMPS) on the hypnotic reversing action of dibutyryl cAMP. Rats were administered Rp-cAMPS, 0.8 ng, dibutyryl cAMP, 0.8 ng, and dexmedetomidine,7 µg, consecutively at 1 minute intervals into the locus coeruleus *via* a chronically-sited indwelling cannula. The number of animals exhibiting loss of righting reflex was determined and expressed as a percentage of the total number of animals tested in each group (n = 8–10/group); * $p < 0.005$ when compared to the "control" group which only received dexmedetomidine.

in the transduction of the hypnotic response to an α_2 agonist in the LC of the rat (Correa-Sales *et al.*, 1992*c*).

To establish the mediating role of various species of calcium (Ca^{2+}) and potassium (K^+) ion channels in the hypnotic response, rats were pretreated with the following drugs prior to the administration of dexmedetomidine LC: S(+)202791 (L-type Ca^{2+} channel activator), nifedipine and R(–)202791 (L-type Ca^{2+} channel blocker), SNX 111 (N-type Ca^{2+} channel blocker), SNX 230 (P-type Ca^{2+} channel blocker), quinine (Ca^{2+}-activated and voltage gated K^+ channel blocker), charybdotoxin (Ca^{2+}-activated K^+ channel blocker), dendrotoxin (voltage-gated K^+ channel blocker) and glybenclamide (ATP-sensitive K^+ channel blocker). The drug doses used were devoid of any behavioral effects which may confound the interpretation of LORR. SNX 230 and the dihydropyridines nifedipine and R(–)202791 produced LORR in the presence of a subhypnotic dose of dexmedetomidine. The hypnotic-enhancing effects of the dihydropyridines could be blocked with S(+)202791, which also diminished LORR in response to dexmedetomidine 7.0 μg LC. Quinine, dendrotoxin, and charybdotoxin attenuated the hypnotic response to dexmedetomidine 7.0 mg LC. The hypnotic response to dexmedetomidine was not significantly altered by SNX 111 and glybenclamide. Both inhibition of ion conductance through either the L- or P-type Ca^{2+} channels and facilitation of conductance through either the voltage-gated or the Ca^{2+}-activated K^+ channels may be involved in the mechanism for the hypnotic action of α_2 adrenergic agonists (Nacif-Coelho *et al.*, 1994). These changes in ion conductance are capable of producing membrane hyperpolarization and decreasing neuronal excitability. There was no evidence for the involvement of ATP-sensitive K^+ channels or N-type Ca^{2+} channels in the hypnotic response to dexmedetomidine.

Finally, the binding affinities (K_i) of several α_2 adrenergic antagonists, RX821002, WB 4101, chlorpromazine, prazosin, and yohimbine, were correlated with their ability to block a hypnotic dose of LC-administered dexmedetomidine. The hypnotic blocking potency of the antagonists correlated highly with the K_i values for the α_{2A} adrenoceptor but not for the two other subtypes.

This suggests that the α_{2A} adrenoceptor subtype in the LC is responsible for the hypnotic effect of dexmedetomidine which is consistent with molecular genetic studies that have found only the mRNA for the α_{2A} subtype to be present in the LC of the rat.

MECHANISM OF ANALGESIC ACTION

The α_2 adrenergic agonists act in a multiplicative fashion at different sites in the nociceptive pathways to produce their analgesic action. These sites include the primary afferent nerve (including the dorsal root ganglion), the dorsal horn of the spinal cord, and the brainstem. The relative importance of each site to the overall analgesic effect of the α_2 agonists may depend on the type of pain.

Both binding (Lawhead *et al.*, 1992) and pharmacologic (Millan, 1992) studies suggest that the α_{2A} adrenoceptor subtype is the predominant receptor subtype in the mammalian spinal cord. In the dorsal horn, where the primary afferent neuron synapses, α_2 adrenergic agonists inhibit the release of substance P (SP), the peptide tachykinin involved in the mediation/modulation of nociceptive transmission (Kuraishi *et al.*, 1985). The molecular mechanism for the analgesic action in the spinal cord does not involve the inhibition of adenylate cyclase (Uhlen *et al.*, 1990); an alternative effector mechanism has not been defined.

The mechanisms involved in the analgesic action of α_2 adrenergic agonists in the brainstem are more obscure. It does appear that some of the analgesic effects of centrally-administered α_2 agonists may depend on descending monoaminergic (including noradrenergic) pathways which influence nociceptive processing at the level of the spinal cord.

CONCLUSION

The clinical application of α_2 adrenergic agonists into the anesthetic paradigm has no precedent in our discipline. More is known about the underlying mechanism of action of these compounds than is known about all other anesthetic agents. The initial promise of this class of agent is now being fulfilled with the development of more selective compounds. These novel compounds will help define, more precisely, the clinical utility of these agents. The recent technologic revolution in molecular biology should facilitate the development of even more selective agents for use in anesthesia.

ACKNOWLEDGMENTS

This work was supported, in part, by grants from the NIH (GM30232) and the Department of Veterans Affairs.

References

Correa-Sales, C., Rabin, B.C. and Maze, M. (1992*a*) A hypnotic response to dexmedetomidine, an α_2 agonist is mediated in the locus coeruleus in rats. *Anesthesiology* **76,** 948–952.

Correa-Sales, C., Reid, K. and Maze, M. (1992*b*) Pertussis toxin-mediated ribosylation of G proteins blocks the hypnotic response to an α_2 agonist in the locus coeruleus of the rat. *Pharmacology, Biochemistry, Behavior* **43,** 723–727.

Correa-Sales, C., Nasif-Coelho, C., Reid, K. and Maze, M. (1992*c*) Inhibition of adenylate cyclase in the locus coeruleus mediates the hypnotic response to an α_2 agonist in the rat. *J Pharmacology and Experimental Therapeutics* **263,** 1046–1049.

Dorman, B.H., Zucker, J.R. Verrier, E.D., Gartman, D.M. and Slachman, F.N. (1993) Clonidine improves perioperative myocardial ischemia, reduces anesthetic requirement, and alters hemodynamic parameters in patients undergoing coronary artery bypass surgery. *J Cardiothoracic and Vascular Anesthesia* **7,** 386–395.

Eisenach, J., Lysak, S.Z. and Viscomi, C.M. (1989) Epidural clonidine analgesia following surgery: Phase 1. *Anesthesiology* **71,** 640–646.

Filos, K.S., Goudas, L.C., Patroni, O. and Polyzou, V. (1992) Intrathecal clonidine as a sole analgesic for pain relief after cesarean section. *Anesthesiology* **77,** 174–267.

Jarvis, D.A., Duncan, S.R., Segan, I.S. and Maze, M. (1992) Ventilatory effects of clonidine alone, and the presence of alfentanil, in human volunteers. *Anesthesiology* **76,** 899–905.

Joris, J., Banache, M., Bonnet, F., Sessler, D.I. and Lamy, M. (1993) Clonidine and ketanserin both are effective treatment for postanesthetic shivering. *Anesthesiology* **79,** 532–539.

Kuraishi, Y., Hirota, N., Sato, Y., Kaneko, S., Satoh, M. and Takagi, H. (1985) Noradrenergic inhibition of the release of substance P from the primary afferents in the rabbit spinal dorsal horn. Brain Res **359,** 177–182.

Lawhead, R.G., Blaxall, H.S. and Bylund, D.B. (1992) α-2A Is the predominant α-2 adrenergic receptor subtype in human spinal cord. *Anesthesiology* **77,** 983–991.

Maier, C., Steinberg, G.K., Sun, G.H., Tian Zhi, G. and Maze, M. (1993) Neuroprotection by the alpha-2 adrenoceptor agonist, dexmedetomidine, in a focal model of cerebral ischemia. *Anesthesiology* **79,** 306–12.

Millan, M.J. (1992) Evidence that an α_{2A}—adrenoceptor subtype mediates antinociception in mice. *European J Pharmacology* **215,** 355–356.

Nacif-Coelho, C., Correa-Sales, C., Chang, L.L. and Maze, M. (1994) Perturbation of ion channel conductance alters the hypnotic response to the α_2 adrenergic agonist dexmedetomidine in rats. *Anesthesiology in Press.*

Nicholas, A.P., Pieribone, V. and Hokfelt, T. (1993) Distribution of mRNAs for alpha-2 adrenergic receptor subtypes in rat brain: an *in situ* hybridization study. *J Comparative Neurology* **328,** 575–594.

Rauck, R.L., Eiseach, J.C., Jackson, K., Young, L.D. and Southern, J. (1993). Epidural clonidine for refractory reflex sympathetic dystrophy. *Anesthesiology* **79,** 1163–1169.

Rostaing, S., Bonnet, F., Levron, J.C., Vodinh, J., Pluskwa, F. and Saada, M. (1991) Effects of epidural clonidine on analgesia and pharmacokinetics of epidural fentanyl in postoperative patients. *Anesthesiology* **75,** 420–425.

Salonen, M., Onaivi, E.S. and Maze, M. (1992*a*) Dexmedetomidine synergism with midazolam in the elevated plus—Maze test in rats. *Psychopharmacology* **108,** 229–234.

Salonen, M., Reid, K. and Maze, M. (1992*b*) Synergistic interaction between alpha$_2$-adrenergic agonists and benzodiazepines in rats. *Anesthesiology* **76,** 1004–1011.

Segal, I.S., Javis, D.J., Duncan, S.R., White, P.F. and Maze, M. (1991) Clinical efficacy of oral-transdermal clonidine combinations during the perioperative period. *Anesthesiology* **74,** 220–225.

Uhlen, S., Persson, M.-L., Alari, L., Post, C., Axelsson, K.L. and Wikberg, J.E.S. (1990) Antinociceptive actions of α_2-adrenoceptor agonists in the rat spinal cord: evidence for antinociceptive α_2-adrenoceptor subtypes and dissociation of antinociceptive α_2-adrenoceptors from cyclic AMP. *J Neurochemistry* **55,** 1905–1914.

Vickery, R.G., Sheridan, B.C., Segal, I.S. and Maze, M. (1988) Anesthetic and hemodynamic effects of the stereoisomers of medetomidine, an alpha$_2$ adrenergic agonist, in halothane-anesthetized dogs. *Anesth Analg* **67,** 611–615.

IV. β–ADRENOCEPTORS

SUBCLASSIFICATION OF THE β-ADRENOCEPTORS

J. PAUL HIEBLE and ROBERT R. RUFFOLO, JR.

SmithKline Beecham Pharmaceuticals, P.O. Box 1539, 709 Swedeland Road, King of Prussia, PA 19406-0939, USA

KEY WORDS: β_1-adrenoceptors, β_2-adrenoceptors, β_3-adrenoceptors, receptor classification

1. INTRODUCTION

For nearly one century, it has been known that certain responses to sympathetic nerve stimulation or to various adrenergic agonists are insensitive to inhibition by the classical adrenoceptor antagonists, such as the ergot alkaloids or β-haloalkylamines (Dale, 1906; Nickerson, 1949). In a manner analogous to the subclassification of the effects of acetylcholine as being either nicotinic or muscarinic, there were attempts to classify the effects of sympathetic nerve stimulation or sympathomimetic amines as being either "excitatory" or "inhibitory". This qualitative distinction was not particularly helpful in identifying or classifying adrenoceptor subtypes, and even led to proposals for the existence of different neurotransmitters that mediated these excitatory and inhibitory responses. In contrast to this qualitative analysis, Ahlquist (1948) used a more quantitative approach to propose the existence of two adrenoceptor subtypes, based on different rank orders of potency within a series of structurally-related natural and synthetic sympathomimetic amines whose responses were evaluated in different tissues. These two subtypes of adrenoceptors designated by Ahlquist (1948) subsequently became known as α-adrenoceptors and β-adrenoceptors. This subclassification of adrenoceptors was consistent with observations based on antagonist sensitivity, such that those responses that were designated by Ahlquist as being mediated by β-adrenoceptors were insensitive to inhibition by the ergot alkaloids or β-haloalkylamines, whereas those responses mediated by α-adrenoceptors were highly sensitive to these antagonists. Final proof for this subclassification scheme came in 1957 with the description of the partial-agonist, dichloroisoproterenol, which was the first agent capable of antagonizing β-adrenoceptor- but not α-adrenoceptor-mediated responses (Powell and Slater, 1957; Moran and Perkins, 1958).

Subsequently, Lands and co-workers (1967), again comparing rank orders of potency of agonists in a manner similar to that performed by Ahlquist (1948), concluded that there were two subtypes of the β-adrenoceptor. The β_1-adrenoceptor, the dominant receptor in heart and adipose tissue, was stimulated equally by norepinephrine and epinephrine, whereas the β_2-adrenoceptor, which produced relaxation of vascular, uterine and airway smooth muscle, was much less sensitive to norepinephrine *vis-a-vis* epinephrine. Highly selective antago-

Correspondence to: Dr. J. Paul Hieble, SmithKline Beecham Pharmaceuticals, 709 Swedeland Road, P.O. Box 1539, King of Prussia, PA 19406-0939 USA.

nists for both β_1- and β_2-adrenoceptors have been identified, as well as many potent and selective β_2-adrenoceptor agonists which are used as bronchodilators in the treatment of asthma.

Evidence has also accumulated over the years for the existence of yet another β-adrenoceptor that is insensitive to both β_1-adrenoceptor and β_2-adrenoceptor antagonists. This receptor has often been referred to as the "atypical" β-adrenoceptor, but with the subsequent identification of selective agonists (Arch *et al.*, 1984; Wilson *et al.*, 1984) and the recent cloning of a recombinant receptor having identical characteristics, this receptor has now been designated as the β_3-adrenoceptor.

More recently, the rapidly developing advances in molecular biology have had a major impact on β-adrenoceptor subclassification. Genes have been cloned for all three of the β-adrenoceptor subtypes that have been identified using functional and/or radioligand binding techniques in native tissues. Transfection of genomic or cDNA clones of these β-adrenoceptors into suitable mammalian cells has allowed for a detailed pharmacological characterization to be performed on these pure populations of the recombinant β-adrenoceptor subtypes, thereby confirming their identity with the corresponding β-adrenoceptor subtypes present in native tissues.

All three β-adrenoceptor subtypes can be activated by norepinephrine and epinephrine. However, in contrast to the α-adrenoceptors, these endogenous catecholamines do have differential affinities for the β-adrenoceptor subtypes. A primary distinction between β_1- and β_2-adrenoceptors is the relative potency of epinephrine and norepinephrine, with these two catecholamines being equipotent at the β_1-adrenoceptor, and epinephrine having up to 100-fold higher affinity than norepinephrine for the β_2-adrenoceptor. Conversely, at least in some systems, norepinephrine is more potent than epinephrine as a β_3-adrenoceptor agonist. The synthetic catecholamine, isoproterenol, is a potent agonist at all three β-adrenoceptor subtypes, with no consistent selectivity for any of the β-adrenoceptor subtypes being observed. Propranolol and its many analogs are potent antagonists at β_1-adrenoceptors and β_2-adrenoceptors; however β_3-adrenoceptor-mediated responses are much less sensitive to inhibition by these antagonists. All three β-adrenoceptor subtypes activate adenylyl cyclase as the primary mechanism for signal transduction.

2. FUNCTIONAL SUBCLASSIFICATION OF B-ADRENOCEPTORS

Many of the functional studies, typically physiological, aimed at subclassifying β-adrenoceptors have compared β_1-adrenoceptor-mediated inotropic and/or chronotropic responses in cardiac (commonly atrial) preparations with β_2-adrenoceptor-mediated relaxation of pulmonary (commonly tracheal) smooth muscle. Although at one time it was assumed that these responses were mediated by only one β-adrenoceptor subtype, it is now clear that this is not the case. Careful analysis of agonist/antagonist interactions have been used to show that both β_1- and β_2-adrenoceptor activation can contribute to both pulmonary and cardiac responses to βadrenoceptor agonists (Kaumann and Lemoine, 1985; Lemoine *et al.*, 1985). Nevertheless, these tissues are still useful for identifying agents showing β-adrenoceptor subtype selectivity. As is the case for the subclassification of most receptors, selective agonists and antagonists are used as pharmacological tools to subclassify the β-adrenoceptors. The most commonly used pharmacological tools to study the β-adrenoceptor subtypes are presented in Table 1.

TABLE 1
Pharmacological Tools Used to Subclassify and Characterize the β-Adrenoceptor Subtypes

Compound	*Index*[1]	β_1	β_2	β_3	*Ref*
		AGONISTS			
NONSELECTIVE					
Isoproterenol	K_i[2]	14	21	ND	Naito *et al.*, 1985
	EC_{50}[3]	1.3	2.2	9.3	Arch *et al.*, 1984
	K_{act}[4]	ND	ND	3.9	Emorine *et al.*, 1992
β_1-SELECTIVE					
Xamoterol	K_i	135	4533	ND	Brodde *et al.*, 1990
Denopamine	K_i	545	2205	ND	Naito *et al.*, 1985
β_2-SELECTIVE					
Salbutamol	EC_{50}	810	26	2300	Arch *et al.*, 1984
	EC_{50}	350	3.3	ND	Yabuuchi, 1977
Salmeterol	EC_{50}	> 30,000	4	ND	Ball *et al.*, 1987
Terbutaline	EC_{50}	18,000	120	ND	Waldeck *et al.*, 1986
Procaterol	EC_{50}	257	1.8	ND	Yabuuchi, 1977
	EC_{50}	12,800	1.7	ND	Waldeck *et al.*, 1986
	K_i	4100	41	ND	Dickinson *et al.*, 1981
β_3-SELECTIVE					
BRL 37344	EC_{50}	680	35	1.7	Arch *et al.*, 1984
		5600	190	8.4	Bloom *et al.*, 1992
	K_{act}	ND	ND	5.9	Emorine *et al.*, 1992
CGP 12177	K_{act}	ND	ND	300	Emorine *et al.*, 1992
CL 316,243	EC_{50}	> 10,000	30,000	3	Bloom *et al.*, 1992
		ANTAGONISTS			
NONSELECTIVE					
Propranolol	K_B[5]	1	0.7	95	Harms *et al.*, 1977
	K_B	2	5	250	Wilson *et al.*, 1984
	K_i	ND	ND	1500	Muzzin *et al.*, 1991
	K_i	ND	ND	450	Nahimas *et al.*, 1991
[^{3}H] CGP 12177	K_D[6]	0.3	0.9	ND	Nanoff *et al.*, 1987
[^{125}I] ICYP[7]	K_D	0.01	0.01	1	Sillence *et al.*, 1993
β_1-SELECTIVE					
Betaxolol	K_i	3	610	ND	Mauriege *et al.*, 1988
	K_i	ND	ND	16,700	Muzzin *et al.*, 1991
Atenolol	K_i	125	1250	ND	Daemmgen *et al.*, 1985
Metoprolol	K_i	3	4200	ND	Mauriege *et al.*, 1988
Bisoprolol	K_i	0.2	1020	ND	Mauriege *et al.*, 1988
	K_B	ND	ND	630	Langin *et al.*, 1991
CGP 20712	K_i	0.6	1500	ND	Mauriege *et al.*, 1988
	K_i	ND	ND	6700	Emorine *et al.*, 1989
β_2-SELECTIVE					
ICI 118551	K_i	710	4.6	ND	Mauriege *et al.*, 1988
	K_i	ND	ND	770	Emorine *et al.*, 1989
	K_B	ND	ND	1600	Wilson *et al.*, 1984

[1]Parameter used to express affinity for the β-adrenoceptor. All dissociation constants and EC_{50} values are expressed in nM.
[2]Dissociation equilibrium constant (K_i) for the inhibition of radioligand binding to the β-adrenoceptor. Inhibition of the binding of a nonselective radioligand in membrane homogenates from a tissue containing both β_1 and β_2 adrenoceptor subtypes is biphasic, and can be used to calculate the K_i value at each subtype. Alternatively, binding can be studied in tissues containing a relatively pure population of β_1- (*e.g.*, human atrium, rat ventricle) or β_2- (*e.g.*, human or rat lung) adrenoceptors.
[3]EC_{50} for induction of a functional response in a tissue containing a relatively pure population of β_1- (*e.g.*, rat atrium), β_2- (*e.g.*, guinea pig trachea) or β_3- (*e.g.*, rat white or brown adipocytes) adrenoceptors.
[4]Affinity constant (K_{act}) for stimulation of adenylate cyclase in CHO cells expressing the β_3-adrenoceptor.

TABLE 1–*continued*

[5]Receptor dissociation constant (K_B) for blockade of a functional response to a β-adrenoceptor agonist in a tissue having a relatively pure population of a particular β-adrenoceptor subtype.
[6]Dissociation constant (K_D) for the interaction of [^{3}H] CGP 12177 with rat cardiac membranes in the presence of subtype-saturating concentrations of a selective β_1- or β_2-adrenoceptor antagonist.
[7]Dissociation constants (K_D) for the interaction of [^{125}I] iodocyanopindolol at β_1-, β_2- and β_3-adrenoceptors in rat brown adipose tissue and soleus muscle.

Several classes of agonists have been identified that will produce relaxation of tracheal smooth muscle at concentrations that have no effect on atrial rate. For example, sulfonterol has an EC_{50} of 17 nM for relaxing guinea pig trachea, whereas concentrations that are more than 1,000-fold higher (*i.e.*, EC_{50} = 28 μM) are required for stimulating the rate of contraction of isolated guinea pig right atrium (Kaiser *et al.*, 1975). However, one must be cautious in interpreting agonist selectivity, since the functional response produced by an agonist is dependent on both its affinity and efficacy at the receptor, and as such, EC_{50} values may not accurately reflect relative affinities or selectivities (Ruffolo, 1982). Further investigation of sulfonterol showed that the compound was a partial agonist at the β_2-adrenoceptor, as reflected by the fact that the compound produced less maximal relaxation than did isoproterenol in carbachol-precontracted guinea pig trachea (Waldeck *et al.*, 1986). This also corresponded to the observation that sulfonterol produced a lower maximal stimulation of adenylyl cyclase in cultured muscle cells (which contain β_2-adrenoceptors) than did isoproterenol (Pittman *et al.*, 1984), and with the failure of sulfonterol to produce activation of adenylyl cyclase in these cells following partial β_2-adrenoceptor desensitization (Pittman *et al.*, 1984). In addition, sulfonterol has moderate affinity, but low efficacy, at the β_1-adrenoceptor, as evidenced by its ability to block isoproterenol-induced increases in the force of contraction of guinea pig papillary muscle. In fact, comparison of the pA_2 values for sulfonterol as an antagonist of isoproterenol-induced tracheal relaxation (6.15) and the increases in papillary muscle contraction (5.82) shows that the affinity of sulfonterol for β_1- and β_2-adrenoceptors is actually quite similar (Waldeck *et al.*, 1986).

However, agonists demonstrating selectivity for the β_2-adrenoceptor have, in fact, been identified. One example of these compounds is procaterol (OPC-2009) which has high intrinsic activity at the β_2-adrenoceptor (Pittman and Molinoff, 1983; Pittman *et al.*, 1984; Waldeck *et al.*, 1986) and does not interact with β_1-adrenoceptors until concentrations that are approximately 100-fold higher than the EC_{50} at β_2-adrenoceptors are attained (Hedberg and Mattsson, 1981; Waldeck *et al.*, 1986). The selectivity of procaterol for β_2-adrenoceptors has been confirmed in radioligand binding studies (Dickinson *et al.*, 1981).

While many functionally selective β_2-adrenoceptor agonists have been identified by comparing potencies in atrial and tracheal preparations, it has been difficult to demonstrate the converse (*i.e.*, an agonist that will stimulate the atrial, but not the tracheal, β-adrenoceptor). This may be due, at least in part, to the presence of a functional β_1-adrenoceptor population in guinea pig trachea, the most commonly used preparation for such studies. Although still less extensively studied than the selective β_2-adrenoceptor agonists, several agents are available that show selectivities for β_1- versus β_2-adrenoceptors that are comparable to the selectivity observed for norepinephrine, but without having the α-adrenoceptor agonist activity associated with the endogenous catecholamine. Among these compounds are Ro-363 (Iakovidis *et al.*, 1980), denopamine (Naito *et al.*, 1985) and xamoterol (Nuttall and Snow, 1982). However, all of these agents are partial agonists at the β_1-adrenoceptor, and some of the apparent *in vivo* β_1-adrenoceptor selectivity may result from their low intrinsic activity at β_2-adrenoceptors (Kenakin and Beck, 1982).

The availability of highly selective β-adrenoceptor antagonists has facilitated the analysis of the functional roles of β_1- and β_2-adrenoceptors. Soon after the initial subclassification of the β-adrenoceptors, several antagonists having moderate selectivity for either β-adrenoceptor subtype, such as practolol (β_1) and butoxamine (β_2), were identified (Levy, 1966; Dunlop and Shanks, 1968; Levy and Wilkenfeld, 1970). However, these compounds had neither high potency or sufficient selectivity to be extremely useful as pharmacological tools. Currently, the selective antagonists most commonly used as pharmacological tools are CGP 20712A, which has a dissociation constant (K_B) at the β_1-adrenoceptor that is below 1 nM, and a β_1/β_2-adrenoceptor selectivity ratio of 10,000-fold (Dooley *et al.*, 1986; Schliep and Harting, 1984), as well as (–)-bisoprolol, which has a K_B of 2–3 nM at the β_1-adrenoceptor and a β_1/β_2-adrenoceptor selectivity of approximately 100-fold (Kaumann and Lemoine, 1985). Conversely, ICI 118,551 is a potent β_2-adrenoceptor antagonist, with a K_B that is below 1 nM, and with a β_2/β_1-adrenoceptor selectivity of over 100-fold (Bilski *et al.*, 1983; Lemoine *et al.*, 1985).

Over one decade ago, functional evidence for the existence of a β-adrenoceptor that was distinct from either the β_1- or β_2-adrenoceptors was provided by studies in rat adipose tissue (Wilson *et al.*, 1984; Arch *et al.*, 1984). Isoproterenol-induced lipolysis in rat adipocytes is relatively insensitive to blockade by either propranolol ($pA_2 = 6.6$), ICI 188,551 ($pA_2 =$ 5.5, 5.8), practolol ($pA_2 = 4.9$) or CGP 20712A ($pA_2 = 4.8$) (Wilson *et al.*, 1984; Hollenga and Zaagsma, 1989). Furthermore, propranolol and several other β-adrenoceptor antagonists demonstrated less stereoselectivity between their respective enantiomers for blocking the lipolytic response in fat cells than was observed for the blockade of other β-adrenoceptor-mediated responses in different test systems (Harms *et al.*, 1977). This novel β-adrenoceptor has now been clearly shown to represent a third β-adrenoceptor subtype, designated as the β_3-adrenoceptor. In contrast to rat adipocytes, the human adipocyte also appears to have β_1-adrenoceptors with the expected antagonist potencies and stereoselectivity (Harms, 1976), although the β_3-adrenoceptor may also mediate a component of the lipolytic response in human fat cells as well (Zaagsma and Nahorski, 1990).

The identification of novel β_3-adrenoceptor agonists, such as BRL 28410, BRL 35135 and BRL 37344, which are potent agonists at the rat adipocyte β-adrenoceptor, with substantially less activity in producing β_1- and β_2-adrenoceptor-mediated responses in rat atria, uterus or trachea (Wilson *et al.*, 1984), provides further support for the existence of a novel β-adrenoceptor in adipocytes. Furthermore, certain phenoxypropanolamines that are potent antagonists of both β_1- and β_2-adrenoceptors act as partial agonists at the β_3-adrenoceptor.

A similar situation exists with respect to the β-adrenoceptors in the rat brown adipocyte. Activation of β-adrenoceptors on these specialized adipose cells also stimulates lipolysis, which results in a thermogenic response (Nicholls and Locke, 1984). These receptors are likely to play an important role in diet-induced heat generation (Rothwell and Stock, 1979). As in the rat white adipocyte, stimulation of lipolysis in the rat brown adipocyte is less sensitive to inhibition by propranolol than are β_1- and β_2-adrenoceptor-mediated responses, and the selective β_3-adrenoceptor agonists, BRL 37344 and BRL 35135, can activate lipolysis in these cells and stimulate metabolic rate (Arch *et al.*, 1984).

Direct measurement of metabolic rate *in vitro* confirms that BRL 37344 is a potent agonist at β_3-adrenoceptors, and that this response is relatively resistant to blockade by nonselective (propranolol), β_1-adrenoceptor-selective (atenolol) or β_2-adrenoceptor-selective (ICI 118,551) antagonists (Sudera and Stock, 1988).

There is a growing body of experimental evidence to support the presence of an atypical β-adrenoceptor in other tissues, perhaps with similar characteristics to the β_3-adrenoceptor found in adipocytes. Thus, in guinea pig right atrium, in addition to a β_2-adrenoceptor-mediated component, high concentrations of (–)-pindolol will produce a chronotropic response that is insensitive to either β_1-adrenoceptor blockade with bisoprolol or β_2-adrenoceptor blockade with ICI 118,551. This response can be antagonized by high concentrations of the nonselective β-adrenoceptor antagonist, bupranolol, and may correspond to a binding site for [^{3}H]-bupranolol that has low affinity ($pK_D = 4.9$) for (–)-pindolol (Walter *et al.*, 1984). A detailed analysis of the cardiac-stimulant action (chronotropic, inotropic and/or adenylyl cyclase activation) of several partial β-adrenoceptor agonists supports the notion that an atypical cardiac β-adrenoceptor may exist that has similar characteristics to the β_3-adrenoceptor of the rat adipocyte (Kaumann, 1989).

Lower concentrations of BRL 37344 produce relaxation of histamine-contracted longitudinal strips of guinea pig ileum (Bond and Clarke, 1988). Concentrations of propranolol and nadolol that are sufficient to produce profound blockade of β_1- and β_2-adrenoceptor-mediated responses in other tissues produce only limited blockade of the relaxation induced by either BRL 37344 or isoproterenol in guinea pig ileum. In order to produce additional blockade of these responses, higher concentrations of the antagonists (up to 1 mM) are required. These results suggest the existence of a minor β_1-adrenoceptor-mediated, propranolol-sensitive component, with the majority of the response being mediated by the β_3-adrenoceptor. In the presence of propranolol at concentrations that saturate the β_1- and β_2-adrenoceptor populations, atenolol produces competitive blockade of the inhibitory effect of isoproterenol in transmurally-stimulated guinea pig ileum ($pA_2 = 6.5$) (Blue *et al.*, 1988). Under these conditions, cyanopindolol is an even more potent antagonist of the propranolol-insensitive response to isoproterenol in this tissue ($pA_2 = 7.63$) (Blue *et al.*, 1989).

In rat colon, another series of novel β-adrenoceptor agonists, which can be considered as hybrids of the phenylethanolamine and aminotetralin structures, produces relaxation of spontaneous contractions, and these responses are sensitive to inhibition by higher concentrations of nonselective β-adrenoceptor antagonists (alprenolol, pindolol and propranolol), but are resistant to either selective β_1- (atenolol) or β_2- (ICI 118,551) adrenoceptor blockade (Croci *et al.*, 1988; Giudice *et al.*, 1989). The pA_2 values for pindolol (6.5) and propranolol (6.2) against these selective agonists are similar to those obtained against isoproterenol, BRL 28140 or BRL 35135 at the β_3-adrenoceptor of rat adipocyte ($pA_2 = 6.2$–7.0) (Harms, 1976; Wilson *et al.*, 1984). Isoproterenol is active in this preparation, and analysis of the effects of β-adrenoceptor antagonists of the isoproterenol response suggests an action on both typical (*i.e.*, β_2 and β_2) and atypical (*i.e.*, β_3) β-adrenoceptors (Croci *et al.*, 1988). In rat distal colon contracted with a high concentration of potassium, both isoproterenol and BRL 37344 induce concentration-dependent relaxation (McLaughlin and MacDonald, 1989). The inhibition by propranolol of the isoproterenol responses suggest an action on a mixed population of β-adrenoceptors, whereas the response to BRL 37344 was insensitive to inhibition by propranolol (at a concentration of 1 μM).

Hence, although there are quantitative differences between the various responses described above, there is sufficient evidence to infer that the rat adipocyte (both white and brown), guinea pig ileum, rat colon and cardiac tissue from several species may all possess β_3-adrenoceptors, probably as a component of a mixed population of β-adrenoceptors. A different proportion of β_3-adrenoceptors *vis-a-vis* β_1- and β_2-adrenoceptors may provide the explanation for the differences observed between these tissues.

3. RECOMBINANT β-ADRENOCEPTOR SUBTYPES

β_2-adrenoceptors Screening of a hamster genomic library with oligonucleotides complementary to peptide fragments of the purified hamster lung β_2-adrenoceptor yielded a clone which, when expressed in a variety of systems, had functional and radioligand binding characteristics that were consistent with those of the β_2-adrenoceptor (Dixon *et al.*, 1986). Using probes derived from this cDNA, β_2-adrenoceptors from mouse, rat and human sources have been cloned. There exist only minor species differences between these clones, which show an overall amino acid identity of between 87 and 93%.

β_1-adrenoceptors It proved more difficult to clone the β_1-adrenoceptor because the human β_2-adrenoceptor cDNA did not cross-hybridize with the β_1-adrenoceptor, even when the full length coding sequence was used. Using the β_2-adrenoceptor as a probe, a related receptor was isolated (Kobilka *et al.*, 1987b), which proved to be the 5-HT_{1A} receptor (Fargin *et al.*, 1988). Using the coding region of the 5-HT_{1A} receptor DNA to probe a human placental cDNA library, the β_1-adrenoceptor clone was eventually identified (Frielle *et al.*, 1987). The overall amino acid identity between the human β_1- and β_2-adrenoceptors is only 54%, although the amino acid identity between these receptors increases to 71% in the hydrophobic regions that are postulated to represent the seven membrane spanning domains.

β_1-Adrenoceptors have also been cloned from the rat (Machida *et al.*, 1990; Shimomura and Terada, 1990) and the mouse (Cohen *et al.*, 1993). Rat and mouse β_1-adrenoceptors show 98% overall amino acid identity with each other, and 92% overall identity to the human β_1-adrenoceptor. Species-related differences between rodent and human receptors are concentrated in the third cytoplasmic loop (Cohen *et al.*, 1993).

Again using purified receptor protein as a source of oligonucleotide probes, the avian β-adrenoceptor of the turkey erythrocyte was cloned (Yarden *et al.*, 1986). This clone resembles the mammalian β_1-adrenoceptor closely (69% overall amino acid identity with the human β_1-adrenoceptor, with 84% amino acid identity in the transmembrane spanning regions).

β_3-adrenoceptors Screening of a human genomic library with probes prepared from the avian β_1-adrenoceptor and the human β_2-adrenoceptor cDNAs resulted in the identification of a novel β-adrenoceptor clone having the predicted sequence for a G-protein-coupled receptor (Emorine *et al.*, 1989). This novel receptor is the β_3-adrenoceptor, although the identity of primary amino acid sequence was only 40–50% *vis-a-vis* the β_1- or β_2-adrenoceptors (64–69% in the putative transmembrane spanning regions). A homologous β_3-adrenoceptor was cloned from rat brown adipose tissue (Muzzin *et al.*, 1991; Granneman *et al.*, 1992).

This rat β_3-adrenoceptor has 80% overall amino acid identity to the human β_3-adrenoceptor. Probing a murine genomic library with the entire coding region of the human β_3-adrenoceptor led to the identification of the mouse β_3-adrenoceptor, which shows 82% overall amino acid identity to the human β_3-adrenoceptor (Nahimas *et al.*, 1992). Although, by analogy to the β_1- and β_2-adrenoceptors, the β_3-adrenoceptor gene was assumed to have no introns, mapping studies of the rat gene have shown the presence of an intron near the 3′ end of the coding block (Granneman *et al.*, 1992, 1993; Bensaid *et al.*, 1993). The existence of similar splicing sites in the rat, mouse and human β_3-adrenoceptor genes suggest that the mouse and human genes also contain introns, and that the complete β_3-adrenoceptor protein contains 6 (human) or 12 (rat and mouse) additional amino acids at the carboxy terminus (Granneman *et al.*, 1992, 1993). It is possible that two forms of the β_3-adrenoceptor could be present, due to splicing variants at this intron.

The properties of the recombinant β_3-adrenoceptor, expressed in mammalian cells, appears to correspond to those previously defined for the atypical β-adrenoceptor. The selective agonist, BRL 37344, produces activation of adenylyl cyclase, with a maximum response that is equivalent to that produced by norepinephrine or isoproterenol (Emorine *et al.*, 1989; Muzzin *et al.*, 1991; Nahimas *et al.*, 1991; Granneman *et al.*, 1992). The pharmacological characteristics of rodent and human β_3-adrenoceptors appear to be qualitatively similar, although the rodent β_3-adrenoceptor appears to be less sensitive to activation by the endogenous catecholamines (Muzzin *et al.*, 1991; Hadjilambris *et al.*, 1994), and the human receptor is less sensitive to activation by the selective β_3-adrenoceptor agonist, BRL 37344 (Liggett, 1992). Recent studies show that both agonists and antagonists can show substantial selectivity between the mouse and human β_3-adrenoceptors (Hadjilambris *et al.*, 1994). Comparison of the pharmacological properties of the truncated (402 amino acids) and full length (408 amino acids) human β_3-adrenoceptors shows no substantial differences (Granneman *et al.*, 1993).

mRNA hybridizing with the recombinant human β_3-adrenoceptor was found to be present in rat tissues where "atypical" β-adrenoceptor-mediated responses have been shown to occur, such as adipose tissue, liver and skeletal muscle (Emorine *et al.*, 1989). In contrast, Muzzin *et al.* (1991) detected β_3-adrenoceptor mRNA only in rat white and brown adipose tissue using a probe derived from rat β_3-adrenoceptor cDNA. Conflicting data have been presented with respect to β_3-adrenoceptor expression in human tissues. Krief *et al.* (1993) found β_3-adrenoceptor mRNA in adipose tissue and gall bladder, and Granneman *et al.* (1993) observed mRNA for β_3-adrenoceptors in adipose and intestinal tissue. In contrast, Thomas and Liggett (1993) could not detect β_3-adrenoceptor mRNA in any human tissue, including adipose tissue, even when using a sensitive reverse transcriptase PCR assay.

4. DO ADDITIONAL β-ADRENOCEPTORS EXIST?

The properties of the avian β-adrenoceptor are sufficiently different from those of the mammalian β_1-adrenoceptor to suggest that these two β-adrenoceptors should be considered as different receptor subtypes. The avian receptor has been classified as a β_1-adrenoceptor because of the nearly equal potencies of norepinephrine and epinephrine. However, correlation between the ability of a series of subtype-selective β-adrenoceptor antagonists to inhibit [^{125}I] iodohydroxypindolol binding to turkey erythrocyte membranes and their potency at either β_1- or β_2-adrenoceptors is low (Minneman *et al.*, 1980).

It is unlikely that all of the "atypical" β-adrenoceptor responses that have been reported to date have characteristics that are entirely consistent with the β_3-adrenoceptor, and hence, the possibility of additional subtypes cannot be excluded at the present time.

References

Ahlquist, R.P. (1948) A study of the adrenotropic receptors. *Am. J. Physiol.*, **153,** 586–600.

Arch, J.R., Ainsworth, A.T., Cawthorne, M.A., Piercy, V., Sennitt, M.V., Thod V.E., Wilson, C. and Wilson, S. (1984) Atypical β-adrenoceptors on brown adipocytes as target for anti-obesity drugs. *Nature*, **309,** 163–165.

Ball, D.I., Coleman, R.A., Denyer, L.H., Nials, A.T. and Sheldrick, K.E. (1987) In vitro characterization of the β_2-adrenoceptor agonist, salmeterol. *Brit. J. Pharmacol.*, **92,** 591P.

Bensaid, M., Kaghad, M., Rodriquez, M., LeFur, G. and Caput, D. (1993) The rat β_3-adrenergic receptor gene contains an intron. *FEBS*, **318,** 223–226.

Bilski, A.J., Halliday, S.E., Fitzgerald, J.D. and Wale, J.L. (1983) The pharmacology of a β_2-selective adrenoceptor antagonist (ICI 118.551). *J. Cardiovasc. Pharmacol.*, **5,** 430–437.

Bloom, J.D., Dutia, M.D., Johnson, B.D., Wissner, A., Burns, M.G., Largis, E.E., Dolan, J.A. and Claus, T.H. (1992) Disodium (R,R)-5-[2-[[2-(3-chlorophenyl)-2-hydroxyethyl]amino]propyl]-1,3-benzodioxole-2,2-dicarboxylate (CL 316,243) A potent β-adrenergic agonist virtually specific for β_3-receptors. A promising antidiabetic and antiobesity agent. *J. Med. Chem.*, **35,** 3081–3084.

Blue, D.R., Bond, R.A., Adham, N., Delmondo, R., Michel, A.D., Eglen, R.M., Whiting, R.L. and Clarke, D.E. (1989) Interaction of dihydroalprenolol and cyanopindolol with atypical β-adrenoceptors in guinea-pig ileum. *Br. J. Pharmacol.*, **96,** 246P.

Blue, D.R., Bond, R.A., and Clarke, D.E. (1988) Blockade of putative adrenoceptor in guinea pig ileum by alprenolol. *Br. J. Pharmacol.*, **95,** 541P.

Bond, R.A. and Clarke, D.E. (1988) Agonist and antagonist characterization of a putative adrenoceptor with distinct pharmacological properties from the alpha- and beta-subtypes. *Brit. J. Pharmacol.*, **95,** 723–734.

Brodde, O.-E., Daul, A.A., Michel-Reher, M., Boomsma, F., Man in't Veld, A.J., Schlieper, P. and Michel, M.C. (1990) Agonist-induced densensitization of β-adrenoceptor function in humans. *Circulation*, **81,** 914–921.

Cohen, J.A., Baggott, L.A., Romano, C., Arai, M., Southerling, T.E., Young, L.H., Kozak, C.A., Molinoff, P.B. and Greene, M.I. (1993) Characterization of a mouse β_1-adrenergic receptor genomic clone. *DNA and Cell Biol.*, **12,** 537–547.

Croci, T., Cecchi, R., Tarantino, A., Aureggi, G., Bianchetti, A., Goigegrain, R. and Manara, L. (1988) Inhibition of rat colon motility by stimulation of atypical β-adrenoceptors with new gut-specific agents. *Pharmacol. Res. Commun.*, **20,** 147–151.

Dale, H.H. (1906) On some physiological actions of ergot. *J. Physiol. (London)*, **34,** 163–206.

Dickinson, K., Richardson, A. and Nahorski, S.R. (1981) Homogeneity of β_2-adrenoceptors on rat erythrocytes and reticulocytes. *Mol. Pharmacol.*, **19,** 194–204.

Dixon, R.A.F., Kobilka, B.K., Strader, D.J., Benovic, J.L., Dohlman, H.G., Frielle, T., Bolanowski, M.A., Bennett, G.D., Rands, E., Diehl, R.E., Mumford, R.A., Slater, E.E., Sigal, I.S., Caron, M.G., Lefkowitz, R.J. and Strader, C.D. (1986) Cloning of the gene and cDNA for mammalian β-adrenergic receptor and homology with rhodopsin. *Nature*, **321,** 75–79.

Dooley, D.J., Bittiger, H. and Reymann, N.C. (1986) CGP 20712A: A useful tool for quantitating β_1- and β_2-adrenoceptors. *Eur. J. Pharmacol.*, **130,** 137–139.

Dunlop, D. and Shanks, R.G. (1968) Selective blockade of adrenotropic beta receptors in the heart. *Br. J. Pharmacol. Chemother.*, **32,** 201–218.

Emorine, L.J., Marullo, S., Briend-Sutren, M.-M., Patey, G., Tate, K., Delavier-Klutchko, C., and Strosberg, A.D. (1989) Molecular characterization of the human β_3-adrenergic receptor. *Science*, **245,** 1118–1121.

Emorine, L.J., Feve, B., Pairault, J., Briend-Sutren, M-M., Nahmias, C., Marullo, S., Delavier-Klutchko, C. and Strosberg, D.A. (1992) The human β_3-adrenergic receptor: Relationship with atypical receptors. *Am. J. Clin. Nutr.*, **55 (Suppl. 1)**, 215s–218s.

Fargin, A., Raymond, S.R., Lohse, M.J., Kobilka, B.K., Caron, M.G., and Lefkowitz, R.J. (1988) The genomic clone G-21 which resembles a β-adrenergic receptor sequence encodes the 5-HT_{1A} receptor. *Nature*, **335,** 358–360.

Frielle, T., Collins, S., Daniel, K.W., Caron, M.G., Lefkowitz, R.J. and Kobilka, B.K. (1987) Cloning of the cDNA for the human β_1-adrenergic receptor. *Proc. Natl. Acad. Sci. USA*, **84,** 7920–7924.

Giudice, A., Croci, T., Bianchetti, A. and Manara, L. (1989) Inhibition of rat colonic motility and cardiovascular effects of new gut-specific β-adrenergic phenylethanolaminotetralines. *Life Sci.*, **44,** 1411–1417.

Granneman, J.G., Lahners, K.N. and Chaudhry, A. (1991) Molecular cloning and expression of the rat β_3-adrenergic receptor. *J. Pharmacol. Exp. Ther.*, **40,** 895–899.

Granneman, J.G., Lahners, K.N. and Rao, D.D. (1992) Rodent and human β_3-adrenergic receptor genes contain an intron within the protein coding block. *J. Pharmacol. Exp. Ther.*, **42,** 964–970.

Granneman, J.G., Lahners, K.N. and Chaudhry, A. (1993) Characterization of the human β_3-adrenergic receptor gene. *J. Pharmacol. Exp. Ther.*, **44,** 264–270.

Hadjilambris, O.W., Skwish, S. and Dickinson, K.E.J. (1994) Different pharmacological characteristics of CHO cells transfected with mouse and human β_3-adrenergic receptors. *FASEB J.*, **8,** A354.

Harms, H.H. (1976) Stereochemical aspects of β-adrenoceptor antagonist-receptor interaction in adipocytes: differentiation of β-adrenoceptors in human and rat adipocytes. *Life Sci.*, **19,** 1447–1452.

Harms, H.H., Zaagsma, J. and de Vente, J. (1977) Differentiation of β-adrenoceptors in right atrium, diaphragm and adipose tissue of the rat, using stereoisomers of propranolol, alprenolol, nifenalol and practolol. *Life Sci.*, **1,** 123–138.

Hedberg, A. and Mattisson, H. (1981) Beta adrenoceptor interaction of full and partial agonists in the cat heart and soleus muscle. *J. Pharmacol. Exp. Ther.*, **219,** 798–808.

Hollenga, C. and Zaagsma, J. (1989) Direct evidence for the atypical nature of functional β-adrenoceptors in rat adipocytes. *Br. J. Pharmacol.*, **98,** 1420–1424.

Iakovidis, D., Malta, E., McPherson, G.A. and Raper, C. (1980) In vitro activity of RO363, a β_1-adrenoceptor selective agonist. *Br. J. Pharmacol.*, **68,** 677–685.

Kaiser, C., Schwartz, M.S., Colella, D.F. and Wardell, J.R. (1975) Adrenergic agents. 3. Synthesis and adrenergic activity of some catecholamine analogs bearing a substituted sulfonyl or sulfonylalkyl group in the meta position. *J. Med. Chem.*, **18,** 674–683.

Kaumann, A.J. (1989) Is there a third heart β-adrenoceptor? *Trends in Pharmacol. Sci.*, **10,** 34–348.

Kaumann, A.J. and Lemoine, H. (1985) Direct labelling of myocardial β_1-adrenoceptors: comparison of binding affinity of ^{3}H-(–)-bisoprolol with its blocking potency. *Naunyn-Schmiedeberg's Arch. Pharmacol.*, **331,** 27–39.

Kenakin, T.P. and Beck, D. (1982) In vitro studies on the cardiac activity of prenalterol with reference to its use in congestive heart failure. *J. Pharmacol. Exp. Ther.*, **220,** 77–85.

Kobilka, B.K., Frielle, T., Collins, S., Yang-Feng, T., Kobilka, T.S., Francke, U., Lefkowitz, R.J. and Caron, M.G. (1987) An intronless gene encodes a potential member of a family of receptors coupled to guanine nucleotide regulatory protein. *Nature*, **329,** 75–79.

Krief, S., Lonqvist, F., Raimbault, S., Baude, B., Van Spronsen, A., Arner, P., Strosberg, A.D., Ricquier, D. and Emorine, L.J. (1993) Tissue distribution of β_3-adrenergic receptor mRNA in man. *J. Clin. Invest.*, **91,** 344–349.

Lands, A.M., Arnold, A., McAuliff, J.P., Luduena, F.P. and Brown, T.G. (1967) Differentiation of receptor systems activated by sympathomimetic amines. *Nature*, **214,** 597–598.

Langin, D., Portillo, M.P., Saulnier-Blanche, J-S. and Lafontan, M. (1991) Coexistence of three β-adrenoceptor subtypes in white fat cells of various mammalian species. *Eur. J. Pharmacol.*, **199,** 291–301.

Lemoine, H., Ehle, B. and Kaumman, A.J. (1985) Direct labelling of β_2-adrenoceptor: comparison of binding potency of ^{3}H-ICI 118,551 and blocking potency of ICI 118,551. *Naunyn-Schmiedeberg's Arch. Pharmacol.*, **331,** 40–51.

Levy, B. (1966) The adrenergic blocking activity of N-tertiary butylmethoxamine (butoxamine). *J. Pharmacol. Exp. Ther.*, **151,** 413–422.

Levy, B. and Wilkenfeld, B.E. (1970) Selective interactions with beta adrenergic receptors. *Fed. Proc.*, **29,** 1362–1364.

Liggett, S.B. (1992) Functional properties of the rat and human β_3-adrenergic receptors: differential agonist activation of recombinant receptors in chinese hamster ovary cells. *J. Pharmacol. Exp. Ther.*, **42,** 634–637.

Machida, C.A., Bunzow, J.R., Searles, R.P., Van Tol, H., Tester, B., Neve, K.A., Teal, P., Nipper, V. and Civelli, O. (1990) Molecular cloning and expression of the rat β_1-adrenergic receptor gene. *J. Biol. Chem.*, **265,** 12960–12965.

Mauriege, P., De Pergole, G., Berlan, M. and Lafontan, M. (1988) Human fat cell beta-adrenergic receptors: Beta-agonist dependent lipolytic responses and characterization of beta-adrenergic binding sites on human fat cell membranes with highly selective beta1-antagonists. *J. Lipid. Res.*, **29,** 587–601.

McLaughlin, D.P. and MacDonald, A. (1989) Responses to catecholamines in rat distal colon may be mediated, in part, by 'atypical' β-adrenoceptors. *Br. J. Pharmacol.*, **98,** 657P.

Minneman, K.P., Weiland, G.A. and Molinoff, P.B. (1980) A comparison of the beta-adrenergic receptor of the turkey erythrocyte with mammalian beta_1 and beta_2 receptors. Mol. *Pharmacol.*, **17,** 1–7.

Moran, N.C. and Perkins, M.E. (1958) Adrenergic blockade of the mammalian heart by a dichloro analogue of isoprenaline. *J. Pharmacol. Exp. Ther.*, **124,** 223–237.

Muzzin, P., Revelli, J-P., Kuhne, F., Gocayne, J.D., McCombie, W.R., Venter, J.C., Giocobino, J-P. and Fraser, C.M. (1991) An adipose tissue-specific β-adrenergic receptor. *J. Biol. Chem.*, **266,** 24053–24058.

Nahmias, C., Blin, N., Elalouf, J.-M., Mattei, M.G., Strosberg, A.D. and Emorine, L.J. (1991) Molecular characterization of the mouse β_3-adrenergic receptor: relationship with the atypical receptor of adipocytes. *EMBO J.*, **10,** 3721–3727.

Naito, K., Nagao, T., Otsuka, M., Harigays, S. and Nakajima, H. (1985) Studies on the affinity and selectivity of denopamine (TA-064), a new cardiotonic agent, for β-adrenergic receptor. *Japan J. Pharmacol.*, **38,** 235–241.

Nakane, T., Tsujimoto, G., Hashimoto, K. and Chiba, S. (1988) Beta adrenoceptors in the canine large coronary arteries: beta-1 adrenoceptors predominate in vasodilation. *J. Pharmacol. Exp. Ther.*, **245,** 936–943.

Nanoff, C., Freissmuth, M. and Schutz, W. (1987) The role of a low β_1-adrenoceptor selectivity of [^{3}H] CGP-12177 for resolving subtype-selectivity of competitive ligands. *Naunyn-Schmiedeberg's Arch Pharmacol.*, **336,** 519–523.

Nicholls, D.G. and Locke, R.M. (1984) Thermogenic mechanisms in brown fat. *Physiol. Rev.*, **64,** 1–64.

Nickerson, M. (1949) The pharmacology of adrenergic blockade. *Pharmacol. Rev.*, **1,** 27–101.

Nuttall, A., and Snow, H.M. (1982) The cardiovascular effects of ICI 118,587: A β_1-adrenoceptor partial agonist. *Br. J. Pharmacol.*, **77,** 381–388.

Pittman, R.N., Reynolds, E.E. and Molinoff, P.B. (1984) Relationship between intrinsic activities of agonists in normal and desensitized tissue and agonist-induced loss of beta adrenergic receptors. *J. Pharmacol. Exp. Ther.*, **230,** 614–618.

Powell, C.E. and Slater, I.H. (1957) Blocking of inhibitory adrenergic receptors by a dichloro analog of isoprenaline. *J. Pharmacol. Exp. Ther.*, **122,** 480–488.

Rothwell, N.J. and Stock, M.J. (1979) A role for brown adipose tissue in diet-induced thermogenesis. *Nature*, **281,** 31–35.

Ruffolo, R.R. Jr. (1982) Important concepts of receptor theory. *J. Auton. Pharmacol.*, **2,** 277–295.

Schliep, H.-J. and Harting, J. (1984) β_1-Selectivity of bisoprolol, a new β-adrenoceptor antagonist, in anesthetized dogs and guinea pigs. *J. Cardiovasc. Pharmacol.*, **6,** 1156–1160.

Shimomura, H. and Terada, A. (1990) Primary structure of the rat beta-1 adrenergic receptor gene. *Nucleic Acids Res.*, **18,** 4591.

Sillence, M.N., Moore, N.G., Pegg, G.G. and Lindsay, D.B. (1993) Ligand binding properties of putative β_3-adrenoceptor compared in brown adipose tissue and skeletal muscle membranes. *Br. J. Pharmacol.*, **109,** 1157–1163.

Sudera, D.K. and Stock, M.J. (1988) Confirmation of an 'atypical' β-adrenoceptor on rat brown adipose tissue. *Proc. 1st Eur. Congr. Obesity, Stockholm*, abstract No. 54.

Thomas, R.F. and Liggett, S.B. (1993) Lack of β_3-adrenergic receptor mRNA expression in adipose and other metabolic tissues in the adult human. *Mol. Pharmacol.* **43,** 343–348.

Waldeck, B., Jeppsson, A.B. and Widmark, E. (1986) Partial agonism and functional selectivity: A study on β-adrenoceptor mediated effects in tracheal, cardiac and skeletal muscle. *Acta Pharmacol. Toxicol.*, **58,** 209–218.

Walter, M., Lemoine, H. and Kaumann, A.J. (1984) Stimulant and blocking effects of optical isomers of pindolol onthe sinoatrial node and trachea of guinea pig: role of β-adrenoceptor subtypes in the dissociation between blockade and stimulation. *Naunyn-Schmiedeberg's Arch. Pharmacol.*, **327,** 159–175.

Wilson, C., Wilson, S., Piercy, V., Sennitt, M.V. and Arch, J.R.S. (1984) The rat lipolytic β-adrenoceptor: studies using novel β-adrenoceptor agonists. *Europ. J. Pharmacol.*, **100,** 309–319.

Yabuuchi, Y. (1977) The β-adrenoceptor stimulant properties of OPC-2009 on guinea pig isolated tracheal, right atrial and left atrial preparations. *Br. J. Pharmacol.*, **61,** 513–521.

Yarden, Y., Rodriguez, H., Wong, S.K.F, Brandt, D.R., May, D.C., Burnier, J., Harkins, R.M., Chen, E.Y., Ramachandran, J., Ulrich, P. and Ross, E.M. (1986) The avian β-adrenergic receptor: Primary structure and membrane topology. *Proc. Natl. Acad. Sci. USA*, **83,** 6795–6799.

Zaagsma, J. and Nahorski, S.R. (1990) Is the adipocyte β-adrenoceptor a prototype for the recently cloned atypical "β_3-adrenoceptor"? *Trends Pharmacol. Sci.*, **11,** 3–7.

ADRENERGIC RECEPTORS AS MODELS FOR THE FAMILY OF G PROTEIN COUPLED RECEPTORS

CATHERINE D. STRADER, TUNG M. FONG, MARI R. CANDELORE, MARGARET A. CASCIERI, DAVID WEINBERG and MARVIN BAYNE

Department of Molecular Pharmacology and Biochemistry, Merck Research Laboratories, Rahway, NJ 07065

KEY WORDS: G proteins, adrenoceptor structure, signal transduction, receptor coupling, second messengers

Adrenergic receptors are members of a large family of cell surface receptor proteins that function by activation of one or more of a family of guanine nucleotide binding regulatory proteins (G proteins). The G proteins consist of heterotrimeric complexes of α,β, and γ subunits: the α subunit contains a guanine nucleotide binding site and interacts with the receptor, while both the α and βγ pair can activate various effector enzymes and ion channels (Conklin *et al.*, 1993). The signal transduction pathway shared by G protein coupled receptors is initiated by the binding of an agonist ligand to the receptor, causing a conformational change that leads to the interaction of the receptor with the GDP liganded form of the G protein α subunit. The nature of this activation step has not yet been elucidated. This ternary agonist-receptor-G protein complex is destabilized by the exchange of GDP for GTP on G_α, leading to activation of the G protein and relaxation of the receptor into a state with lower affinity for the agonist. The decrease in agonist affinity for the receptor after G protein activation allows the receptor to return to its unliganded basal state, while the G protein is recycled to its ground state by an intrinsic GTPase activity. By this mechanism, various receptors can activate a variety of specific G proteins leading to activation or inhibition of adenylyl or guanylyl cyclase, Ca^{++} or K^+ channels, phospholipases C or A_2, or other effectors, resulting in increases or decreases in levels of intracellular second messengers.

Several hundred members of the G protein coupled receptor family have been cloned and expressed, making it the largest family of receptor proteins characterized to date (Strader *et al.*, 1994). In addition, several previously undetected subtypes of receptors were first elucidated by molecular cloning, with the result that most G protein coupled receptors are now known to exist as families of homologous subtypes. These receptor subtypes are characterized by different pharmacological properties and often have different tissue distributions, opening new avenues for the discovery of subtype-selective agonists or antagonists that act therapeutically at these receptors. Conversely, the development of pharmacological ligands has played a key role in the discovery and characterization of subtypes of G protein coupled receptors. These ligands have revealed differences in the affinities measured among

Correspondence to: Catherine D. Strader, Department of Molecular Pharmacology and Biochemistry, Merck Research Laboratories, Rahway, NJ 07065.

tissues in binding studies, as well as between receptors found in native tissues and cloned receptors expressed in cell lines. These pharmacological differences could reflect the cellular environment of the receptor. Alternatively, they may provide a clue to the existence of a yet undiscovered subtype of that receptor family. Having access to all members of a related receptor family is critical for a clear understanding of the biology and pharmacology of the receptor. Therefore, it is important to characterize the subtype pharmacology of a receptor family; molecular cloning techniques have provided a powerful tool for identifying novel family members.

As an example, in the α_1-adrenergic receptor family, three distinct receptor subtypes have been cloned. However, discrepencies between the pharmacological characteristics observed in native tissues and those determined for cloned receptors have led to the suggestion that a fourth subtype of α 1 receptor might exist (Bylund, 1992). The three known human α-1 receptor subtypes, α 1a, 1b, and 1c, share substantial sequence homology (75%) within their transmembrane domains (Forray *et al.*, 1994; Weinberg *et al.*, 1994). Additional members of the α 1 family might be expected to have similar homology. However, homology-based screening of cDNA libraries did not yield a fourth subtype, leaving open the question of whether the entire family of subtypes had been identified. A genomic approach was used, therefore, to address this issue more broadly. Using a probe corresponding to transmembrane domains 2–5 of the human α 1a, α 1b, or α 1c receptors, we established conditions under which a probe from any of these three subtypes would hybridize with both of the other known subtypes in a genomic southern blot. This analysis identified all three known α-1 receptors in human genomic DNA, but did not reveal any additional bands consistent with another receptor subtype (D. Weinberg, unpublished data). In addition, these conditions were used to screen a human genomic cosmid library. The probe derived from the α 1a receptor identified numerous clones representing α 1a, α 1b and α 1c receptors, as well as a large number of clones that hybridized more faintly with the probe. Only one of these weakly hybridizing clones proved to be a G protein coupled receptor; upon sequencing, that clone was found to encode the dopamine D4 receptor. These data indicate that there is not another α-1 receptor subtype present in the human genome that is as homologous to the α 1a, 1b, and 1c receptors as these three subtypes are to each other, although a less homologous fourth subtype cannot be ruled out. Recent studies have shown that the cloned rat α-1 receptor subtypes have pharmacological characteristics similar to those previously described in native tissues, suggesting that the pharmacological differences are the result of species differences, rather than a fourth receptor subtype (Forray *et al.*, 1994). These experiments demonstrate the difficulty in conclusively determining the existence of novel receptor subtypes from discrepancies between the pharmacology measured in animal tissue preparations and that determined using cloned human receptors.

The molecular basis for the pharmacological distinctions among structurally related receptors has been the focus of research in a number of laboratories over the past several years. The endogenous ligands for G protein coupled receptors range from small biogenic amines such as catecholamines, to peptides such as substance P or angiotensin, to proteins such as lutenizing hormone or glucagon. Despite the diversity of their ligands, however, these receptors share characteristic structural features, presumably reflecting their similar mechanisms of action. G protein coupled receptors are characterized by seven stretches of hydrophobic amino acids, postulated to form transmembrane α-helices, connected by more hydrophilic extracellular and intracellular loops. The majority of the primary sequence conservation among these receptors resides within the hydrophobic transmembrane domain, with the hydrophilic loops being more divergent. Deletion mutagenesis of the β_2-adrenergic receptor (βAR) revealed that the ligand binding domain for small adrenergic agonists and antagonists resides within the conserved hydrophobic core of the receptor (Dixon *et al.*, 1994). Most of the hydrophilic loop regions of the receptor could be deleted without affecting the ligand binding properties of the receptor, indicating that these regions are not required for receptor-ligand interactions to occur. This localization of the binding site of the receptor within the transmembrane helices is reminiscent of the interaction of retinal with rhodopsin (Khorana, 1992), suggesting the existance of a conserved transmembrane binding site for this family of receptors (Dixon *et al.*, 1994).

Because of the extensive medicinal chemical experience in developing structure-activity relationships for β-agonists and β-blockers, it was of interest to determine the specific sites of interaction of these classes of compounds with the βAR. β-agonists are catecholamines or catecholamine mimics, containing a 3,4-substituted catechol ring connected to a protonatable amine group via a β-hydroxy ethyl side chain (c.f. isoproterenol in Figure 1) (Main and Tucker, 1985). Receptor affinity and subtype specificity is determined largely by the substituents on the amine, whereas the catechol ring, or a close analog, is required for agonist activation of the receptor. In contrast, antagonists (β-blockers) are characterized by a more hydrophobic aromatic group, often consisting of a fused ring system, linked to the amine by a β-hydroxypropyl ether side chain (c.f. propranolol in Figure 1) (Main and Tucker, 1985).

Genetic analysis of the transmembrane region of the βAR by single residue substitutions indicated that adrenergic agonists and antagonists are linked to the receptor via an ion pair formed between the protonated amine group of the ligand and the carboxylate side chain of Asp113 in transmembrane helix 3 (Strader *et al.*, 1988). Substitution of this Asp residue with a Ser, replacing the anionic side chain with a hydroxyl group, results in a mutant receptor that is activated by catechol esters and catechol ketones, compounds that do not activate the wild type Asp113-containing receptor (Strader *et al.*, 1991). Thus, substitution of a single amino acid side chain at the active site of the βAR is sufficient to alter the binding specificity of the receptor from a catachol*amine* receptor to one that recognizes catechol-*esters*. These data suggest that, although the amine moiety of the ligand is a major contributor to the overall binding energy of the interaction, the position of the catechol ring is critical for receptor activation, consistent with the structure-activity relationships derived from pharmacological characterization of catecholamine analogs.

The site of interaction of the catechol ring of adrenergic agonists with the receptor was mapped by a combination of genetic alteration of the receptor and chemical alteration of the ligand. Substitution of either Ser204 or Ser207 in the fifth transmembrane domain of the βAR with Ala, thereby removing the hydroxyl side chain from the receptor, resulted in a decrease in the affinity of the receptor for catecholamine agonists, with no change in affinity for the more hydrophobic antagonists, suggesting that these residues might interact with the catechol hydroxyl groups of the agonist (Strader *et al.*, 1989). Further evidence for this assignment came from analysis of isoproterenol analogs in which either the *meta* or the *para* hydroxyl group was replaced with a hydrogen atom. These compounds showed a decrease in the affinity and efficacy at the wild type receptor, reflecting the requirement of

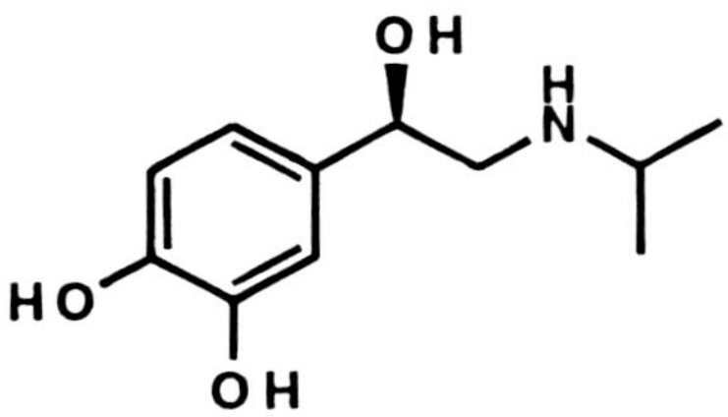

Isoproterenol

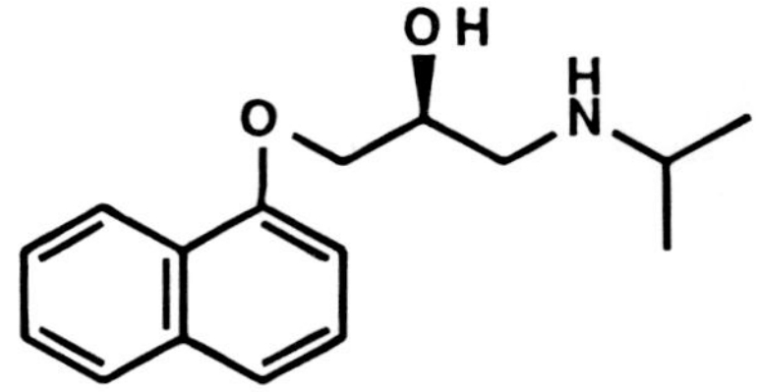

Propranolol

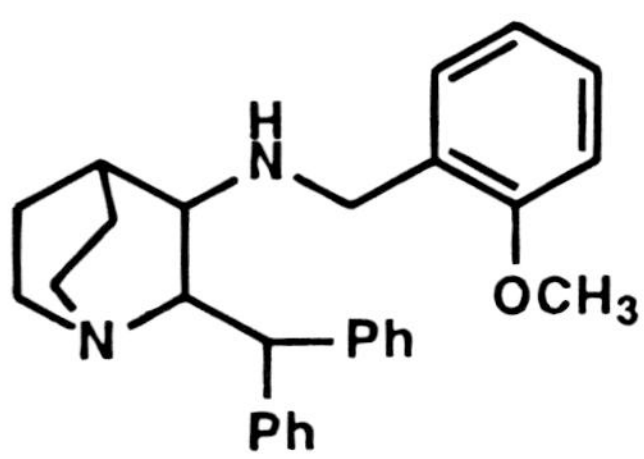

CP-96,345

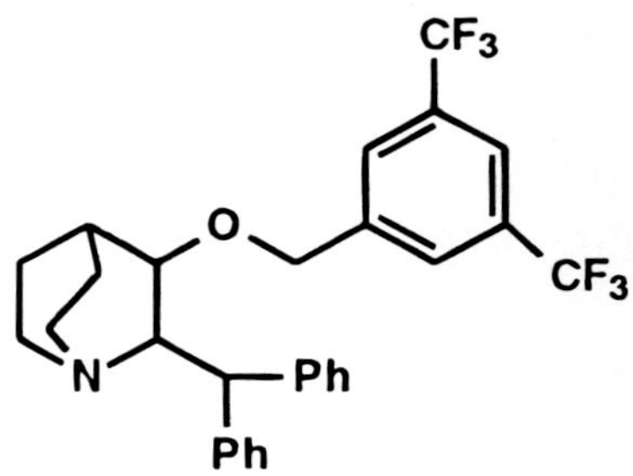

L-703,776

FIGURE 1 Structures of ligands for the adrenergic and neurokinin receptors.

both of the catechol hydroxyl groups for full activation of the wild type receptor. An examination of the additivity of the effects of removing the hydroxyl groups from both the receptor and the ligand showed that the effects of removing the hydroxyl group from position 207 of the receptor were not exacerbated by further removal of the *para*-hydroxyl group from the ligand. In contrast, this Ala207 mutant receptor was sensitive to substitution of the *meta* hydroxyl group of the ligand, suggesting a specific interaction between Ser207 in the receptor and the *para*-hydroxyl group of the ligand. Similar evidence implicates Ser204 of the receptor in a specific interaction with the *meta*-hydroxyl group of catecholamine ligands. Further, these data indicate that these residues are important in determining both the affinity and the efficacy of catecholamine agonists at the receptor. The position of these two Ser residues, predicted from molecular modeling studies to reside one helical turn apart on transmembrane helix 5 of the receptor, is consistent with two hydrogen bonding interactions anchoring the catechol moieties of the agonist to the receptor. Similar experiments have indicated that the aromatic catechol-containing ring interacts with the side chain of Phe290 in transmembrane helix 6 of the receptor (Dixon *et al.*, 1988). These three interactions would serve to specifically orient the catechol ring of the agonist in the binding site of the receptor (Figure 2A). A variety of experimental approaches in several different laboratories have implicated the third cytoplasmic loop of G protein coupled receptors, which

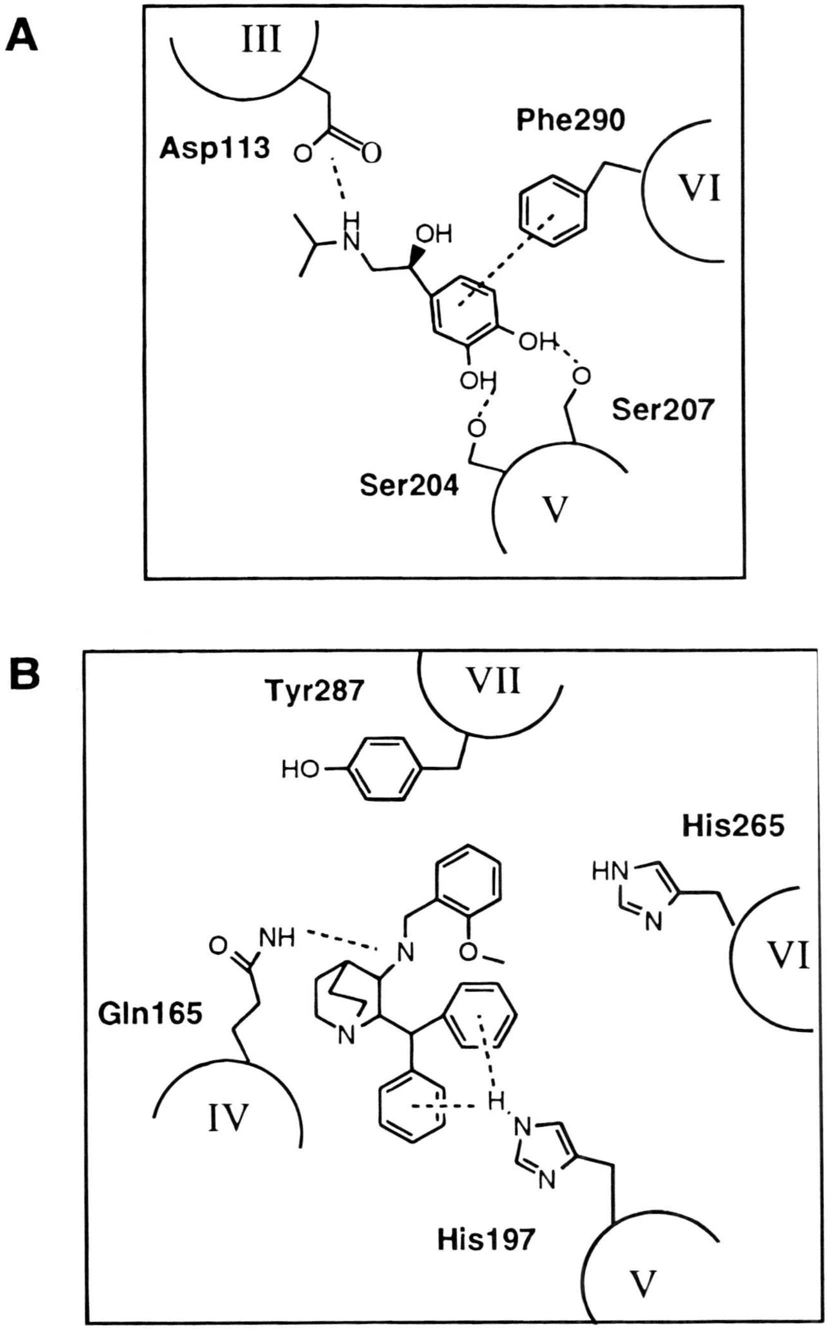

FIGURE 2 Model for the ligand binding sites of (A) β-adrenergic receptor with the agonist isoproterenol and (B) neurokinin-1 receptor with the antagonist CP-96,345. The side chains of key residues are shown.

connects transmembrane helices 5 and 6, as a major determinant of G protein activation (Strader *et al.*, 1994).

Analysis of the primary sequences of the G protein coupled receptors cloned to date reveals that all of the receptors that bind catecholamine ligands (14 different subtypes of adrenergic receptors from several different species) contain an Asp residue at the position analogous to Asp113 in transmembrane helix 3 of the βAR, two Ser residues in transmembrane helix 5, and a Phe in transmembrane helix 6, strongly suggesting that this arrangement of amino acid side chains comprises the catecholamine ligand binding motif for this entire receptor family (Dixon *et al.*, 1988). Biophysical approaches to determine the tertiary arrangement of these critical side chains are limited by the lack of availability of sufficient quantities of receptor protein. Because of the ready availability of rhodopsin as the major protein component of retinal rod outer segment membranes, a variety of biophysical techniques have been used to study its conformation. Recently, a projection map of rhodopsin was determined at 9 Å resolution from electron diffraction of a two-dimensional crystalline array (Schertler *et al.*, 1993). This level of resolution is sufficient to confirm the seven transmembrane helix configuration of the protein, but not to localize individual amino acid side chains. Thus, while the catecholamine binding motif of the adrenergic receptors can be easily recognized from the primary sequence of the receptor, the specific receptor-ligand interactions can only be modeled. The best available models are based on a synthesis of the structural data on rhodopsin and the mutagenesis data on the βAR and related adrenergic receptors, with inter-atom distances constrained by the possible conformations of the catecholamine ligands themselves (Strader *et al.*, 1994). Future biophysical analysis of a variety of G protein coupled receptors will be required to further refine these models and to determine the conformation of the receptor-ligand complex.

The availability of cloned human G protein coupled receptors and the explosion of known receptor subtypes has led to the widespread use of these receptors as tools for drug discovery. Indeed, in recent years, new small molecule antagonists have been discovered for a variety of therapeutically important receptors in this family. It is of interest to identify the binding sites for small molecule antagonists for G protein coupled receptors whose endogenous ligands are peptides, in order to characterize the "binding motifs" responsible for the interactions of these compounds with the target receptors. Recently, a variety of antagonists for the neurokinin-1 (NK1) receptor have been described, typified by the quinuclidines CP96,345 (Snider *et al.*, 1991) and L-703,776 (Figure 1). The NK1 receptor binds the undecapeptide agonist substance P to activate the G protein G_q and stimulate the phospholipase C signal transduction system. Molecular modeling studies have failed to identify a clear relationship between the parameters involved in binding the peptide agonists and those involved in binding the small molecule antagonists.

Site-directed mutagenesis studies have identified His197 in transmembrane helix 5 of the NK1 receptor as a critical element in the interaction with the antagonist (Fong *et al.*, 1993). A combination of different amino acid substitutions at this position in the receptor and modification of various functional groups on the ligand suggests that the imidazole side chain of His197 is involved in an amino-aromatic interaction with the benzhydryl group at the 2-position of the quinuclidine. A similar experimental approach has implicated the side chain of His265 in transmembrane helix 6 of the receptor in binding the 3,5 disubstituted benzyl group at the 3-position of quinuclidines such as L-703,776 (Fong *et al.*, 1994). The heteroatom in the alkyl side chain at the 3-position of the quinuclidine interacts with the amide side chain of Gln165 in transmembrane helix 4 of the NK1 receptor, probably through a hydrogen bond (Fong *et al.*, 1994). Several structural classes of selective NK1 receptor antagonists have been described, most of which have been found to interact with the receptor through these or nearby amino acid residues (Cascieri *et al.*, 1994). Thus, these compounds appear to bind to the receptor through specific interactions with amino acid side chains in the transmembrane domain of the receptor, in particular involving residues in transmembrane helices 4, 5, 6, and 7 (Figure 2B).

Interestingly, most of the residues determined to be involved in the binding of these antagonists are not required for the interaction with peptide ligands: substitution of these residues has little or no effect on the affinity of the receptor for peptide agonists. In contrast, residues in transmembrane helices 2 and 7 have been found to be required for high affinity peptide binding to the receptor (Huang *et al.*, 1994). A comparison of the binding of various analogs of substance P to the wild type and mutant receptors suggests that the side chain of Asn85 is involved in a direct interaction with the C terminal amide of substance P. In addition, other residues in transmembrane helices 2 and 5 have been identified as critical for agonist activation of the NK1 receptor. Although the molecular interactions linking the peptide agonists to the receptor are, for the most part, different from those involved in binding the small molecule antagonists, these compounds can be functionally defined as *competitive* antagonists of receptor activation. This observation suggests that the competitive antagonism of these compounds arises from steric exclusion of the agonist from the binding site, rather than from direct inhibition of specific receptor-agonist interactions (Huang *et al.*, 1994).

Despite the divergence between the endogenous agonists for the βAR and the NK1 receptor, there are striking structural similarities between the small molecule binding sites on these two receptors. In particular, alignment of the primary sequences of these receptors reveals that His265, which is important in binding the substituted benzyl group at the 3-position of quinuclidine antagonists to the NK1 receptor, corresponds to Phe290, which is critical for binding the catechol ring of adrenergic agonists to the βAR. In addition, His197 in the NK1 receptor, which binds the benzhydryl group of the quinuclidines, would be predicted to be located one helical turn above the position corresponding to that of Ser204, which binds the *meta*-hydroxyl group of the catechol moiety, in the βAR. There is not complete overlap of the two binding pockets: Pro112 in the NK1 receptor, which corresponds to Asp113 in the βAR, does not appear to be involved in the binding of agonists or antagonists to the NK1 receptor. However, these data, and similar results from other receptors (recently reviewed in Strader, *et al.*, 1994), suggest the existance of a general small molecule binding site within the transmembrane domains of G protein coupled receptors. Further mutagenesis and biophysical analysis of a variety of these receptor proteins may permit us to recognize the binding motifs that determine the specificity of these receptor-ligand interactions.

References

Bylund, D. (1992) *FASEB J.*, **6**, 832–839.

Cascieri, M.A., MacLeod, A.M., Underwood, D., Shiao, L.L., Ber, E., Sadowski, S., Yu, H., Merchant, K.J., Swain, C.J., Strader, C.D. and Fong, T.M. (1994) *J. Biol. Chem.*, **269**, 6587–6591.

Conklin, B.R., Bourne, H.R. (1993) *Cell*, **73,** 631–41.
Dixon, R.A.F., Sigal, I.S. and Strader, C.D. (1988) *Cold Spring Harbor Symp. Quant. Biol.*, **53,** 487–497.
Dixon, R.A.F., Sigal, I.S., Rands, E., Register, R.B., Candelore, M.R., Blake, A.D. and Strader, C.D. (1987) *Nature*, **326,** 73–77.
Fong, T.M., Cascieri, M.A., Yu, H., Bansal, A., Swain, C. and Strader, C.D. (1993) *Nature*, **362,** 350–353.
Fong, T.M., Yu, H., Cascieri, M.A., Underwood, D., Swain, C.J., Strader, C.D. (1994) *J. Biol. Chem.*, **269,** 2728–2732.
Fong, T.M., Yu, H., Cascieri, M.A., Underwood, D., Swain, C.J., Strader, C.D. (1994) *J. Biol. Chem.*, **269,** 14957–14961.
Forray, C., Bard, J.A., Wetzel, J.M., Chiu, G., Shapiro, E., Tang, R., Lepor, H., Hartig, P.R., Weinshank, R.L., Branchek, T.A. and Gluchowski, C. (1994) *Mol. Pharmacology*, **45,** 703–708.
Huang, R.R., Yu, H., Strader, C.D. and Fong, T.M. (1994) *Biochemistry*, **33,** 3007–3013.
Khorana, H.G. (1992) *J. Biol. Chem.*, **267,** 1–4.
Main, B.G. and Tucker, H. (1985) *Prog. Med. Chem.*, **22,** 122–164.
Schertler, G.F.X., Villa, C. and Henderson, R. (1993) *Nature*, **362,** 770–772.
Snider, R.M., Constantine, J.W., Lowe, J.A., III, Longo, K.P., Lebel, W.S., Woody, H.A., Drozda, S.E., Desai, M.C., Vinick, F.J., Spencer, R.W. and Hess, H.J. (1991) *Science*, **251,** 435–437.
Strader, C.D., Candelore, M.R., Hill, W.S., Sigal, I.S. and Dixon, R.A.F. (1989) *J. Biol. Chem.*, **264,** 13572–13578.
Strader, C.D., Fong, T.M., Tota, M.R., Underwood, D. and Dixon, R.A.F. (1994) *Ann. Rev. Biochem.*, **63,** 101–132.
Strader, C.D., Gaffney, T., Sugg, E.E., Candelore, M.R., Keys, R., Patchett, A.A. and Dixon, R.A.F. (1991) *J. Biol. Chem.*, **266,** 5–8.
Strader, C.D., Sigal, I.S., Candelore, M.R., Rands, E., Hill, W.S. and Dixon, R.A.F. (1988) *J. Biol. Chem.*, **263,** 10267–10271.
Weinberg, D.H., Trivedi, P., Tan, C.P., Mitra, S., Perkins-Barrow, A., Borkowski, D., Strader, C.D. and Bayne, M.L. (1994) *BBRC*, **201,** 1296–1304.

β-ADRENERGIC REGULATION OF ADENYLYLCYCLASE: EFFECT OF RECEPTOR NUMBER

BRENDA S. WHALEY, NANYONG YUAN, ROGER BARBER and RICHARD B. CLARK

Laboratories of Cyclic Nucleotide Research, University of Texas, Graduate School of Biomedical Sciences, Houston, Texas 77225

We recently demonstrated that significant changes are observed in the EC_{50} and V_{max} for the agonist-stimulated activation of adenylylcyclase by the β_2-adrenergic receptor (βAR) as receptor density varies (Whaley *et al.*, 1994). The EC_{50} and V_{max} observed for adenylylcyclase activation by mutant and wild type βAR are commonly used as a direct measure of the ability of the receptors to couple to G_s and activate adenylylcyclase, regardless of differences in receptor densities. Our work indicates that a direct comparison of these parameters, without taking variations in receptor expression levels into account, may lead to mistaken conclusions concerning the ability of a mutant receptor to interact with G_s/adenylylcyclase. In this publication we quantitatively define the ability of βAR to activate adenylylcyclase as the receptor's "coupling efficiency". We present simple equations which allow the calculation of coupling efficiency using the easily measured values of agonist K_d, EC_{50} and V_{max} and the receptor density for both wild type and mutant receptors. This allows a comparison of the ability of βAR to activate adenylylcyclase for receptors expressed at different levels. We demonstrate with our own data, as well as previously published data, that this rigorous approach to the analysis of βAR activity often yields altered conclusions concerning the relative ability of a mutant receptor to interact with G_s/adenylylcyclase when compared to wild type. In addition, we present data which shows that the equations predicting the variation of EC_{50} and V_{max} with receptor number may apply to the M_2 muscarinic receptor which activates phospholipase C through G_q, indicating the principles presented here for the βAR/G_s/adenylylcyclase system may be applicable in other G-protein coupled receptor systems.

KEY WORDS: β-adrenergic receptor, G-protein, adenylylcyclase, coupling

INTRODUCTION

The β_2-adrenergic receptor (βAR) is a widely studied member of a family of receptors which are active in intracellular signal transduction and which exhibit a common structure of seven transmembrane helices (Lefkowitz *et al.*, 1988). The stimulation of adenylylcyclase by hormone-bound βAR through the stimulatory G-protein, G_s, has been well-characterized (Levitzki *et al.*, 1993) and in recent years an intense effort has been directed toward identifying the functional domains of βAR (Birnbaumer *et al.*, 1990). Typically, these studies in an attempt to identify regions which are responsible for ligand binding, coupling to G_s, and receptor desensitization compare the ability of wild-type and mutant βAR to activate adenylylcyclase (Barak *et al.*, 1994; Cheung *et al.*, 1992; Johnson *et al.*, 1990; Moffett *et*

Correspondence to: Brenda S. Whaley, Laboratories of Cyclic Nucleotide Research, University of Texas, Graduate School of Biomedical Sciences, Houston, Texas 77225.

al., 1993). Generally, the mutant and wild type receptors are expressed in a model cell system over a wide range of receptor densities and their activities are often compared without taking these differences in receptor level into account. We have recently shown that varying levels of βAR can significantly affect the EC_{50} and V_{max} observed for adenylylcyclase activation (Whaley *et al.*, 1994). Furthermore, we have demonstrated that these changes in adenylylcyclase stimulation are predicted by equations derived from the currently accepted models (Cassel *et al.*, 1977; Tolkovsky *et al.*, 1981) of the βAR/adenylylcyclase system. Here we will discuss the implications of our studies on the comparison of adenylylcyclase activation curves from mutant and wild type receptors expressed at different densities. We will present simple equations which allow a quantitative calculation of the ability of βAR to activate the G-protein/adenylylcyclase system rather than a qualitative comparison of adenylylcyclase activation curves. Using the values for K_d, EC_{50}, V_{max}, and the receptor density obtained both from our own work and from previously published studies we will calculate receptor coupling efficiencies for wild type and mutant βAR and show that in some cases a more rigorous analysis of adenylylcyclase activation data may result in altered conclusions concerning the ability of a mutant receptor to interact with adenylylcyclase.

METHODS

Transfection of βAR into L cells, membrane preparation and measurement of adenlylcyclase activation and βAR receptor density have been previously described in detail (Whaley *et al.*, 1994).

RESULTS AND DISCUSSION

Coupling efficiency defined

We have previously published equations which predict the EC_{50} and V_{max} for the stimulation of adenylylcyclase by an agonist will vary with the density of βAR present in the membranes (see figure 2 legend). We have demonstrated the validity of these equations for membrane preparations of both hamster and human βAR transfected into L cells. These results demonstrate that EC_{50} and V_{max} values can vary dramatically with receptor density and that measurements from two different preparations cannot be reasonably compared unless receptor numbers are taken into consideration. As mentioned above, the literature indicates studies involving the evaluation of the ability of mutant receptors to activate adenylylcyclase relative to wild-type βAR usually do not attempt to quantitate the effects of receptor density.

We propose that a more quantitative assessment of adenylylcyclase activation is the calculation of k_1/k_{-1}, as it is defined in Table 1. k_1 is a measure of the coupling efficiency between βAR and adenylylcyclase as this rate constant reflects the ability of a receptor to activate adenylylcyclase. k_{-1} is a measure of the intrinsic GTPase activity of G_s. Since this parameter is dependent only on the G-protein population present in the membranes, and not on receptor density, it can be assumed it remains constant for all clones from a given cell line. Therefore, a comparison of the k_1/k_{-1} ratio obtained for two different clones from the same cell line should allow an accurate comparison of the ability of a receptor to activate adenylylcyclase regardless of differences in receptor density.

TABLE 1

Hamster βAR Coupling Efficiency

Receptor Number, fmol/mg protein	*Epinephrine[a] EC_{50}, nM*	k_1/k_{-1}
5	144	0.23
6	197	0.09
6	173	0.13
24	42	0.26
30	49	0.17
34	51	0.15
36	63	0.11
37	42	0.17
40	43	0.15
42	81	0.07
47	35	0.16
51	22	0.25
52	31	0.17
59	31	0.15
65	27	0.16
72	29	0.13
76	48	0.07
91	47	0.06
97	39	0.07
103	30	0.09
120	32	0.07
210	9.3	0.15
218	4.6	0.30
266	5.0	0.23
275	3.7	0.30
281	4.5	0.24
371	8.7	0.09
393	13	0.06
Average		0.153
S.E.		0.003

[a]$K_d = 306 \pm 111$ nM

$$\frac{k_1}{k_{-1}} = \frac{K_d - EC_{50}}{EC_{50}r}$$

This expression was derived from the expression for EC_{50} shown in the legend for figure 2A and can be used to calculate the k_1/k_{-1} for wild type βAR.

Calculations of k_1/k_{-1}

The k_1/k_{-1} ratio for hamster βAR was calculated for twenty-eight EC_{50} measurements made on membranes containing hamster βAR and the results are shown in Table 1. An average k_1/k_{-1} value of 0.15 ± 0.003 for the hamster receptor was obtained after discarding two values which were greater than two standard deviations above the mean. Similar calculations were performed on the twenty EC_{50} measurements made on membranes containing human βAR. Table 2 shows an average k_1/k_{-1} ratio of 0.90 ± 0.03 was obtained for the human receptor after discarding three values.

The usefulness of k_1/k_{-1} in the evaluation of adenylylcyclase activation curves is illustrated in figure 1 which shows the epinephrine dose response curves obtained for two different clones of L cells transfected with human βAR in which receptor is expressed at two

TABLE 2

Human βAR Coupling Efficiency

Receptor Number, fmol/mg protein	*Epinephrine*[a] *EC_{50}, nM*	k_1/k_{-1}
10	102.0	1.37
28	233.0	0.19
236	5.0	1.27
268	26.0	0.21
300	3.0	1.66
306	3.8	1.29
357	18.4	0.23
454	3.0	1.10
749	10.9	0.18
833	1.0	1.8
1459	0.9	1.14
1490	1.6	0.63
2535	0.6	0.99
2535	1.1	0.54
2946	1.0	0.51
3296	0.3	1.52
10546	0.2	0.71
Average		0.90
S.E.		0.031

[a]$K_d = 1.5 \pm 1.0$ μM
k_1/k_{-1} was calculated as in Table 1.

different densities. Although V_{max} is increased five-fold and EC_{50} is decreased thirty-fold when βAR is expressed at a higher level, the calculation of k_1/k_{-1} indicates there is no significant difference in receptor coupling efficiency between the two membrane preparations. This is expected as both membrane preparations contained the same human βAR, merely expressed at different levels. The figure demonstrates that k_1/k_{-1} may be used to evaluate the ability of a mutant receptor to activate adenylylcyclase relative to wild-type. This allows a quantitative assessment of mutant receptor activity that compensates for any difference in the level of receptor expression.

k_1/k_{-1} *for mutant βAR*

Table 3 shows our calculations of k_1/k_{-1} using previously published data from three different studies of mutant βAR. The first two mutants contain a serine to alanine mutation at either position 204 or 207 of the hamster βAR (Strader *et al.*, 1989). As the authors pointed out, these mutations lead to a dramatic decrease in the affinity of the receptor for isoproterenol. The calculation of coupling efficiencies yields even further information from this data. S204A, although it binds poorly to the agonist as indicated by the high K_d, still seems to be interacting normally with G_s as indicated by a k_1/k_{-1} ratio which is comparable to wild type due to a relatively low EC_{50} for agonist. However, S207A, which also binds agonist poorly, seems to be highly uncoupled from G_s as indicated by a k_1/k_{-1} ratio which is only 8% of that of wild type due to a very high EC_{50} for agonist.

The third mutant in Table 3 represents a cysteine to glycine mutation at position 341 of human βAR which is expected to prevent palmitoylation of the receptor at this position (Moffett *et al.*, 1993; O'Dowd *et al.*, 1989). The mutant exhibits a suppressed V_{max} for

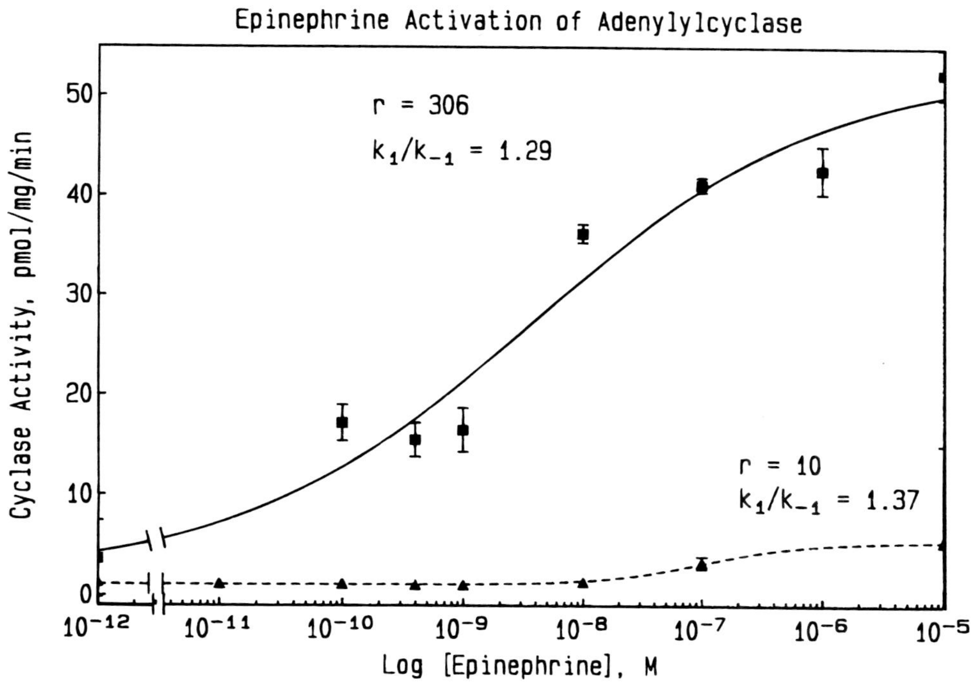

FIGURE 1 Adenylylcyclase activity for human βar at high and low receptor expression denylylcyclase activity was measured for two L cell clones which expressed human βAR at the receptor densities shown; (■) EC_{50} = 3.8 nM, V_{max} = 46.7 pmol/mg/min; (▲) EC_{50} = 102.0 nM, V_{max} = 5.5 pmol/mg/min. k_1/k_{-1} is calculated as described in Table 1.

adenylylcyclase activation, a four-fold increase in basal phosphorylation and a resistance to agonist-induced desensitization. The authors concluded that the mutant receptor is largely uncoupled from G_s and is highly desensitized even in its naive state. Our analysis of the data from the two publications cited above shows the mutant receptor has a coupling efficiency only slightly lower than wild type as reflected by EC_{50} values which are comparable to those of wild type receptor. It appears that in this case the suppressed V_{max} is mainly due to decreased expression levels for the mutant and there is no indication the receptor is extensively desensitized.

The final mutant shown in Table 3 represents a mutation of tyrosine to alanine at position 326 of the human βAR (Barak *et al.*, 1994). The authors show that this mutant abolishes agonist-mediated sequestration and concluded that since the Y326A has the ability to maximally activate adenylylcyclase the mutant receptor was able to "interact effectively with G-proteins". Although the mutant does yield a V_{max} equivalent to wild type, a ten-fold right shift in EC_{50} indicates the receptor is functioning very inefficiently. A calculation of k_1/k_{-1} for the mutant and wild type demonstrates that the Y326A mutant coupling efficiency is only 3% that of wild type indicating the receptor is largely uncoupled from the G-protein. A reduction in efficiency of this magnitude seriously brings into question the conclusion that the mutant specifically blocks internalization without interfering with other receptor functions.

TABLE 3

Mutant βAR Coupling Efficiency

Receptor	*Receptor Number, fmol/mg*	*Kd, nM*	*EC_{50}, nM*	*V_{max}, (norm)*	*k_1/k_{-1}*	*% of Wild type*[f]
hamster						
βAR[a]	500	200	43	100	0.0073[e]	
S204A	650	5200	430	54	0.0079[g]	108%
S207A	490	5100	5800	40	0.00056[g]	8%
human						
βAR[b]	2110	185	54.9	100	0.0011[e]	
C341G	1480	206	37.4	32.4	0.00083[g]	76%
βAR[c]	2400	233	38	100	0.0021[e]	
C341G	1900	287	22	40	0.0018[g]	86%
human						
βAR[d]	600	379	27	100	0.022[e]	
Y326A	400	327	267	100	0.00056[g]	3%

[a](Strader *et al.*, 1989); [b](O'Dowd *et al.*, 1989); [c](Moffett *et al.*, 1993); [d](Barak *et al.*, 1994).
[e]k_1/k_{-1} for wild type βAR was calculated as in Table 1.
[f]Relative k_1 for mutants was calculated as follows:

$$\frac{k_{1M}}{k_{1WT}} = \frac{V_{maxM} K_{dM} EC_{50WT} r_{WT}}{V_{max\,WT} K_{dWT} EC_{50M} r_M}$$

[g]k_1/k_{-1} for mutants is the product of the relative k_1 for the mutant and the k_1/k_{-1} for wild type.

Other G-protein coupled receptors

We have tested our equations which predict a change in V_{max} and EC_{50} for agonist stimulation as receptor density increases with data published for other receptor systems. We have found that we can simulate quite well the results obtained in a 1987 study of the effect of differential expression of the muscarinic receptor on the hydrolysis of phosphoinositides (Ashkenazi *et al.*, 1987). Figure 2 shows the data we interpolated from the four dose response curves Ashkenzai published for phosphoinositide hydrolysis in Chinese hamster ovary cells which had been transfected with porcine M_2 muscarinic receptor. Receptor expression varied over a 10-fold range. The EC_{50} for carbachol activation decreased from 32 to 5 μM as receptor density increased. The dashed line in figure 2A is a simulation of EC_{50} predicted by our equation (shown in the figure legend) when K_d is set to 161 μM (the value reported by the authors) and k_1/k_{-1} is set to 1.5. As EC_{50} decreased, V_{max} was doubled at higher levels of receptor expression. The dashed line in figure 2B is a simulation of V_{max} predicted by our equation (shown in the figure legend) when V_{100} was set to 170 fmol/10^6 cells/30 min and k_1/k_{-1} was set to 1.5 (the same value used to fit the EC_{50} data). The muscarinic receptor is thought to be coupled to phospholipase C through G_q. This demonstrates that our equations describe the activity of at least one other G-protein coupled receptor system as well as that of βAR/adenylylcyclase. However, our attempts to model the M_2 muscarinic receptor stimulated inhibition of adenylylcyclase (Ashkenazi *et al.*, 1987) and the 5-HT_{1A} receptor stimulated inhibition of adenylylcyclase (Varrault *et al.*, 1992; Yocca *et al.*, 1992), which both act through G_i, were unsuccessful. This may be due to a significant difference in the mechanism of G-protein coupling for these receptor systems as compared to that utilized in the βAR/G_s/adenylylcyclase and M_2/G_q/phospholipase C systems.

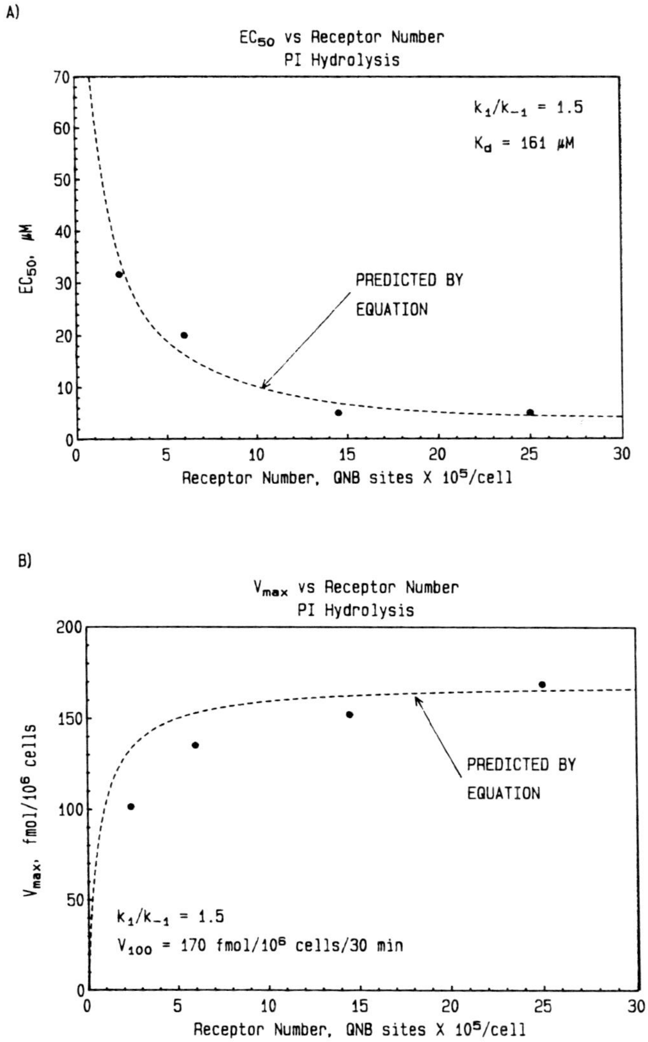

FIGURE 2 Phosphoinositide hydrolysis at varying M_2 muscarinic receptor density. (Ashkenazi *et al.*, 1987)
A) Simulation of EC_{50} as receptor density changes:

$$EC_{50} = \frac{K_d}{\frac{k_1}{k_{-1}} r + 1}$$

B) Simulation of V_{max} as receptor density changes:

$$V_{max} = \frac{V_{100} \frac{k_1}{k_{-1}} r}{\frac{k_1}{k_{-1}} r + 1}$$

where V_{100} is the theoretical V_{max} observed when k_1 approaches infinity (Whaley *et al,* 1994).

In conclusion, this study demonstrates that the comparison of EC_{50} and V_{max} for adenylylcyclase activation between two membrane preparations with different levels of receptor expression may lead to inaccurate conclusions concerning the ability of βAR to activate adenylylcyclase. We have presented equations which allow a quantitative analysis of βAR/G_s/adenylylcyclase coupling using the readily measured parameters of agonist K_d, the EC_{50} and V_{max} of adenylylcyclase activation, and the receptor density of the membrane preparation. The calculation of the k_1/k_{-1} ratio allows a more accurate assessment of receptor coupling efficiency than has been available in the past. This parameter should be very useful in the evaluation of mutants used in the studies of βAR agonist binding, G_s coupling and desensitization. Finally, we have shown that the equations for EC_{50} and V_{max} which we have previously verified for the βAR activation of adenylylcyclase may also apply in the M_2 muscarinic receptor activation of phospholipase C.

ACKNOWLEDGMENTS

We wish to thank Lutz Birnbaumer for his gracious gift of the human βAR clones.

References

Ashkenazi, A., Winslow, J.W., Peralta, E.G., Peterson, G.L., Schimerlik, M.I., Capon, D.J. and Ramachandran, J. (1987) An M2 muscarinic receptor subtype coupled to both adenylyl cyclase and phosphoinositide turnover. *Science (Washington D.C.)*, **238,** 672–675.

Barak, L.S., Tiberi, M., Freedman, N.J., Kwatra, M.M., Lefkowitz, R.J. and Caron, M.G. (1994) A highly conserved tyrosine residue in G protein-coupled receptors is required for agonist-mediated β_2-adrenergic receptor sequestration. *J. Biol. Chem.*, **269,** 2790–2795.

Birnbaumer, L., Abramowitz, J. and Brown, A.M. (1990) Receptor-effector coupling of G proteins. *Biochimica et Biophysica Acta*, **1031,** 163–224.

Cassel, D. and Selinger, Z. (1977) Mechanism of adenylate cyclase activation by cholera toxin: inhibition of GTP hydrolysis at the regulatory site. *Proc. Natl. Acad. Sci. USA*, **74,** 3307–3311.

Cheung, A.H., Huang, R.C. and Strader, C.D. (1992) Involvement of specific hydrophobic, but not hydrophilic, amino acids in the third intracellular loop of the β-adrenergic receptor in the activation of G_s. *Mol. Pharmacol.*, **41,** 1061–1065.

Johnson, J.A., Clark, R.B., Friedman, J., Dixon, R.A.F. and Strader, C.D. (1990) Identification of a specific domain in the β-adrenergic receptor required for phorbol ester-induced inhibition of catecholamine-stimulated adenylylcyclase. *Mol. Pharmacol.*, **38,** 289–293.

Lefkowitz, R.J. and Caron, M.G. (1988) Adrenergic receptors. *J. Biol. Chem.*, **263,** 4993–4996.

Levitzki, A., Marback, I. and Bar-Sinai, A. (1993) The signal transduction between β-receptors and adenylylcyclase. *Life Sci.*, **52,** 2093–2100.

Moffett, S., Mouillac, B., Bonin, H. and Bouvier, M. (1993) Altered phosphorylation and desensitization patterns of a human β_2-adrenergic receptor lacking the palmitoylated cys341. *EMBO J.*, **12,** 349–356.

O'Dowd, B.F., Hnatowich, M., Caron, M.G., Lefkowitz, R.J. and Bouvier, M. (1989) Palmitoylation of the human β_2-adrenergic receptor. *J. Biol. Chem.*, **264,** 7564–7569.

Strader, C.D., Candelore, M.R., Hill, W.S., Sigal, I.S. and Dixon, R.A.F. (1989) Identification of two serine residues involved in agonist activation of the β-adrenergic receptor. *J. Biol. Chem.*, **264,** 13572–13578.

Tolkovsky, A.M. and Levitzki, A. (1981) Theories and predictions of models describing sequential interactions between the receptor, the GTP regulatory unit, and the catalytic unit of hormone dependent adenylate cyclase. *J. Cyclic Nucleotide Res.*, **7,** 139–150.

Varrault, A., Journot, L., Audigier, Y. and Bockaert, J. (1992) Transfection of human 5-hydroxytryptamine$_{1A}$ receptors in NIH-3T3 fibroblasts: effects of increasing receptor density on the coupling of 5-hydroxytryptamine$_{1A}$ receptors to adenylyl cyclase. *Mol. Pharmacol.*, **41,** 999–1007.

Whaley, B.S., Nanyong, Y., Birnbaumer, L., Clark, R.B. and Barber, R. (1994) Differential expression of the β-adrenergic receptor modifies agonist stimulation of adenylylcyclase: a quantitative evaluation. *Mol. Pharmacol.*, **45,** 481–489.

Yocca, F.D., Iben, L. and Meller, E. (1992) Lack of apparent receptor reserve at postsynaptic 5-hydroxytryptamine$_{1A}$ receptors negatively coupled to adenylylcyclase activity in rat hippocampal membranes. *Mol. Pharmacol.*, **41,** 1066–1072.

β-ADRENERGIC RECEPTORS ARE REGULATED TRANSCRIPTIONALLY, POST-TRANSCRIPTIONALLY AND POST-TRANSLATIONALLY

BABY G. THOLANIKUNNEL, VIJAYA KAROOR, MEILING SHIH and CRAIG C. MALBON[1]

Department of Pharmacology, Diabetes and Metabolic Diseases Research Group, School of Medicine, Health Sciences Center, Suny/Stony Brook, Stony Brook, NY 11794-8651

β-Adrenergic receptors (βAR) are members of a populous (> 300) family of G-protein-linked receptors that transduce binding of ligand (or photoactivation for photopigment receptors) into a conformational change(s) that activates effector units (ion channels and enzymes) via G-protein intermediaries. Regulation of receptor expression and function includes several motifs. Receptor expression may be modulated transcriptionally (both via activation or repression) or post-transcriptionally (via destabilization of receptor mRNA), whereas receptor function is modulated post-translationally (via protein phosphorylation). The β-AR family (β_1AR, β_2AR, β_3AR) display regulation at all three levels. Glucocorticoids activate transcription of the β_2AR gene, while repressing transcription of the β_1AR gene. Agonist-induced down-regulation of β_2AR up-regulates a receptor mRNA-binding protein (βARB) and the destabilization of pre-existing receptor message. Receptor (β_2AR) function can be attenuated by agonist as well as by counter-regulatory tyrosine kinase receptors through protein phosphorylation.

KEY WORDS: β-adrenoceptors, receptor expression, receptor regulation, MRNA, receptor synthesis

INTRODUCTION

Unraveling the molecular basis of transmembrane signaling remains a central target for modern biology. Catecholamine stimulation of intracellular cyclic AMP provides a paradigm for G-protein-linked signal transduction from β-ARs to adenylylcyclase via G_s. Equally captivating is the question of how receptor function and expression are regulated *in vivo*. Recent studies reveal a rich, dynamic regulation of β-AR subtypes. The molecular basis for receptor regulation, where known, is summarized below in the context of transcriptional, post-transcriptional, and post-translational mechanisms (Hausdorf *et al.*, 1990; Hadcock & Malbon, 1991).

TRANSCRIPTIONAL REGULATION

Thyroid status regulates β-adrenergic sensitivity, hyperthyroidism enhancing and hypothyroidism blunting (except in liver) β-adrenergic responses. In the heart, thyroid hormones increase the expression of β_1AR and β_2AR. Transcriptional activation of the β_1AR

Correspondence to: Baby G. Tholanikunnel, Department of Pharmacology, Diabetes and Metabolic Diseases Research Group, School of Medicine, Health Sciences Center, Suny/Stony Brook, Stony Brook, NY 11794-8651.

gene by thyroid hormones provides a basis for the increase in receptor mRNA and a putative response element has been identified in the rat gene that extends from +118 to +139 relative to the start-site for transcription (Bahouth, 1991). In contrast to heart, liver displays blunted responses in hyper- and enhanced responses in hypothyroidism (Malbon, 1980). Hepatic β_2AR expression increases in hypothyroidism, but the molecular basis for this control remains to be established.

Steroids (like glucocorticoids) also display the ability to regulate the physiological responses mediated by β-adrenergic receptors. Glucocorticoids stimulate expression of β_2ARs, increasing receptor levels several-fold within hours of administration (Hadcock *et al.*, 1989b,c; Wang *et al.*, 1989). Receptor mRNA increased 3–4 fold in response to dexamethasone and actinomycin D effectively blocked this response. Analysis of transcription rates confirmed that glucocorticoids induced expression of β_2AR by transcription activation (Collins *et al.*, 1988; Hadcock & Malbon, 1988). Elimination of a glucocorticoid-response element (GRE) at position –1031 in the 5′ UTR of the hamster gene abolishes the transcriptional activation by steroid (Malbon & Hadcock, 1988). For β_1ARs, glucocorticoids down-regulate receptor expression by suppression of transcription (Kiely *et al.*, 1994). Promoter analysis has revealed a functional GRE mediating glucocorticoid-induced transcriptional suppression of the β_1AR gene. Non-steroidal agents like the morphogen retinoic acid also regulates expression of β-ARs (Galvin-Parton *et al.*, 1990). The molecular basis for retinoic acid-induced regulation of these genes is unknown. Short-term challenge with agonist increases transcription of the β_2AR gene via a cyclic AMP response element (Collins *et al.*, 1989), although the complement of receptor remains largely unchanged (Wang *et al.*, 1989a–c).

POST-TRANSCRIPTIONAL REGULATION

Agonist-induced down-regulation of receptor expression provides an explanation for the attenuation in a signaling response that results from chronic stimulation of the receptor (Hadcock *et al.*, 1989c; Hadcock & Malbon, 1991). Chronic use of β-catecholamines *in vivo* promotes adaptation, requiring alternative therapies or higher doses of agent to overcome this loss in responsiveness. The molecular basis for agonist-induced down-regulation of receptors was illuminated through studies in which the mRNA levels for these low-abundance proteins could be defined. Agonist was found to induce a rapid decline in both β_1AR and β_2AR mRNA levels, which was the basis for the decline in receptor synthesis and decline in steady-state receptor levels in adaptation (Hadcock & Malbon, 1988a; Bouvier *et al.*, 1989). Further work revealed that the effects were not on transcription, but rather on the stability ($t_{1/2}$) of pre-existing receptor mRNA or post-transcriptional regulation (Hadcock *et al.*, 1989b). A receptor mRNA-binding protein (M_r 35,000, termed β-ARB protein) was discovered (Port *et al.*, 1992) which specifically bound mRNA that displayed agonist-induced destabilization and whose levels are increased in response to challenge with a β-adrenergic agonist. This M_r 35,000 β-ARB protein induced by agonists requires both an AUUUA pentamer and U-rich domains on the 3′ and 5′ flanking regions for mRNA recognition (Huang *et al.*, 1993). Furthermore, this RNA-binding protein distinguishes among βAR subtype mRNAs and recognizes mRNAs for other G-protein-linked receptors that undergo agonist-induced destabilization (Tholanikunnel *et al.*, 1994). Whether β-ARB protein displays intrinsic nuclease activity, is a modulator of a nuclease, or is itself a stabilization factor for RNA requires purification and molecular cloning of the gene.

POST-TRANSLATIONAL REGULATION

Protein phosphorylation is the biochemical basis for most short-term regulation in signaling. Lefkowitz and colleagues have demonstrated the central role of phosphorylation of adrenergic receptors by protein kinase -A, -C, and a novel class of receptor-specific kinases termed "β-ARKs" (Hausdorff *et al.*, 1990). Recently counter-regulation of β-catecholamines by insulin has been shown to result from insulin-stimulated phosphorylation of βARs on tyrosyl residues *in vivo* (*Hadcock et al.*, 1992), reflecting the ability of G-protein-linked receptors to act as substrates for tyrosine kinase growth factor receptors such as the insulin and IGF-I receptors (Baltensperger *et al.*, 1994). Although βARs are clearly multi-phosphorylated by distinct classes of kinases, the precise location(s), stoichiometries, and temporal sequences of phosphorylations remains unknown. Utilizing antisense RNA expression *in vivo*, cells deficient in specific families of kinase activities have been created in which functional studies and structural studies can be performed to answer this most fundamental question about phosphorylation and cell signaling (Shih & Malbon, 1994). Crosstalk among G-protein-linked pathways highlights the integration of transcriptional, post-transcriptional and post-translational mechanisms in the biology of βARs (Ros *et al.*, 1988; Hadcock *et al.*, 1991; Morris *et al.*, 1991, Port *et al.*, 1992).

ACKNOWLEDGMENTS

This work was supported by United States Health Services Grant DK25410 from the National Institutes of Health. C.C.M. was a recipient of career development award K04 AM00786 from the NIH.

References

Bahouth, S.W. (1991) Thyroid hormones transcriptionally regulate the beta-1 adrenergic receptor gene in cultured ventricular myocytes. *J. Biol. Chem.*, **266,** 15863–15869.

Bouvier, M., Collins, S., O'Dowd, B.F., Campbell, P.T., De Blasi, A., Kobilka, B.K., MacGregor, G.P., Caron, M.G. and Lefkowitz, R.L. (1989) Two distinct pathways of cyclic AMP mediated down-regulation of the β_2-adrenergic receptor: phosphorylation of the receptor and regulation of its mRNA. *J. Biol. Chem.*, **264,** 16786–16792.

Collins, S., Caron, M.G. and Lefkowitz, R.J. (1988) β_2-Adrenergic receptors in hamster smooth muscle cells are transcriptionally regulated by glucocorticoids. *J. Biol. Chem.*, **263,** 9067–9070.

Collins, S., Altschmied, J., Herbsman, O., Caron, M.G., Mellon, P.L. and Lefkowitz, R.J. (1990) A cyclic AMP response element in the β_2-adrenergic receptor gene confers transcriptional autoregulation by cyclic AMP. *J. Biol. Chem.*, **265,** 19330–19335.

Galvin-Parton, P.A., Watkins, D.C. and Malbon, C.C. (1990) Retinoic acid modulation of transmembrane signaling. *J. Biol. Chem.*, **265,** 17771–17779.

Hadcock, J.R. and Malbon, C.C. (1988a) Down-regulation of β-adrenergic receptors: agonist-induced reduction in receptor mRNA levels. *Proc. Natl. Acad. Sci. U.S.A.*, **85,** 5021–5025.

Hadcock, J.R. and Malbon, C.C. (1988b) Regulation of β-adrenergic receptors by "permissive" hormones: glucocorticoids increase steady-state levels of receptor mRNA. *Proc. Natl. Acad. Sci. U.S.A.*, **85,** 8415–8419.

Hadcock, J.R. and Malbon, C.C. (1991) Regulation of receptor expession by agonists: transcriptional and post-transcriptional controls. *Trends in Neurological Sciences*, **14,** 242–247.

Hadcock, J.R., Port, J.D., Gelman, M.S. and Malbon, C.C. (1992) Cross-talk between tyrosine kinase and G-protein-linked receptors: phosphorylation of β_2-adrenergic receptors in response to insulin. *J. Biol. Chem.*, **267,** 26017–26022.

Hadcock, J.R., Port, J.D. and Malbon, C.C. (1991) Cross-regulation between G-protein-mediated pathways:

Activation of the inhibitory pathway of adenylyl cyclase increases expression of β_2-adrenergic receptors. *J. Biol. Chem.*, **266,** 11915–11922.

Hadcock, J.R., Ros, M. and Malbon, C.C. (1989a) Agonist regulation of β-adrenergic receptor mRNA: analysis in S49 mouse lymphoma mutants. *J. Biol. Chem.*, **264,** 13956–13961.

Hadcock, J.R., Wang, H.-Y. and Malbon, C.C. (1989b) Agonist-induced destabilization of β-adrenergic receptor mRNA. Attenuation of glucocorticoid-induced up-regulation of β-adrenergic receptors. *J. Biol. Chem.*, **264,** 19928–19933.

Hadcock, J.R., Williams, D.W. and Malbon, C.C. (1989c) Physiological regulation at the level of mRNA. *Am. J. Physiol.*, **257,** C457-C465.

Hausdorff, W.P., Caron, M.G. and Lefkowitz, R.J. (1990) Turning off the signal: desensitization of β-adrenergic rceptor function. *FASEB J.*, **4,** 2881–2889.

Huang, L.-y., Tholanikunnel, B.G., Vakalopoulou, E. and Malbon, C.C. (1993) The M_r 35,000 β-adrenergic receptor mRNA-binding (βARB) protein induced by agonist requires both an AUUUA pentamer and U-rich domains for RNA recognition. *J. Biol. Chem.*, **268,** 25769–25775.

Kiely, J., Hadcock, J.R., Bahouth, S.W. and Malbon, C.C. (1994) Glucocoticoids down-regulate β_1-adrenergic receptor expression by suppressing transcription of the receptor gene. *Biochem. J.*, **000,** in press.

Malbon, C.C. (1980) Liver cell adenylate cyclase and β-adrenergic receptors: increased β-adrenergic receptor number and responsiveness in the hypothyroid rat. *J. Biol. Chem.*, **255,** 8692–8699.

Malbon, C.C. and Hadcock, J.R. (1988) Evidence that the GRE in the 5′non-coding region of the hamster β_2-adrenergic receptor gene are obligate for glucocorticoid regulation of hamster mRNA levels. *Biochem. Biophys. Res. Commun.*, **154,** 676–681.

Morris, G.M., Hadcock, J.R. and Malbon, C.C. (1991) Cross-regulation between G-protein coupled receptors: Activation of β_2-adrenergic receptors increases a_1-adrenergic receptor mRNA levels. *J. Biol. Chem.*, **266,** 2233–2238.

Port, J.D., Hadcock, J.R. and Malbon, C.C. (1992) Cross-regulation between G-protein-mediated pathways: Acute activation of the inhibitory pathway of adenylyl cyclase reduces β_2-adrenergic receptor phosphorylation and increases β-adrenergic receptor responsiveness. *J. Biol. Chem.*, **267,** 8468–8472.

Port, J.D., Huang, L.-y. and Malbon, C.C. (1992) β-Adrenergic agonists that down-regulate receptor mRNA up-regulate a 35,000-M_r protein(s) that selectively bind to β-adrenergic receptor mRNAs. *J. Biol. Chem.*, **267,** 8468–8472.

Ros, M., Northup, J.K. and Malbon, C.C. (1988) Steady-state levels of G-proteins and β-adrenergic receptors in rat fat cells: permissive effects of thyroid hormones. *J. Biol. Chem.*, **263,** 4362–4368.

Shih, M. and Malbon, C.C. (1994) Oligodeoxynucleotides to protein kinase-A, -C, and βARK reveal novel, cell type-specific roles in agonist-induced desensitization. *Proc. Natl. Acad. Sci. U.S.A.*, in press.

Wang, H.-Y., Berrios, M. and Malbon, C.C. (1989b) Indirect immunofluorescence localization of β-adrenergic receptors and G-proteins in human A431 cells. *Biochem. J.*, **263,** 519–532.

Wang, H.-Y., Berrios, M. and Malbon, C.C. (1989c) Localization of β-adrenergic receptors in A431 cells *in situ*: effect of chronic exposure to agonist. *Biochem. J.*, **263,** 533–538.

Wang, H.-Y., Lipfert, L., Malbon, C.C. and Bahouth, S.W. (1989a) Site-directed antibodies define the the topography of the b-adrenergic receptor. *J. Biol. Chem.*, **264,** 14424–14434.

CONTRIBUTION OF DIFFERENTIAL EFFICACY TO THE PHARMACOLOGY OF HUMAN β_1- AND β_2-ADRENOCEPTORS

A. J. KAUMANN, J. A. LYNHAM, LOUISE SANDERS, A. M. BROWN and P. MOLENAAR

The Babraham Institute, Cambridge CB2 4AT, UK, SmithKline Beecham Pharmaceuticals, Harlow, Essex CM19 5AW, UK, Department of Pharmacology, University of Melbourne, Victoria 3052, Australia

In human heart maximal activation by a catecholamine of an individual β_1- and β_2-adrenoceptor can cause the production of 500 and 2000 molecules of cyclic AMP per minute, respectively. Selective coupling of β_2-adrenoceptors to adenylyl cyclase has been confirmed with cloned human β_1- and β_2-adrenoceptors. Binding assays in human heart membranes reveal a greater ratio of high and low affinity receptor isomers (the latter state stabilized by GTP) for β_2- compared to β_1-adrenoceptors, consistent with selective coupling of β_2-adrenoceptors to G_s protein. In failing human left ventricle, fewer β_2- than β_1-adrenoceptors can mediate matching inotropic and lusitropic effects of epinephrine. In myocytes from isolated human atrium, a lower β_2- than β_1-adrenoceptor density (assessed by radioautography) mediates a matching incidence of arrhythmias by physiological catecholamines. These results are consistent with a greater efficacy of catecholamines at β_2- than β_1-adrenoceptors.

KEY WORDS: β_1 and β_2-adrenoceptors, human heart, arrhythmias, ventricular relaxation, receptor-selective coupling, efficacy

INTRODUCTION

The coexistence of two β-adrenoceptor (βAR) populations, β_1 and β_2, in human tissues has raised questions. Do both receptors function in the same cell and is there cross-talk between the 2 receptor populations? Are there differences between β_1AR and β_2AR in their coupling to effectors and if so, what are the functional consequences? Is the functional contribution of coexisting β_1AR and β_2AR proportional to the corresponding population densities? Answers to these questions are beginning to emerge from studies with human heart muscle and cells as well as with cloned and transfected β_1AR and β_2AR.

Studies with human single ventricular myocytes have shown that epinephrine enhances contraction and hastens relaxation through both β_1AR and β_2AR (Del Monte *el al.*, 1993), thus demonstrating that the 2 receptor subtypes function in the same cell and suggesting the possibility of intracellular cross-talk between the 2 receptor populations. Two observations suggest intracellular cross-talk in atrial myocytes. Chronic treatment of patients with blockers selective for β_1AR causes atrial inotropic β_2AR hyperresponsiveness (Hall *et al.*, 1990) but predominantly atrial arrhythmic β_1AR hyperresponsiveness (Kaumann and

Correspondence to: Dr. A. J. Kaumann, The Babraham Institute, Cambridge CB2 4AT, UK.

Sanders, 1993). These divergent results suggest that the intracellular cross-talk between β_1AR and β_2AR leading to inotropic β_2AR hyperresponsiveness occurs in a different cell compartment than the selective β_1AR sensitization for the mediation of arrhythmias.

Differences in the coupling of β_1AR and β_2AR to adenylyl cyclase were first reported in 1985. Gille *et al.*, (1985) estimated in human atrial membranes that 2/3 of the epinephrine-stimulated and norepinephrine-stimulated adenylyl cyclase occurred through β_2AR and only 1/3 through β_1AR, despite the higher density of β_1AR compared to β_2AR (reviewed by Brodde, 1991). The physiological catecholamines also stimulated human ventricular adenylyl cyclase selectively through β_2AR despite the lower density of β_2AR compared to β_1AR (Kaumann and Lemoine, 1987; Bristow *et al.*, 1989). When the specific activity of catecholamine-stimulated adenylyl cyclase in human ventricular membranes was divided by the corresponding βAR density it was estimated that one β_1AR induces the generation of 500 molecules cyclic AMP per minute while one β_2AR molecule induces the production of 2000 molecules cyclic AMP per minute (Kaumann and Lemoine, 1987).

Subsequently, the selective coupling of β_2AR to adenylyl cyclase has been verified with human cloned β_1AR and β_2AR, transfected either separately (Green *et al.*, 1992) or jointly (Levy *et al.*, 1993; Birnbaumer *et al.*, 1994) into host cells. Joint transfections of human β_1AR and β_2AR into murine fibroblasts as a function of βAR density showed that at relatively low receptor concentrations that are equivalent to physiological levels in human heart (41 or 113 fmol.mg^{-1} protein β_1AR plus 16 or 28 fmol.mg^{-1} β_2AR) stimulation of adenylyl cyclase through β_1AR was only about 1/3 of that observed through β_2AR (Levy *et al.*, 1993). When receptors were overexpressed, β_2AR caused a ceiling in adenylyl cyclase stimulation (at 1000 fmol.mg^{-1}) while β_1AR continued stimulating cyclase as a function of receptor density to reach 80% (at 3200 fmol.mg^{-1}) of that caused by β_2AR (Levy *et al.*, 1993).

As expected from more efficient coupling of β_2AR, Green *et al.* (1992) reported a greater GTP-induced decrease in affinity for catecholamine binding to β_2AR than for β_1AR with the cloned human βAR, transfected separately. We have now asked whether the selective decrease of catecholamine binding to β_2AR by GTP happens also with the *native* receptors in human heart. We have also wondered whether a finite Gs protein pool limits the interaction of catecholamines with β_1AR and β_2AR in human heart. The results are consistent with a more efficient coupling of β_2AR than of β_1AR to G_s protein in human heart. We did not detect evidence that a finite G_s protein pool limits catecholamine-receptor interactions with either subtype. Further, we have addressed the question of whether selective coupling of β_2AR to G_s protein and adenylyl cyclase is relevant to β_2AR-evoked effects distal to the adenylyl cyclase such as positive inotropic effects (Kaumann and Lemoine, 1987; Lemoine *et al.*, 1988; Hall *et al.*, 1990) and atrial arrhythmias (Kaumann and Sanders, 1993).

METHODS

Radioligand binding Human heart ventricle was dissected in the operating theater from freshly explanted hearts, quickly freed of fat and connective tissue and frozen in liquid nitrogen. Tissue powder was prepared in a mortar under liquid nitrogen and homogenized in buffer with a Polytron PT7. The incubation buffer contained (mM): Tris HCl 50 pH 7.4 (37°C), EGTA 5, EDTA 1, $MgCl_2$ 2, ascorbate 1, phenyl methyl sulfonyl fluoride 0.5, and 0.10% bovine serum albumin. (–)^{125}I-cyanopindolol (ICYP) 1–300 pM was used to assess saturation kinetics and around 50 pM for inhibition of binding by competing ligands.

Binding removed by 200 μM (–) isoproterenol was considered specific. After 2 h incubation at 37°C 200 μl samples were diluted with ice cold buffer and filtered through Whatman GF/B filters with a Brandel-Harvester.

Radioautography and arrhythmias Atrial tissues from patients undergoing coronary artery bypass surgery were obtained as described and used for studies of frequency-dependent arrhythmias (Kaumann and Sanders, 1993). To assess myocyte β_1AR and β_2AR, segments of these atrial tissues were studied with the densitometric analysis of Molenaar *et al.* (1990).

Adenylyl cyclase Assays were as by Kaumann and Lemoine (1987).

Materials. ICYP, [α-^{32}P]-ATP and [^{3}H]cAMP were from NEN, guanyl nucleotides were from Sigma. All other materials were those of Kaumann and Lemoine (1987).

RESULTS AND DISCUSSION

β_2AR couple more tightly to G_s protein than β_1AR in human heart We determined the total βAR density from saturation binding and then assessed the proportions of β_1AR and β_2AR using the β_1-selective antagonist CGP 20712A (Kaumann and Lemoine, 1987) as shown with left ventricle in panels a and b of Fig. 1. Binding of both epinephrine and norepinephrine to β_1AR was decreased to a small extent by the non-hydrolysable GTP analogue GTPγS (Fig. 1 panels c and d) compared to a large effect of the nucleotide on β_2AR (β_1AR being blocked) for both agonists (Fig. 1 panels e and f). Estimation of low and high reciprocal affinity constants K_L and K_H revealed that the ratio of K_L/K_H was 12 and 13 for β_1AR and 387 and 219 for β_2AR with epinephrine and norepinephrine respectively, consistent with a greater coupling of β_2AR than β_1AR to G_s protein (Kent *et al.*, 1980). Similar results were obtained on membranes from human right atrial appendage and right ventricle (assays not shown) in which binding of epinephrine and norepinephrine to β_2AR were equally affected by GTPγS, in spite of the lower affinity of norepinephrine for this receptor (Fig. 1), suggesting that both amines cause a similar conformational change of the β_2AR that leads to its coupling to the G_s protein. As expected, stimulation of adenylyl cyclase by epinephrine in left ventricular membranes of the patient of Fig. 1 occurred predominantly (66%) through β_2AR (Fig. 2A).

Competition for a finite G_s protein pool does not limit the effects of GTPγS on catecholamine binding to β_1AR and β_2AR in human heart membranes. This is shown in panels g and h of Fig. 1. The marked effect of GTPγS on the catecholamine binding to β_2AR (when β_1AR are blocked by CGP 20712A) compared to β_1AR (when β_2AR are blocked by ICI 118551) raised the question whether in the absence of blocking agents confiscation of G_s protein by the more numerous β_1AR might prevent the marked effect of GTPγS on β_2AR. However, the smaller decrease in catecholamine affinity caused by GTPγS observed on βAR in the absence of blocking agents was consistent with the sum of the large effects on the smaller β_2AR population and small effects on the larger β_1AR population.

Selective G_s protein coupling and adenylyl cyclase stimulation of β_2AR is conveyed to effectors. Epinephrine causes the same maximal positive inotropic effects through β_2AR as through β_1AR in isolated human right atrial appendages and is actually more potent on β_2AR than on β_1AR (Lemoine *et al.*, 1988; Hall *et al.*, 1990; Kaumann, 1991) despite the lower

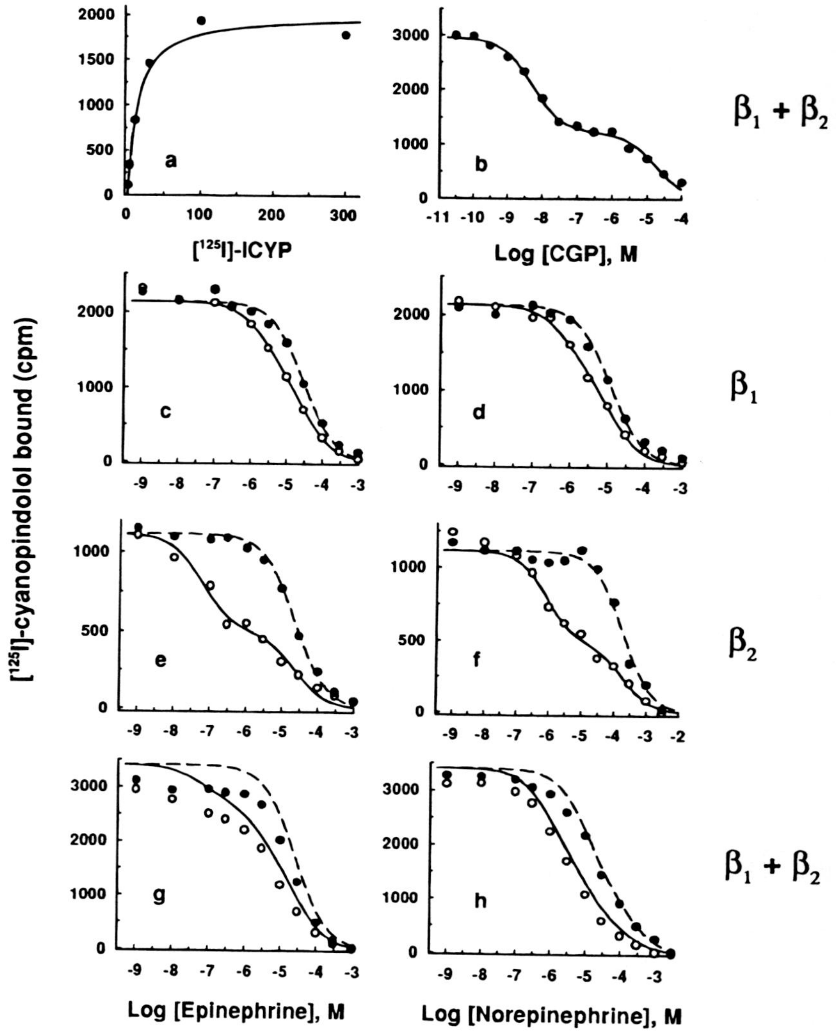

FIGURE 1 Binding of ICYP to membranes from left ventricle of a freshly explanted heart of a 58 y old male patient with dilated cardiomyopathy. Panel a shows specific binding to βAR (β_1AR + β_2AR = 22 fmol.mg^{-1}) with an overall equilibrium dissociation constant of 12.9 pM. Panel b shows inhibition of ICYP (37.5 pM) binding by CGP 20712A; 58% of the receptors were β_1AR, 42% β_2AR. Panels c, e, g and d, f, h show inhibition of ICYP binding by epinephrine and norepinephrine respectively, measured in a single experiment. Open and closed symbols represent data obtained in the absence and presence of 100 μM GTPγS respectively. Data in panels c and d were obtained in the presence of ICI 118551 50 nM to saturate β_2AR; data in panels e and f were measured in the presence of CGP 20712A 300 nM to saturate β_1AR. Data in panels g and h are in the absence of antagonist. 39% of β_1AR (panels c and d) and 56% of β_2AR (panels e and f) were in a high affinity state. K_L and K_H for epinephrine were 7.0 μM and 0.58 μM for β_1AR and 5.8 μM and 0.015 μM for β_2AR. K_L and K_H for norepinephrine were 2.8 μM and 0.22 μM for β_1AR and 46 μM and 0.21 μM for β_2AR. The K_Ds of CGP 20712A were 0.86 nM for β_1AR and 4.4 μM for β_2AR. All curves are theoretical. The data in panels c-f were fitted by non-linear regression and the parameters thus obtained were then used to produce the curves shown in these panels and in g and h. The small discrepancy between the data and fit in panel g probably reflects experimental variation.

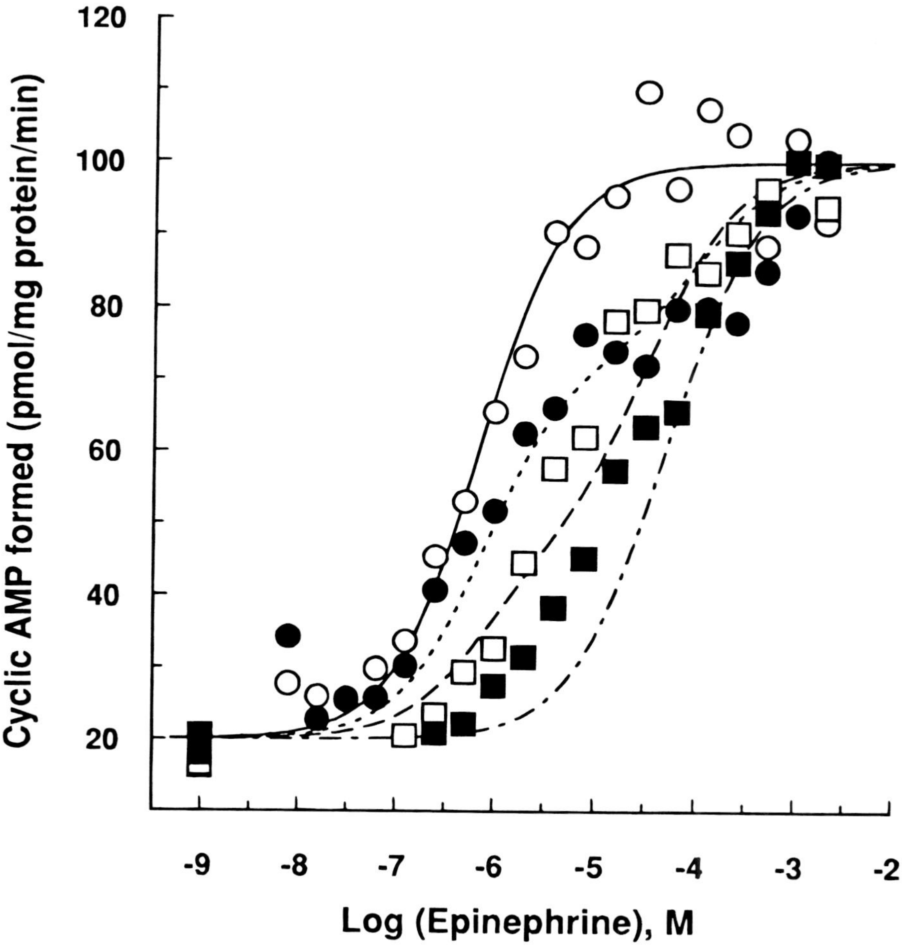

FIGURE 2A Left ventricular tissues obtained from the patient of Fig.1. **A**. Adenylyl cyclase stimulation by epinephrine in the absence of antagonists (○), presence of 50 nM ICI 118551 (□), 300 nM CGP 20712A (●) and both antagonists (■). The fractions of adenylyl cyclase activity stimulated through β_1AR and β_2AR, estimated as by Gille *et al.* (1985), were 0.34 and 0.66 respectively.

density of the former than the latter receptors in this tissue (reviewed in Brodde, 1991). These findings are consistent with a greater efficacy of epinephrine on β_2AR than on β_1AR.

Similarly, epinephrine causes at least as pronounced inotropic and lusitropic effects through β_2AR as norepinephrine through β_1AR in left ventricular trabeculae of the patient of Figs. 1 and 2A, despite the smaller β_2AR population (Fig.2B).

Another example of the selective β_2AR involvement is the production of atrial arrhythmias (Fig. 3). Epinephrine elicits a greater incidence of frequency-dependent arrhythmias through β_2AR than norepinephrine through β_1AR, despite the smaller β_2AR compared to the β_1AR population.

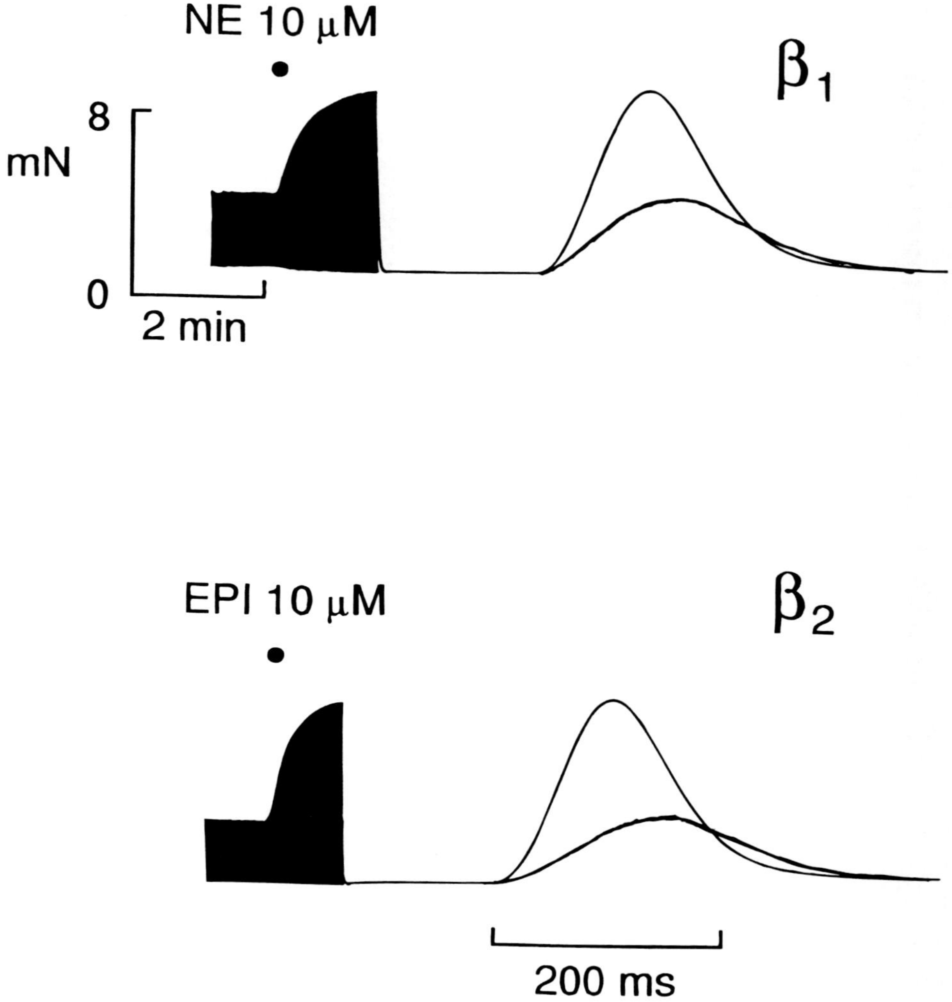

FIGURE 2B Inotropic and lusitropic effects of norepinephrine (in the presence of 50 nM ICI 118551) and epinephrine (in the presence of 300 nM CGP 20712A). The right hand tracings are fast speed tracings before and 4 min after the administration of catecholamine.

Interestingly, chronic treatment of patients with β blockers enhances arrhythmias mediated through β_1AR more than through β_2AR with only marginal increases of the corresponding receptor populations, suggesting selective facilitation of β_1AR coupling to ion channels.

CONCLUSION

Selective G_S-adenylyl cyclase coupling of β_2AR compared to β_1AR contributes to pronounced β_2AR-mediated inotropic, lusitropic and arrhythmic effects of physiological catecholamines in human heart.

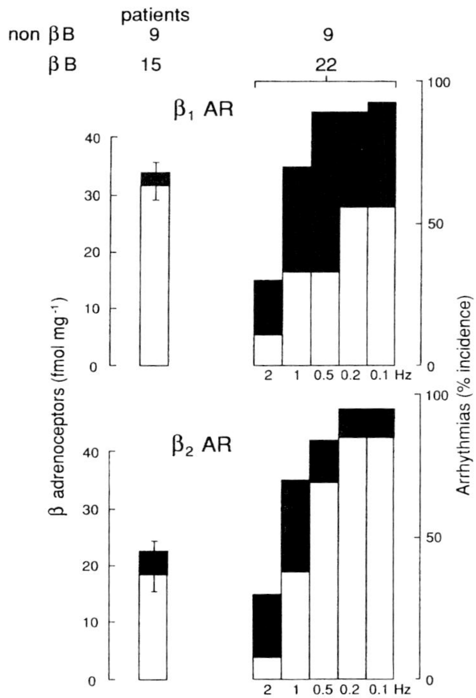

FIGURE 3 Relationship between the incidence of frequency-dependent arrhythmias (right hand panels) and myocyte receptor densities (left hand columns) in human right atrial appendages. Open and closed columns represent data from non-β blocker-treated (non βB) and β blocker-treated (βB) patients. Vertical bars through left hand columns are SEM values.

References

Birnbaumer, L., Levy, F.O., Zhu, X. and Kaumann, A.J. (1994) Studies on the intrinsic activity (efficacy) of human adrenergic receptors. *Texas Heart Institute Journal*, **21,** 16–21.

Bristow, M.R., Hershberger, R.E., Port, J.D., Minobe, W. and Rasmussen, R. (1989) β_1- and β_2-adrenergic receptor-mediated adenylate cyclase stimulation in nonfailing and failing human ventricular myocardium. *Mol. Pharmacol.*, **35,** 295–303.

Brodde, O.E. (1991) β1- and β2-adrenoceptors in the human heart: properties, function and alterations in chronic heart failure. *Pharmacol. Rev.*, **43,** 203–242.

DelMonte, F., Kaumann, A.J., Poole-Wilson, P.A., Wynne, D.G., Pepper, J. and Harding, S.E. (1993) Coexistence of functional β_1- and β_2-adrenoceptors in single myocytes from human ventricle. *Circulation*, **88,** 889–893.

Gille, E., Lemoine, H., Ehle, B. and Kaumann, A.J. (1985) The affinity of (-)-propranolol for β_1- and β_2-adrenoceptors of human heart. Differential antagonism of the positive inotropic effects and adenylate cyclase stimulation by (-)-noradrenaline and (-)-adrenaline. *Naunyn-Schmiedeberg's Arch. Pharmacol.*, **331,** 60–70.

Green, S.A., Holt, B.D. and Liggett, S.B. (1992) β_1- and β_2-adrenergic receptors display subtype-selective coupling to Gs. *Mol. Pharmacol.*, **41,** 889–893.

Hall, J.A., Kaumann, A.J. and Brown, M.J. (1990) Selective β_1-adrenoceptor blockade enhances positive inotropic effects to endogenous catecholamines through β_2-adrenoceptors in human atrial myocardium. *Circ. Res.*, **66,** 1610–1623.

Kaumann, A.J. (1991) Some aspects of heart beta adrenoceptor function. *Cardiovasc Drugs Ther.*, **5,** 549–560.

Kaumann, A.J. and Lemoine, H. (1987) β_2-adrenoceptor-mediated positive inotropic effect of adrenaline in human ventricular myocardium. Quantitative discrepancies with binding and adenylate cyclase stimulation. *Naunyn-Schmiedeberg's Arch. Pharmacol.*, **335,** 403–411.

Kaumann, A.J. and Sanders, L. (1993) Both β_1- and β_2-adrenoceptors mediate catecholamine-evoked arrhythmias in isolated human right atrium. *Naunyn-Schmiedeberg's Arch. Pharmacol.*, **348,** 536–540.

Kent, R.S., DeLean, A. and Lefkowitz, R.J. (1980) A quantitative analysis of beta-adrenergic receptor interactions: resolution of high and low affinity binding states of the receptor by computer modelling of ligand binding data. *Mol. Pharmacol.*, **17,** 14–23.

Lemoine, H., Schoenell, H. and Kaumann, A.J. (1988) Contribution of β_1- and β_2-adrenoceptors of human atrium and ventricle to the effects of noradrenaline and adrenaline as assessed with (-)-atenolol. *Br.J.Pharmacol.*, **95,** 55–66.

Levy, F.O., Zhu, X., Kaumann, A.J. and Birnbaumer, L. (1993) Efficacy of β_1-adrenergic receptors is lower than that of β_2-adrenergic receptors. *Proc. Natl. Acad. Sci. USA*, **90,** 10798–10802.

Molenaar, P., Russell, F.D., Shimada, T. and Summers, R.J. (1990) Densitometric analysis of β_1- and β_2-adrenoceptors in guinea-pig atrioventricular conducting system. *J. Mol. Cell. Cardiol.*, **22,** 483–495.

SUBCLASSIFICATION OF β-ADRENOCEPTORS: THE PHARMACOLOGY OF β_3-ADRENOCEPTORS IN TISSUES

JONATHAN R. S. ARCH

SmithKline Beecham Pharmaceuticals The Frythe, Welwyn, Herts AL6 9AR, UK

β_1- and β_2-adrenoceptors (β_1 and β_2AR) were originally identified by functional studies using catecholamine agonists and subsequently using antagonists, receptor binding studies and by cloning of their respective genes. β_3AR proved more difficult to identify. Their main features are low affinity for standard antagonists of β_1 and β_2AR, low ability to distinguish the enantiomers of these antagonists, selective stimulation by certain non-catecholamine agonists, and stimulation by high concentrations of certain β_1 and β_2AR antagonists. Secondary features are that the relative potencies of agonists can vary according to the response measured, and tachyphylaxis to selective β_3-agonists occurs in the gut. Pindolol analogues appear to bind in two ways to β_3AR, only one of which—the lower affinity binding mode—can mediate a functional response. The pharmacology of β_3AR varies between tissues. This can mostly, but not entirely, be explained in terms of human and rat β_3AR types, and variations in receptor number or coupling to response.

KEY WORDS: β_3-Adrenoceptors, BRL 37344, CGP 12177, pindolol, adipocytes

INTRODUCTION

β-Adrenoceptors (βAR) were first subdivided into β_1 and β_2AR on the basis of the relative potencies of a series of catecholamine agonists in various *in vitro* and *in vivo* systems (Lands *et al.*, 1967). It is now realised that this approach can be misleading, especially if agonists of differing and poor efficacy are used (Kenakin *et al.*, 1992). Nevertheless, the β_1/β_2 classification received support in the ensuing years from evaluation of the potencies of antagonists in functional studies, receptor binding experiments and eventually the elucidation of the sequences of the β_1AR and β_2AR genes.

Two points were not appreciated when βAR were originally subclassified. First, subtypes of βAR can coexist within one tissue, so that comparison of agonist potencies merely indicated the *predominant* functional subtype. Secondly, the tissues that were classified as containing β_1AR in fact included one tissue (rabbit heart) that contained mainly β_1AR and two tissues (the rat white adipocyte and rabbit jejunum) that contained mainly what are now called β_3AR. β_3AR could not be distinguished from β_1AR with the range of agonists available at that time, and although they were subsequently quite clearly distinguished from β_1AR by the low potencies of β-blockers in tissues that contain β_3AR (Harms *et al.*, 1974), some workers seemed reluctant to accept this evidence in the absence of a selective β_3AR

Correspondence to: Jonathan R. S. Arch, SmithKline Beecham Pharmaceuticals The Frythe, Welwyn, Herts AL6 9AR, UK.

antagonist. Moreover, the low potencies of the available radiolabelled βAR ligands at β_3AR made it impossible to detect β_3AR by receptor binding techniques. It was only when selective β_3AR agonists were described (Arch *et al.*, 1984) that the possible existence of these receptors was fully accepted, and this work ultimately led to the cloning of the human β_3AR receptor (Emorine *et al.*, 1989). For a more comprehensive review of the classification of β_3AR the reader is referred to Arch and Kaumann (1993).

FEATURES OF β_3-ADRENOCEPTOR PHARMACOLOGY

Low antagonist potencies

The most important feature of the pharmacology of β_3AR is that they have a low affinity for the standard β-blockers. For example, Harms *et al.* (1977) showed that the (-)-enantiomers of propranolol, alprenolol and practolol have much higher pA_2 values versus responses to isoprenaline in rat right atrium than in rat white adipocytes, reflecting differences in affinity of the antagonists for βAR in the two tissues of about 100–600-fold.

These early findings have been extended to other tissues, and to other antagonists and agonists, including selective β_3-agonists, such as BRL 37344, and "non-conventional partial agonists", such as CGP 12177 (see below). An intriguing finding is that antagonists usually have lower pA_2 values against responses to BRL 37344 or CGP 12177 than against the same responses to isoprenaline (eg Stock and Sudera, 1988; Van Liefde *et al.*, 1992). This is partly because β_3AR occur in tissues alongside β_1 and β_2AR, so that pA_2 values against responses to isoprenaline reflect an amalgam of antagonist affinities at two or three receptors. In rat colon, however, pA_2 values for antagonists have been determined in the presence of other antagonists at concentrations that block β_1 and β_2 but not β_3AR, and yet pA_2 values are again higher with isoprenaline than with BRL 37344 as the agonist (McLaughlin and MacDonald, 1990). These results might be explained if different agonists bind to somewhat different conformations of the β_3AR.

Eudismic ratios for antagonists

Harms *et al.* (1977) also showed that rat white adipocyte β_3AR are less able than β_1 or β_2AR to distinguish between the (−)- and (+)- enantiomers of β-blockers, ie eudismic ratios are low at β_3AR. This feature of β_3AR appears to have been largely ignored by other workers and deserves more attention. Since Harms *et al.* (1977) used isoprenaline as their agonist, which would act via both β_1 and β_3AR, their eudismic ratios may not accurately reflect those at β_3AR. Studies using agonists that are acting only via β_3AR or on cloned β_3AR expressed in the absence of other subtypes would help to define the receptor better.

Selective β_3AR agonists

Agonists that are more potent at β_3 than at β_1 or β_3AR, such as BRL 37344, were first identified as agents that stimulate metabolic rate and cause weight loss in rats and mice (Arch *et al.*, 1984). BRL 37344 is a selective stimulant of rat and mouse β_3AR relative to β_1 or β_3AR, but is less selective in absolute terms at human β_3AR (Tate *et al.*, 1991). Its profile contrasts markedly, however, with the *low* potency of isoprenaline or adrenaline for human β_3AR relative to β_1 or β_3AR. BRL 37344 stimulates cyclic AMP accumulation in cells that express human β_3AR in relatively high numbers, but it generally has low efficacy when stimulating downstream responses in human tissues, even though β_3AR mRNA is known to be present (Lönnqvist *et al.*, 1993). Calcium uptake by human platelets, which is remarkably sensitive to isoprenaline, is the only example of a human tissue response where BRL 37344 has been reported to respond well to BRL 37344 (Gill *et al.*, 1991).

Non-conventional partial agonists

Certain compounds that were originally identified as antagonists (or weak partial agonists) at β_1 and β_3AR have been found, at significantly higher concentrations, to stimulate β_3AR. These compounds, which have been called "non-conventional partial agonists", include pindolol and its analogues, carazolol, oxprenolol and CGP 12177. Only the (−)-enantiomers of these compounds appear to show this property (Kaumann, 1989). In contrast to BRL 37344, CGP 12177 is more effective at human than rat β_3AR (Fève *et al.*, 1991; Liggett, 1992). This, coupled with its lack of agonist activity at β_1 and β_3AR, makes it a particularly useful tool for the identification of β_3AR in human tissues (Lönnqvist *et al.*, 1993).

Relative potencies of agonists vary with response

The characterisation and identification of β_3AR are complicated by findings that the relative potencies of agonists can vary according to the response measured. In rat white adipocytes, BRL 37344 is as potent or more potent than isoprenaline as a stimulant of lipolysis, but isoprenaline is more potent than BRL 37344 as a stimulant of adenylyl cyclase in adipocyte membranes (Hollenga *et al.*, 1991; Murphy *et al.*, 1993). Potencies for activation of protein kinase A parallel those for activation of lipolysis (Murphy *et al.*, 1993). Hollenga *et al.* (1991) suggested that cyclic AMP generated by β_3AR is better coupled to lipolysis than that generated by β_1AR, and that the relative potencies of BRL 37344 and isoprenaline as stimulants of adenylyl cyclase and lipolysis differ because BRL 37344 acts exclusively via β_3AR and isoprenaline by both β_1 and β_3AR. The difference between the relative potencies was not entirely removed, however, when experiments were conducted in the presence of CGP 20712A to block β_1AR. Whilst not directly relevant to the interpretation of these results, it is also intriguing that the relative potencies of agonists as stimulants of cloned β_3AR vary according to whether the response measured is adenylyl cyclase or cyclic AMP accumulation (Table 1). It would not be surprising if the relative potencies of agonists at other AR and non-adrenoceptors also varied according to the response measured.

TABLE 1
Stimulation of adenylyl cyclase and cyclic AMP accumulation in human cloned β_3AR

	EC_{50} (nM)		
	Adenylyl cyclase[1]	*cyclic AMP accumulation*[2]	*EC_{50} ratio cyclase/cAMP*
Isoprenaline	495	3.9	127
BRL 37344	78	5.9	13
CGP 12177	330	38.7	8
Pindolol	1100	153.0	7

[1] Fève *et al.*, 1991
[2] Emorine *et al.*, 1989

Tachyphylaxis To Agonists

The gut develops tachyphylaxis to βAR agonists, including isoprenaline, when it is treated with selective β_3AR agonists (Coleman *et al.*, 1987; McLaughlin and MacDonald, 1990). Isoprenaline does not cause tachyphylaxis and it has not been reported whether non-conventional partial agonists have this effect. Tachyphylaxis to selective β_3AR agonists has been described only for gut responses and surprisingly, occurs not only for smooth muscle relaxation, but also for acid and bicarbonate secretion (Canfield and Paraskeva, 1992). This raises the possibility that it is due to β_3AR agonists "sticking" to the gut, a suggestion supported by the finding that, in the presence of propranolol (0.1 μM) to block β_1 and β_2AR, cyanopindolol (10 nM) prevented the development of tachyphylaxis to BRL 37344 in rat colon (J. Kelly, unpublished).

Receptor binding studies

β_3AR are difficult to detect by binding studies in tissues, because they are present together with β_1 and β_2AR, which have higher affinity for the available radioligands. Nevertheless, they have been successfully detected using [^{3}H]-CGP 12177 and [^{125}I]-iodocyanopindolol. In most studies these radioligands have been used at higher concentrations than would be employed for studies on β_1 or β_2AR but it is, nevertheless, surprising what low concentrations have been successfully used and what low K_D values for β_3AR have been obtained. Thus EC_{50} and pA_2 values for iodocyanopindolol and cyanopindolol suggest affinities for β_3AR in the range 8 to 76 nM whereas [^{125}I]-iodocyanopindolol has been used at concentrations ranging from 0.06 to 0.3 nM to detect β_3AR, and K_D values of about 1 nM have been reported for binding to both native and cloned β_3AR (see Arch and Kaumann, 1993; Emorine *et al.*, 1994).

Part of the explanation for this paradox may be that iodocyanopindolol is actually more potent than cyanopindolol in functional studies, the one reported EC_{50} value for iodocyanopindolol being misleading. An alternative explanation is suggested, however, by studies on cloned β_3AR which show that, compared to other β-agonists, pindolol is more potent in binding than in functional studies. Thus in the original paper on cloned β_3AR (Emorine *et al.*, 1989), isoprenaline, salbutamol and BRL 37344 had far higher K_i values for displacement of iodocyanopindolol binding than their EC_{50} values for stimulation of cyclic AMP accumulation. Pindolol, by contrast, had a lower K_i than EC_{50} value (Table 2). It must be emphasised that even isoprenaline was a partial agonist relative to noradrenaline

TABLE 2
Ratios of EC_{50} values (cyclic AMP accumulation) and K_i values (inhibition of iodocyanopindolol binding) in human cloned β_3AR[1]

	Accumulation of cyclic AMP		*Binding*	K_i/
	% noradrenaline max ± S.D., n = 4	EC_{50} *(nM)*	K_i*(nM)*	EC_{50}
Isoprenaline	90 ± 2	3.9	620	159
Salbutamol	77 ± 1	266	53,000	199
BRL 37344	87 ± 15	5.9	183	31
Pindolol	55 ± 5	153	11	0.07

[1] Emorine *et al.* (1989)

in these studies, suggesting that it had no spare receptors and its EC_{50} was roughly equal to its K_A value at β_3AR. These results suggest that the discrepancy between EC_{50} and pA_2 values on the one hand, and K_i values on the other, might be explained if the pindolol derivatives can bind to β_3AR in two ways, only one of which is of functional relevance.

TISSUE DISTRIBUTION AND VARIABLE PHARMACOLOGY OF β_3-ADRENOCEPTORS

β_3AR clearly play a functional role in white and/or brown adipocytes of the rat, mouse, hamster, dog and rabbit, in guinea-pig ileum and in rat gastric fundus. Thus, the pharmacology of βAR in these tissues displays the three main features of β_3AR: low antagonist potencies, selective stimulation by agonists such as BRL 37344, and stimulation by high concentrations of non-conventional partial agonists (see Arch and Kaumann, 1993). In other rat and guinea-pig gut preparations, and in whole animal studies of insulin secretion and anti-depressant activity in rats and mice, the first two of these features have been demonstrated but responses to non-conventional partial agonists have not been investigated. Rat colon shows the first two features but does not respond to partial agonists, possibly because β_3AR number or coupling to muscle relaxation is poor in this tissue.

In contrast to the rat colon, human adipocytes have been found by some, though not all workers to respond to non-conventional partial agonists, but any response to BRL 37344 has at most a minor β_3AR-mediated component (Lönnqvist *et al.*, 1993). Human taenia coli similarly responds to CGP 12177 but not to BRL 37344 (J. Kelly, unpublished). These findings are consistent with studies on cloned β_3AR which show that the efficacies of CGP 12177 and BRL 37344 are reversed for the human and rat receptors. It is, however, more difficult to explain why rat right atrium responds to non-conventional partial agonists but no β_3AR-mediated component has been detected in responses to BRL 37344, unless this rat tissue expresses a β_3AR of the "human" type. Perhaps the β_3AR-mediated response to BRL 37344 is simply undetectable in the presence of a large β_1AR-mediated response.

It is interesting that in tissues that respond to non-conventional partial agonists but not to selective β_3-agonists via β_3AR, antagonists have low pA_2 values against responses to the partial agonists but not against responses to isoprenaline. The human cloned β_3AR responds to isoprenaline, so why can a component of β_3AR stimulation not be detected in the action of isoprenaline? This may be because the human β_3AR is generally expressed in much lower amounts than β_1 or β_2AR or because it competes poorly for G proteins.

ACKNOWLEDGMENTS

I thank my many colleagues at SmithKline Beecham whose active research and informed discussion has contributed to this article. In particular I thank Lee Beeley, Mike Cawthorne, Jon Chambers, Axel Kaumann, John Kelly, Greg Murphy, Matthew Sennitt, Shelagh Wilson and Paul Young.

References

Arch, J.R.S., Ainsworth, A.T., Cawthorne, M.A., Piercy, V., Sennitt, M.V., Thody, V.E., Wilson, C. and Wilson S. (1984) Atypical β-adrenoceptor on brown adipocytes as target for anti-obesity drugs. *Nature*, **309**, 163–165.

Arch, J.R.S. and Kaumann, A.J. (1993) β_3 and atypical β-adrenoceptors. *Medicinal Research Reviews*, **13**, 663–729.

Canfield, P. and Paraskeva, P. (1992) β-adrenoceptor agonist stimulation of acid secretion by rat stomach *in vitro* is mediated by 'atypical' β-adrenoceptors *British Journal of Pharmacology*, **106**, 583–586.

Coleman, R.A., Denyer, L.H. and Sheldrick, K.E. (1987) β-adrenoceptors in guinea-pig gastric fundus - are they the same as the 'atypical' β-adrenoceptors in rat adipocytes?. *British Journal of Pharmacology*, **90,** 40P.

Emorine, L., Blin, N. and Strosberg, A.D. (1994) The human β_3-adrenoceptor: the search for a physiological function. *Trends in Pharmacological Sciences*, **15,** 3–7.

Emorine, L.J., Marullo, S., Briend-Sutren, M-M., Patey, G., Tate, K., Delavier-Klutchko, C. and Strosberg, D. (1989) Molecular characterisation of the human β_3-adrenergic receptor. *Science*, **245,** 1118–1121.

Fève, B., Emorine, L.J., Lasnier, F., Blin, N., Baude, B., Nahmias, C., Strosberg, A.D. and Pairault, J. (1991) Atypical β-adrenergic receptor in 3T3-F442A adipocytes. Pharmacological and molecular relationship with the human β_3-adrenergic receptor. *The Journal of Biological Chemistry*, **266,** 20329–20336.

Gill, J., Dandona, P. and Jeremy, J.Y. (1991) Atypical β- and α_2-adrenoceptor activation, dibutyryl cyclic AMP and iloprost stimulate [$^{45}Ca^{2+}$] uptake by human platelets. *Platelets*, **2,** 93–98.

Harms, H.H., Zaagsma, J. and de Vente, J. (1977) Differentiation of β_3-adrenoceptors in right atrium diaphragm and adipose tissue of the rat using stereoisomers of propranolol, alprenolol, nifenalol and practolol. *Life Sciences*, **21,** 123–128.

Harms, H.H., Zaagsma, J. and Van der Wal (1974) β-Adrenoceptor studies on the β-adrenoceptors in rat adipose tissue. *European Journal of Pharmacology*, **25,** 87–91.

Hollenga, Ch., Brouwer, F. and Zaagsma, J. (1991) Relationship between lipolysis and cyclic AMP generation mediated by atypical β-adrenoceptors in rat adipocytes. *British Journal of Pharmacology*, **102,** 577–580.

Kaumann, A.J. (1989) Is there a third heart β-adrenoceptor? *Trends in Pharmacological Sciences*, **10,** 316–320.

Kenakin, T.P., Bond, R.A. and Bonner T.I. (1992) II. Definition of pharmacological receptors. *Pharmacological Reviews*, **44,** 351–362.

Lands, A.M., Arnold, A., McAuliff, J.P., Luduena, F.P. and Brown, T.G. (1967) Differentiation of receptor systems activated by sympathomimetic amines. *Nature*, **214,** 597–598.

Langin, D., Portillo, M.P., Saulnier-Blache, J-S. and Lafontan, M. (1991) Coexistence of three β-adrenoceptor subtypes in white fat cells of various mammalian species. *European Journal of Pharmacology*, **199,** 291–301.

Liggett, S.B. (1992) Functional properties of the rat and human β_3-adrenergic receptors: differential agonist activation of recombinant receptors in Chinese hamster ovary cells. *Molecular Pharmacology*, **42,** 634–637.

Lönnqvist, F., Krief, S., Strosberg, A.D., Nyberg, B., Emorine, L.J. and Arner P. (1993) Evidence for a functional β_3-adrenergic receptor in man. *British Journal of Pharmacology*, **110,** 929–936.

McLaughlin, D.P. and MacDonald, A. (1990) Evidence for the existence of 'atypical' β-adrenoceptors (β_3-adrenoceptors) mediating relaxation in the rat distal colon *in vitro*. *British Journal of Pharmacology*, **101,** 569–574.

Murphy, G.J., Kirkham, D.M., Cawthorne, M.A. and Young P. (1993) Correlation of β_3-adrenoceptor-induced activation of cyclic AMP-dependent protein kinase with activation of lipolysis in rat white adipocytes. *Biochemical Pharmacology*, **46,** 575–581.

Stock, M.J. and Sudera, D.K. (1988) Schild analysis of β-adrenoceptor antagonists inhibition of brown adipocyte respiratory responses to novel β-adrenoceptor agonists. *British Journal of Pharmacology*, **95,** 741P.

Tate, K.M., Briend-Sutren, M-M., Emorine, L.J., Delavier-Klutchko, C., Marullo, S. and Strosberg, A.D. (1991) Expression of three human β-adrenergic-receptor subtypes in transfected Chinese hamster ovary cells. *European Journal of Biochemistry*, **196,** 356–361.

Van Liefde, I. Van Witzenburg, A. and Vauquelin, G. (1992) Multiple beta adrenergic receptor subclasses mediate the L-isoprenaline-induced lipolytic response in rat adipocytes. *The Journal of Pharmacology and Experimental Therapeutics*, **262,** 552–558.

PHARMACOLOGICAL CHARACTERISTICS OF β_3-ADRENERGIC RECEPTOR

F. PIETRI-ROUXEL, M. F. DRUMARE and A. D. STROSBERG

Institut Cochin de Génétique Moléculaire, Molecular Immuno-Pharmacology Laboratory, CNRS-UPR 0415, 22, rue Méchain F-75014 Paris, France

A large number of β-AR ligands has in the past been developed using animal tissues in which several receptor subtypes may be expressed simultaneously, albeit in varying proportions. The availability of the cloned receptor genes has now allowed for a thorough revision of the properties of these ligands by studying their interaction with receptors individually expressed in model systems.

Much of the work delineating the precise pharmacological comparison of the three β-AR receptors has come from investigations with stably transfected Chinese hamster ovary cells (CHO cells). This review discusses recent findings on β3-AR and the properties of a number of β3-AR ligands toward the cloned receptors. Carazolol, tertatolol and Cl 316,243 were thus recently described as selective and potent β3-AR agonists, and ZD 2079 and salmeterol appear to display a full but non-selective agonistic activity on the β3-AR.

With the ongoing characterization of the β3-AR, at the molecular and cellular level, and with the advent of computer-assisted molecular modeling to aid in the determination of the 3-dimensional structure of the receptor, it is envisaged that novel β3-AR compounds will become available with improved selectivity and potency. In time, these should facilitate the discovery of new therapeutic agents.

KEY WORDS: β-adrenergic-receptor, β-AR agonists, pharmacology, adipose tissue

INTRODUCTION

Beta-adrenergic receptors (β-ARs) belong to the superfamily of receptors that contain seven transmembrane helices, are coupled to guanine nucleotide regulatory proteins (G proteins) and are involved in the transmission of signals across cell membranes. β-ARs mediate the physiological actions of adrenaline and noradrenaline by activation of adenylyl cyclase via the stimulatory Gs-protein, leading to the production of the second messenger cAMP. In 1967, Lands *et al.*, established a pharmacological classification of β-ARs into two subtypes: the receptor mediating responses in the heart was designated β1 and the receptor modulating vasodepressor activity and bronchodilatation was designated β2. This classification in two subtypes was later recognized to be insufficient to account for non-β1- and non-β2-adrenergic responses of rat adipose tissues (reviewed in Zaagsma and Nahorski, 1990). This raised the hypothesis of the existence of a third, "atypical", β-AR subtype mediating thermogenesis in brown adipose tissue and lipolysis in white adipose tissue. An important step was made by Emorine *et al.*, in 1989 through the cloning, from a human genomic library, of a gene coding for this third β-AR designated as the β3-AR. This β3-AR, after transfec-

Correspondence to: A.Donny Strosberg, Laboratoire d'Immuno-Pharmacologie Moléculaire, Institut Cochin de Génétique Moléculaire, 22, rue Méchain—75014 Paris (France).

tion of the gene in CHO cells, displayed ligand binding and adenylyl cyclase activating properties, clearly distinct from those of β1- and β2-ARs (Blin *et al.*, 1993, 1994). While the β3-AR showed the highest degree of pharmacological similarity with the atypical receptor described earlier, it also appeared to correspond to the β3-AR expressed in ileum, colon, and gastric fundus and skeletal muscle of rodents, and cat heart (reviewed in Arch and Kaumann 1993).

Here we discuss some of the most recent findings on the β3-AR and, in particular, review the properties of a number of β3 ligands toward the cloned human and rodent receptors.

PHARMACOLOGICAL CHARACTERIZATION OF β3-AR

Main properties of the β3-AR

Much of the work delineating the precise pharmacological comparison of three β-ARs has come from investigations with stably transfected Chinese hamster ovary cells (CHO cells) lines that express high copy numbers of each subtype (Tate *et al.*, 1991). The methods used for these studies were the stimulation of adenylyl cyclase and quantitative determination of cAMP accumulation and direct binding or competition of [^{125}I]-iodocyanopindolol binding. From these studies, the following properties of the β3-AR were recognized: (1) atypically low affinity for conventional β-AR antagonists including radioligands, (2) atypically low stereoselectivity indices for agonist and antagonist enantiomers, (3) atypically low potencies of reference agonists, (4) high potency of a novel class of compounds initially described as potent activators of lipolysis and thermogenesis in adipose tissues of rodents, (5) partial agonistic activity properties of several antagonistic of β1- and β2-ARs, reflecting intrinsic sympathomimetic activities in certain tissues. These properties, similar to the atypical profile (Blin *et al.*, 1994) were further confirmed when the mouse (Nahmias *et al.*, 1991) and rat (Granneman *et al.*, 1991) β3-AR were described.

Three classes of β3 compounds

Three classes of compounds were thus distinguished: (1) β1/β2/β3 agonists; (2) β1/β2 antagonists displaying β3 agonistic properties and (3) β1/β2/β3 antagonists.

β1/β2/β3 agonists These ligands were initially defined as potent and selective activators of lipolysis and thermogenesis for the atypical receptors in white and brown adipocyte tissues. BRL37344, which is a parent acid of BRL 35135, is the most potent ligand of this class of arylethanolamine: it exhibited a ten-fold β3-AR selectivity relative to β1-AR and β2-AR (Table 1). This reference ligand is 11-fold more potent than (–)isoproterenol as a stimulant of rat white adipocytes lipolysis (Arch and Kaumann, 1993), and 38-fold more potent on mouse β3-AR than on human β3-AR. LY79711 is also a potent β3-AR agonist which exhibited 5- and 17-fold greater potency towards β3-AR than on β1-AR and β2-AR respectively (Table 1) (Blin, et al., 1993).

SR58611A and SM-11044 were described as β2/β3-AR selective agonists relative to β1-AR, (Table 1) (Blin *et al.*, 1993). In common with other phenylethanolaminotetralines, SR58611A induced lipolysis in rat white adipocytes (Manara and Bianchetti, 1990) and stimulated fatty acid synthesis in brown adipose tissue of obese diabetic mice. The 5-lipoxygenase inhibitor, SM-11044, has been shown to stimulate atypical β-AR. This ligand also

TABLE 1

Comparison of pharmacological properties of ligands on human β1-, β2-, β3-AR expressed in CHO cells.
Ligands were classified as β1/β2/β3 antagonists; β1/β2 anta. β3 ago: β1-/β2-AR antagonist displaying β3 agonistic properties; β1/β2/β3 antagonists. All data were obtained using similar experimental conditions.
Binding competition assays were carried out with intact cells for 45 min at 37°C in the presence of [^{125}I]CYP. Adenylyl cyclase stimulation assays were performed with intact cells preincubated or not with 10 nM isoproterenolol for 10 min and incubated with drugs for 30 min at 37°C. Concentration-response curves were fitted using least square regression analysis, and binding competition (Ki: inhibition constant) and adenylyl cyclase stimulation (Kact: activation constant) were deduced. IA was calculated for each drug relative to isoproterenol-induced maximal cAMP accumulation. Values are means of at least three or two independent experiments performed in duplicate.

	Human β1-AR			***Human β2-AR***			***Human β3-AR***		
	binding	*adenylyl cyclase stimulation*		*binding*	*adenylyl cyclase stimulation*		*binding*	*adenylyl cyclase stimulation*	
	Ki (nM)	*Kact (nM)*	*IA*	*Ki (nM)*	*Kact (nM)*	*IA*	*Ki (nM)*	*Kact (nM)*	*IA*
β1/β2/β3 agonists									
BRL 37344	1,750 ± 310[a]	112 ± 28[a]	1.30 ± 0.1[a]	1,120 ± 380[a]	177 ± 47[a]	0.80 ± 0.04[a]	287 ± 92[a]	15 ± 3[a]	1.11 ± 0.1[a]
LY 79771	—	86 ± 8[a]	1.42 ± 0.3[a]	—	325 ± 121[a]	0.22 ± 0.03[a]	555 ± 71[a]	18 ± 3[a]	1.06 ± 0.04[a]
SR 58611A	38,500 ± 13,400[a]	12,000 ± 600[a]	0.96 ± 0.07[a]	187 ± 26[a]	36 ± 19[a]	0.87 ± 0.07[a]	6,640 ± 960[a]	25 ± 5[a]	1.23 ± 0.2[a]
SM 11044	18,100 ± 1,700[a]	190 ± 20[a]	1.50 ± 0.2[a]	4,100 ± 200[a]	62 ± 6[a]	1.03 ± 0.08[a]	1,300 ± 200[a]	84 ± 10[a]	0.98 ± 0.1[a]
ZD 2079	—	—	—	—	agonist	—	—	191	0.91
SALMETEROL	—	—	—	—	agonist	—	—	45	0.85
β1/β2 anta. β3 ago.									
BUCINDOLOL	0.20 ± 0.04[a]	antagonist		0.10 ± 0.03[a]	antagonist		23 ± 10[a]	7.0 ± 1.2[a]	1.01 ± 0.1[a]
ICI 201651	549 ± 200[a]	antagonist		2,860 ± 750[a]	antagonist		85 ± 12[a]	20 ± 9[a]	1.14 ± 0.1[a]
CGP 12177A	0.9 ± 0.1[a]	antagonist		4 ± 2[a]	antagonist		88 ± 22[a]	139 ± 44[a]	0.68 ± 0.02[a]
PINDOLOL	3.4 ± 0.7[a]	antagonist		2.3 ± 0.9[a]	antagonist		11 ± 2[a]	153 ± 12[a]	0.55 ± 0.05[a]
cyano-PINDOLOL	—	antagonist		—	antagonist		—	174 ± 58[a]	0.82 ± 0.04[a]
PROPRANOLOL	6.3 ± 1.0[a]	antagonist		0.7 ± 0.3[a]	antagonist		145 ± 8[a]	1,490 ± 550[a]	0.51 ± 0.1[a]
CARAZOLOL	—	antagonist		—	antagonist		2[b]	10[b]	1.07[b]
(±)TERTATOLOL	—	antagonist		—	antagonist		15	413	0.92
(–)TERTATOLOL	—	antagonist		—	antagonist		9	173	0.91
Cl 315,514	—	antagonist		—	antagonist		—	161	1
Cl 316,243	—	antagonist		—	antagonist		14,000	68	1.6
β1/β2/β3 antagonists									
(–)BUPRANOLOL	1.7 ± 0.3[a]	antagonist		0.4 ± 0.1[a]	antagonist		50 ± 14[a]	antagonist	
ICI 118551	120 ± 3[a]	antagonist		1.2 ± 0.2[a]	antagonist		257 ± 34[a]	antagonist	
CGP 20712A	1.5 ± 0.2[a]	antagonist		1,800 ± 400[a]	antagonist		2,300 ± 450[a]	antagonist	

[a] results reported by Blin *et al.*, 1993
[b] results reported by Mejean *et al.*, 1994

stimulated the relaxation of guinea pig trachea and lung parenchyma, and increased right atrial rate (Sugasawa *et al.*, 1992).

ZD 2079 is a new ligand which belongs to the family of arylethanolamine compound with a structure close to BRL 37344 and ICI 201651 structure (figure 1). Clinical studies on animals, showed a low acute oral toxicity and revealed that a repeated exposure to high concentrations of the compound resulted in cardiovascular changes and flushing of the skin

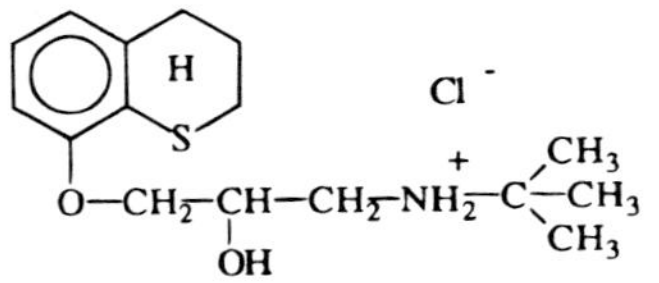

(D,L) Tertatolol

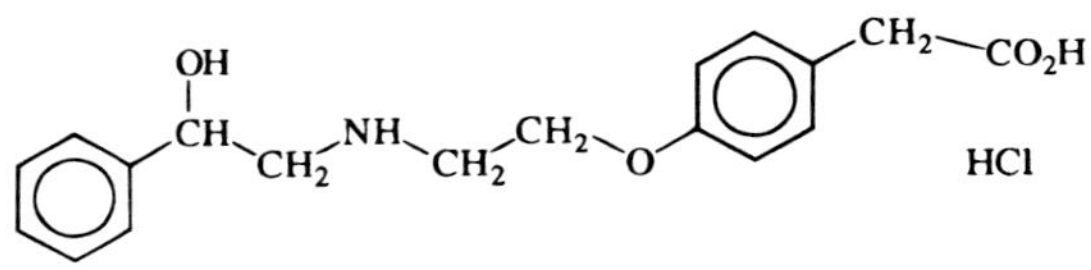

ZD 2079

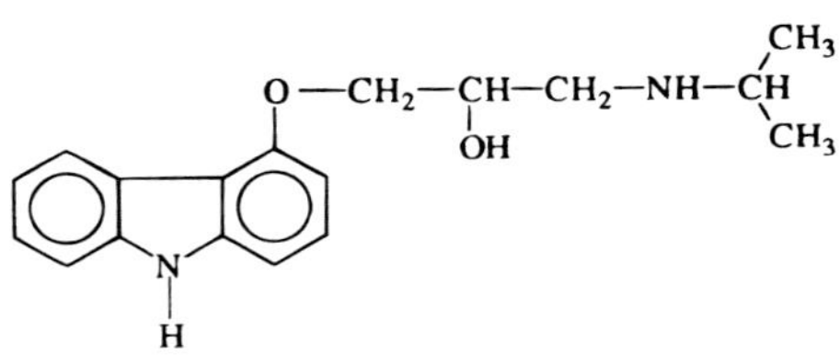

Carazolol

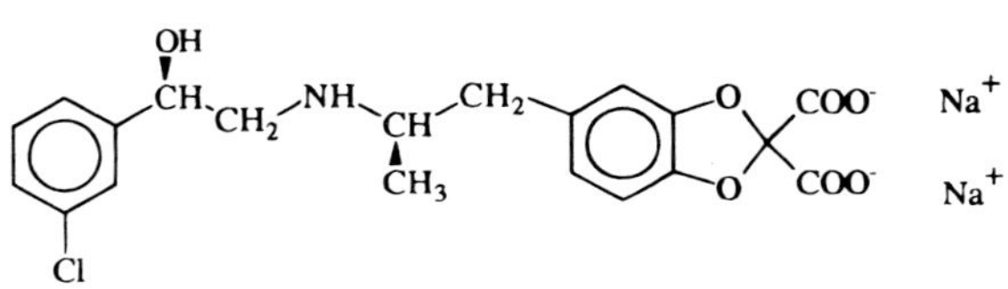

(D,L) CI 314,514

(L) CI 316,243

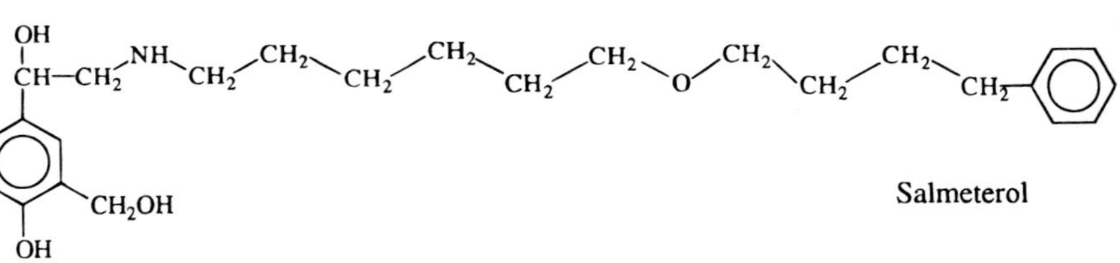

Salmeterol

FIGURE 1 Structural formulas for β3-AR ligands.

and mucous membranes. ZD 2079 exhibited a potent and a full agonistic activity on β3-AR expressed in CHO cells (Table 1). ZD 2079 was not as selective for β3-AR since it displayed a full agonistic activity on β2-AR expressed in CHO cells. Salmeterol is a β2 adrenoceptor agonist that has been shown to produce long acting bronchodilatation in both, *in vitro* and *in vivo*, animal models (Ball *et al.*, 1987; Bradshaw *et al.*, 1987). This molecule was characterized by its long duration of action and its highly selectivity for β2-AR in bronchial smooth muscle; it is 2 to 15 times more potent than the potent β2-AR agonist salbutamol. Salmeterol displays non-competitive membrane binding and a slow dissociation with an effect for many hours. Because of the structure of salmeterol with two side chains that are linked by a long hydroxycarbon chain (figure 1), its mechanism of action was supposed to be through drug-lipid interactions rather than through drug-receptor interactions through an exo-site which is distinct from the β2-AR (Johnson *et al.*, 1993). Experiments carried out on transfected CHO cells confirmed that this ligand is a potent β2-AR agonist (Table 1) and revealed potent and full agonistic activity of salmeterol on β3-AR. Because of its non-selectivity and its apparent non-conventional mechanism of action, the relevance of its action remains to be established in humans.

β1-/β2 antagonists displaying β3 agonistic properties Among the ligands described for their β1/β2-AR antagonistic properties, some revealed high binding affinities and agonistic activity on β3-AR. ICI 201651 is one of the more important compounds of this class, it exhibits a high β3-AR selectivity in binding affinity, relative to β1- and β2-AR (Table 1). This ligand is the in vivo metabolized form of ICI D7114 which was able to stimulate the reappearance of brown adipose tissue in adult dogs treated with this compound for a month or more (Champigny et al., 1991).

Bucindolol was described as a high affinity non-selective β-AR antagonist in human ventricular myocardium. Studies on β-AR expressed in CHO cells showed that bucindolol displayed the same high affinity for β1- and β2-AR and possessed the best full potent and selective agonistic effect described on β3-AR up to this date (Table 1). Functional properties of the mouse and human β3-AR, evaluated in transfected CHO cells, showed that CGP 12177A possessed a partial agonistic activity with a high potency in stimulating adenylyl cyclase, and did not distinguish between the two species (Nahmias *et al.*, 1991; Blin *et al.*, 1993; Lönnqvist *et al.*, 1993). CGP 12177A, pindolol and cyanopindolol were 10 to 100-fold less potent in stimulating β3-AR than bucindolol and ICI 201651 (Table 1). Furthermore, cyanopindolol and iodocyanopindolol were described as partial agonists in rat and hamster white adipocytes, and bound to the β3-AR at 10–100-fold higher concentrations than those used for β1- and β2-AR. These compounds have been termed "non conventional partial agonists" on the basis of their similar properties in cardiac tissues (Blin *et al.*, 1993).

Carazolol is an arylethanolamine characterized by a carbazole ring (figure 1). This compound, developed in 1975, was known as a β1/β2-AR antagonist and is still used currently in veterinary medecine. Its β-AR blocking activity consists in reducing cardiovascular responses to an excessive release of catecholamines with an efficacy on hundred times greater than the classical β-AR blocker propranolol (Bartsch *et al.*, 1977). The β3-AR activity of carazolol was evaluated on mouse β3-AR and on the three human subtypes of β-AR expressed in CHO cells (Mejean *et al.*, 1994). The results showed that carazolol displayed the highest affinity for binding on β3-AR of all compounds studied so far. Its potency is 12-fold greater than the most selective β3-AR: bucindolol (Table 1). As a blocker, carazolol antagonized the β1-/β2-AR mediated response to isoproterenol but acted as a full and potent agonist with potency comparable to that of BRL 37344 on human and mouse β3-AR (Table 1), (Mejean *et al.*, 1994). Carazolol displayed full and potent lipolytic activity in murine 3T3-F442A cells in which the β3-AR is functionally coupled to lipolysis (Mejean *et al.*, 1994). Clinical investigations earlier reported that drug-treated rats showed a significant reduction in body mass in comparison with control animals; this change possibly reflected an increase in the responsiveness of the brown adipose tissue (Hartig *et al.*, 1986). Taken together, the properties of carazolol make this ligand interesting for development of selective thermogenic β-AR agonists for the treatment of obesity.

Tertatolol an aryloxoarylamine characterized by a thiochromane ring (see figure 1), was initially described for its β-AR-blockers effects and its renal vasodilatating properties demonstrated in animals and confirmed in man. Recently, clinical investigations showed a decrease of arterial pressure in man, a significant reduction of diastolic blood pressure and an antihypertensive efficacy of tertatolol, 20 times greater than the potent β1-AR blocker atenolol (Vivet, 1993). Extensive pharmacological studies showed that (–)-tertatolol is an antagonist of postsynaptic 5-HT1A receptors mediating hypothermia in the rat (Millan *et al.*, 1993) and possesses a potent antagonist effect on pre- and post-synaptic serotonin 5-HT1A receptors in the rat brain (Jolas *et al.*, 1993). Effects of (±)-tertatolol and (–)-tertatolol were evaluated on β1-, β2-, and β3-ARs expressed on CHO cells (Table 1). The results confirmed the antagonistic activity of tertatolol on β1- and β2-ARs but revealed a potent and full agonist on β3-AR. Moreover, (–)-tertatolol displayed a 2.5- and 7.5-fold better agonistic activity than (+)-tertatolol and the racemic form, respectively. This compound possessed an affinity for the β3-AR comparable to bucindolol affinity (Table 1).

CL 316,243, the pure (–)-enantiomer and CL 514,214, the racemic form, are benzodioxole-containing phenethalonamines (Figure 1). These two compounds were recently described as highly selective agonists for the β3-AR (Table 1), (Dolan, *et al.*, 1994). CL 316,243 and CL 314,514 stimulated rat adipocytes lipolysis while having no effect on the rate of contraction of guinea pig atria (β1-AR effect) and little or no ability to prevent the insulin-stimulated incorporation of [^{14}C]glucose into rat soleus muscle glycogen (β2-AR effect) at high concentrations (> 100μM). Pharmacological properties of mouse and human β3-AR carried out on CHO transfected cells showed that CL 316,243 was highly mouse selective versus human (Table 1), (Dolan, *et al.*, 1994). These compounds were described for their potent thermogenic, antidiabetic and antiobesity properties in animal models of non insulino-dependent diabetes mellitus and obesity (Largis *et al.*, 1994). CL 316,243 is now in Phase II clinical trials.

Propranolol has established a clear distinction between the human and the mouse β3-AR: studies achieved on transfected CHO cells showed that this ligand behaves as a weak and partial agonist on human β3-AR and as an antagonist on mouse β3-AR (Table 1), (Blin *et al.*, 1994). This result thus revealed a species-specific difference for the β3-AR which may point to a key direction for the development of selective therapeutic agents.

β1-/β2-/β3-AR antagonists The third class of ligands includes the selective antagonists, CGP 20712A for β1-AR and ICI 118551 for β2-AR and the non selective antagonist bupranolol (Table 1). For bupranolol, the receptor binding order of selectivity in CHO transfected cells with human β-ARs, was β2- > β1- > β3-AR. In CHO cells expressing mouse β3-AR, bupranolol exhibited a potency similar to that measured on human β3-AR, whereas ICI

118551 and CGP 20712A were one to two orders less potent (Blin et al., 1994). Bupranolol appears, today, to be the best antagonist available to characterize the β3-AR.

III FUTURE DIRECTIONS

During the past 5 years, scores of publications have described the emergence of compounds useful in the study of the pharmacological characteristics and physiological implications of β3-AR in tissues as well as in model systems. With the now available pharmacological profile and the molecular characterisation of the three β-AR subtypes, the comparison of the agonist or antagonist properties of synthetic ligands will allow exquisitely precise structure-activity relationship studies. Site-directed mutagenesis, affinity labelling analysis and ultimately X-ray diffraction of the crystallized proteins should soon help establish a well-defined picture of the ligand binding site and contribute to understand the mechanism involved in the intracellular signal transmission. These studies will contribute to define a pharmacophore adapted to each receptor subtype and selective drugs will become available to serve as therapeutic agents.

ACKNOWLEDGMENTS

We are grateful to Emmanuelle Jubin for helpful technical cell culture assistance. This work was supported by grants from the Centre National de la Recherche Scientifique, the Institut National de la Santé et de la Recherche Médicale, the Ministère de l'Enseignement Supérieur et de la Recherche, the Université Paris VII, Bristol-Myers-Squibb Company, the Fondation pour la Recherche Médicale, the Association pour le Développement de la Recherche sur le Cancer, and the Ligue Nationale Française contre le Cancer.

References

Arch, J.R.S. and Kaumann, A.J. (1993) β3 and atypical β-adrenoceptors. *Med. Res. Rev.*, **14,** 663–729.

Ball, D.I., Coleman, R.A., Denyer, L.H., Nials, A.T. and Sheldrick, K.E. (1987) In vitro characterization of the β2-adrenoceptor agonist salmeterol [abstract]. *Br. J. Pharmacol.*, **92,** 591P.

Bartsch, W., Dietmann, K., Leinert, H. and Sponer, G. (1977) Cardiac action of carazolol and methypranol in comparison with other β-receptor blockers. *Arzneim. Forsch.*, **27,** 1022–1026.

Blin, N., Camoin, L., Maigret, B. and Strosberg, A.D. (1993) The β3-adrenergic receptor: structure-activity relationship of β-adrenergic ligands. *Mol. Pharmacol.*, **44,** 1094–1104.

Blin, N., Nahmias, C., Drumare, M.F. and Strosberg, A.D. (1994) Mediation of most atypical effects by species homologues of the β3-adrenoceptor. *Br. J. Pharmacol.*, **112,** 911–919.

Bradshaw, J., Brittain, R.T., Coleman, R.A. and al, e. (1987). The design of salmeterol, a long acting selective β2-adrenoceptor agonist [abstract]. *Br. J. Pharmacol.*, **92,** 590P.

Champigny, O., Ricquier, D., Blondel, O., Mayers, R.M., Briscoe, M.G. and Holloway, B.R. (1991) β3-adrenergic receptor stimulation restores message and expression of brown-fat mitochondrial uncoupling protein in adult dogs. *Proc. Natl. Acad. Sci. USA*, **88,** 10774–10777.

Dolan, J.A., Muenkel, H.A., Burns, M.G., Pellegrino, S.M., Fraser, C.M., Piétri, F., Strosberg, A.D., Largis, E.E., Dutia, M.D., Bloom, J.D., Bass, A.S., Tanikella, T.K., Cobuzzi, A., Lai, F.M. and Claus, T.H. (1994) Beta-3 Adrenoceptor of the Dioxolane Dicarboxylate phenethanolamines. *J. Pharmacol. Exper. Ther.*, **269,** 1000–1006.

Emorine, L.J., Marullo, S., Briend-Sutren, M.-M., Patey, G., Tate, K., Delavier-Klutchko, C. and Strosberg, A.D. (1989) Molecular characterization of the human β3-adrenergic receptor. *Science*, **245,** 1118–1121.

Granneman, J.G., Lahners, K.N. and Chaudhry, A. (1991) Molecular cloning and expression of the rat β3-adrenergic receptor. *Mol. Pharmacol.*, **40,** 895–899.

Hartig, F., Vollmar, J., Bode, G., Hebold, G. and Czerwek, H. (1986) Investigations on the exogenous influencing of the spontaneous tumour rate in rats. *Exp. Pathol.*, **30,** 157–164.

Johnson, M., Butchers, P.R., Coleman, R.A., T., N.A., Strong, P., Summer, M.J., Vardey, C.J. and Whelan, C.J. (1993) The pharmacology of salmeterol. *Life Sci.*, **52,** 2131–2143.

Jolas, T., Haj-Dahmane, S., Lanfumey, L., Fattaccini, C.M., Kidd, E.J., Adrien, J., Gozlan, H., Guardiola-Lemaitre, B. and Hamon, M. (1993) (–)Tertatolol is a potent antagonist at pre- and postsynaptic serotonin 5-HT 1A receptors in the rat brain. *Naunyn-Schmiedebergs-Arch-Pharmacol.*, **347,** 453–463.

Lands, A.M., Arnold, A., Mc Auliff, J.P., Luduena, F.P. and Brown, T.G. (1967) Differentiation of receptor systems activated by sympathomimetic amines. *Nature*, **214,** 597–598.

Largis, E.E., Burns, M.G., A., M.H., Dolan, J.A. and Claus, T.H. (1994) Anti-diabetic and anti-obesity effects of a highly Selective β3-Adrenoceptor Agonist (CL 316,243). *Drug Devel. Res.*, **32,** in press.

Lönnqvist, F., Krief, S., Strosberg, A.D., Nyberg, B., Emorine, L.J. and Arner, P. (1993) Evidence for a functional β3-adrenoceptor in man. *Br. J. Pharmacol.*, **110,** 929–936.

Manara, L. and Bianchetti, A. (1990) The phenylethanolaminotetralines: New selective agonists for atypical β-adrenoceptors. *Tr. in Pharmacol. Sci.*, **11,** 229–230.

Mejean, A., Guillaume, J.L. and Strosberg, A., D. (1994) Carazolol: A potent, selective β3-adrenergic receptor agonist. *submitted,*

Millan, M.J., Rivet, J.M., Canton, H., Le Marouille-Girardon, S. and Gobert, A. (1993) Induction of Hypothermia as a Model of 5-Hydroxytryptamine1A Receptor-Mediated Activity in the rat: A pharmacological characterization of the actions of novel agonists and antagonists. *J. of Pharmacol. and Experi. Ther.*, **264,** 1364–1376.

Nahmias, C., Blin, N., Elalouf, J.-M., Mattei, M.G., Strosberg, A.D. and Emorine, L.J. (1991) Molecular characterization of the mouse β3-adrenergic receptor: relationship with the atypical receptor of adipocytes. *EMBO J.*, **10,** 3721–3727.

Sugasawa, T., Matsuzaki, M., Morooka, S., Foignant-Chaverot, N., Blin, N. and Strosberg, A.D. (1992) In vitro study of a novel atypical β-adrenergic agonist, SM-11044. *Eur. J. Pharmacol.*, **216,** 207–215.

Tate, K.M., Briend-Sutren, M.-M., Emorine, L.J., Delavier-Klutchko, C., Marullo, S. and Strosberg, A.D. (1991) Expression of three human β-adrenergic-receptor subtypes in transfected chinese hamster ovary cells. *Eur. J. Biochem.*, **196,** 357–361.

Vivet, P. (1993) Overview of Clinical Safety and efficacy of Tertatolol. *Cardiology*, **83 suppl.,** 41–50.

LOCALISATION AND CHARACTERISATION OF ATYPICAL β-ADRENOCEPTORS IN SKELETAL MUSCLE AND GUT

R. J. SUMMERS*, F. D. RUSSELL, S. J. ROBERTS, V. R. BONAZZI, A. SHARKEY, B. A. EVANS and P. MOLENAAR

Department of Pharmacology, University of Melbourne, Parkville, Vic 3052, AUSTRALIA

(-)-[^{125}I]-CYP binds to high affinity and low affinity sites in rat soleus muscle (RS) and human colon (HC). High affinity sites (K_D 30.5 and 12pM respectively) correspond to conventional β-adrenoceptors while low affinity (K_D 522.5 and 691pM) sites have pharmacological and regulatory properties appropriate for atypical β-adrenoceptors. Competition for low affinity sites in RS and HC showed that antagonists competed for binding with a low degree of stereoselectivity (stereoselectivity indices tertatolol HC 9.3, RS 2.6; alprenolol HC 6.8, RS 3.1). Comparison of the sites in RS and HC showed them to be similar to each other and to atypical β-adrenoceptors in gastrointestinal studies but clearly different from β_1- or β_2-adrenoceptors. Autoradiographic studies of atypical β-adrenoceptors in rat hind limb muscle showed localisation to soleus and other oxidative muscle fibres and co-localisation of β_2-adrenoceptors. In HC autoradiographic studies showed high concentrations of atypical β-adrenoceptors in mucosa, submucosal arteries and veins, myenteric plexus and longitudinal smooth muscle. In contrast, β_1-adrenoceptors were found in muscularis mucosae, submucosal arteries and longitudinal smooth muscle and β_2-adrenoceptors at low levels in mucosa, submucosa and in localised areas in smooth muscle. Molecular studies detected β_3-mRNA in soleus muscle by RT/PCR and localised it to mucosa by *in situ* hybridisation in human colon. These studies provide further evidence for the presence of β_3-adrenoceptors in rat soleus and human colon localised to areas where they are likely to have metabolic, secretory or absorptive roles.

KEY WORDS: β_3-adrenergic receptor, rat soleus, human colon, receptor binding, autoradiography

INTRODUCTION

There is convincing evidence from functional, receptor binding and molecular studies for the existence of atypical β-adrenoceptors with pharmacology distinct from that of β_1- or β_2-adrenoceptors. Atypical β-adrenoceptors are resistant to blockade by (-) propranolol and other conventional β-adrenoceptor antagonists, display low stereoselectivity indices, are activated by novel lipolytic β-adrenoceptor agonists such as BRL 28410, BRL 37344 and ICI D7114, and resemble cloned β_3-adrenoceptors pharmacologically (Arch & Kaumann, 1993; Emorine *et al.*, 1989; Granneman *et al.*, 1991; Nahmias *et al.*, 1991; Granneman *et al.*, 1992; Granneman *et al.*, 1993). Atypical β-adrenoceptors have been described in white and brown adipose tissue (BAT & WAT), areas of the gastrointestinal tract including oesophagus, gastric fundus, jejunum, ileum, and colon as well as heart, skeletal muscle,

*Correspondence to: R. J. Summers, Department of Pharmacology, University of Melbourne, Parkville, Vic 3052, Australia.

airways and blood vessels (see Arch & Kaumann, 1993). There is also evidence for roles for atypical β-adrenoceptors in control of insulin secretion and sensitivity (Arch *et al.*, 1991), glucose uptake (Abe *et al.*, 1993) and thermogenesis in skeletal muscle (Challiss *et al.*, 1988), bicarbonate secretion in caecum (Canfield & Abdul-Ghaffar, 1992), acid secretion in stomach (Canfield & Paraskeva, 1992), and blood pressure by central (Peres-Polon & Correa, 1984) and peripheral mechanisms (Wilson *et al.*, 1992). mRNA coding for β_3-adrenoceptors has been detected in BAT and WAT in rat (Emorine *et al.*, 1989; Granneman *et al.*, 1991) and human (Krief *et al.*, 1993), rat ileum (Emorine *et al.*, 1989), human ileum, colon and gall bladder (Krief *et al.*, 1993; Granneman *et al.*, 1993). Although binding studies have not demonstrated atypical β-adrenoceptors in SK-N-MC neuroblastoma cells (Esbenschade *et al.*, 1992), rat jejunum (van der Vliet *et al.*, 1990) and rat colon (Landi *et al.*, 1992), they are clearly identified in BAT (Muzzin *et al.*, 1992; Sillence *et al.*, 1993) and rat soleus muscle (Roberts *et al.*, 1993; Sillence *et al.*, 1993) and in cDNA expression systems (Tate *et al.*, 1991; Feve *et al.*, 1991; Muzzin *et al.*, 1991). The aims of the present study were to use receptor binding techniques to characterise atypical β-adrenoceptors in rat soleus and human colon, to localise the receptors by autoradiography and to detect and localise mRNA coding for the β_3-adrenoceptor in these tissues.

Receptor binding

Characterisation of β-adrenoceptors in rat soleus muscle. Atypical β-adrenoceptors have proved difficult to detect in intact tissues due to the lack of a high affinity, selective radioligand. Current β-adrenoceptor ligands have high affinity for β_1- and β_2-adrenoceptors compared with the β_3-adrenoceptor and mixtures of β-adrenoceptors are often found in the same tissues. Several recent studies have however identified sites with binding characteristics appropriate for atypical β-adrenoceptors in rat soleus muscle using the strategy of high ligand concentration (500 pM) and assays conducted in the presence of (-) propranolol to block β_1- and β_2-adrenoceptor binding (Roberts *et al.*, 1993; Sillence *et al.*, 1993). These assays involved dissection of soleus muscles from male Sprague-Dawley rats (280–320g) that had been stunned by a blow to the head and exsanguinated. Tissues were weighed and homogenised in Krebs buffer containing phenyl methyl sulphonyl fluoride (PMSF, 10μM), filtered through nylon mesh (212 μm) and membranes prepared as previously described (Roberts *et al.*, 1993). Protein was measured using the Lowry *et al.*, (1951) method using BSA as the standard. Homogenate (50μl) was incubated for 60 min at 37°C with (-)-[^{125}I]-CYP (5–1000 pM) and (±)-alprenolol (500 μM) used to define non-specific binding. Determinations were conducted in duplicate or triplicate. Assays were terminated by addition of 4ml ice cold buffer and rapid vacuum filtration (Brandel M30-R) over Whatman GF/B filters presoaked in 2% polethyleneimine (PEI) and washed 3 times with 4ml of buffer. Radioactivity was determined using a Packard gamma counter (B5424) at an efficiency of 79%. Results were analysed using EBDA (McPherson, 1983) and LIGAND (Munson & Rodbard, 1980). When studied under these conditions (-)-[^{125}I]-CYP binding equilibrated and dissociated more rapidly than at β_1- or β_2-adrenoceptors (Roberts *et al.*, 1993). Saturation curves were biphasic (Fig.1), and two sites were identified, a high affinity site (K_D 30.5 ± 16.3 pM; Bmax 9.4 ± 1.4 fmol/mg protein) and a low affinity site (K_D 522.5 ± 29.1 pM; Bmax 62.19 ± 11.8 fmol/mg protein). It is likely that the high affinity site represents binding to β_2-adrenoceptors since competition curves with (-) propranolol (pK_{D1} 8.3, pK_{D2} 5.33) and the β_2-adrenoceptor selective antagonist ICI 118,551 (pK_{D1} 9.78; pK_{D2} 5.45) were bipha-

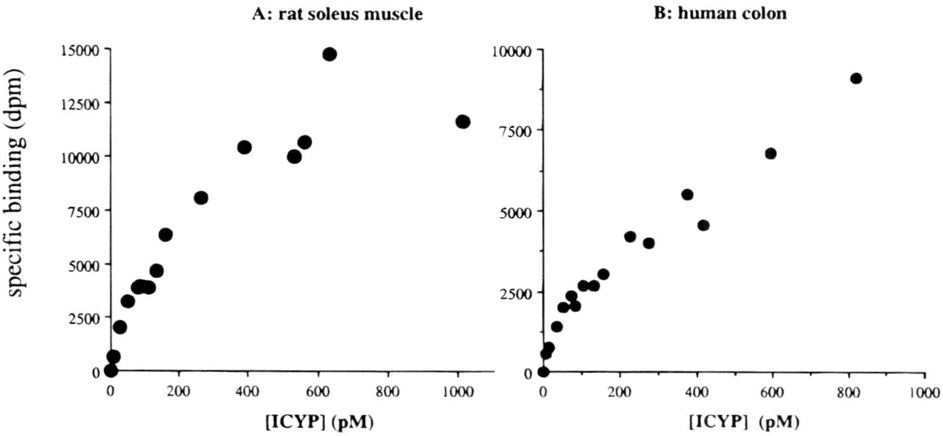

FIGURE 1 Saturation of specific (-)-[^{125}I]-CYP binding in A, rat soleus (Roberts *et al.*, 1993): K_{D1} 30.5 ± 16.3pM; $Bmax_1$ 9.4 ± 1.4fmol/mg protein; K_{D2} 522.5 ± 29.1pM; $Bmax_2$ 62.19 ± 11.8fmol/mg protein (n = 4) and B, human colon; K_{D1} 12 ± 5pM; $Bmax_1$ 8.2 ± 2.1fmol/mg protein; K_{D2} 691 ± 385pM; $Bmax_2$ 53 ± 8fmol/mg protein (n = 4). Non-specific binding was defined by alprenolol (500μM).

sic (Roberts *et al.*, 1993; Fig 2). The β_1-adrenoceptor selective antagonist CGP 20712A competed at a single site with low affinity (Fig. 2). Comparison of a series of β-adrenoceptor antagonists shows that all have low affinity and low stereoselectivity indices at this site compared to binding at β_1- or β_2-adrenoceptors. (-)-[^{125}I]-CYP, (±)-cyanopindolol and tertatolol had the highest affinities (Table 1) and the values compare with those determined in functional assays in ileum (Bond & Vanhoutte, 1992). The results are compatible with the hypothesis that in homogenates of rat soleus, (-)-[^{125}I]-CYP binds with high affinity to β_2-adrenoceptors (pK_D 10.52) and with low affinity to atypical β-adrenoceptors (pK_D 9.28).

Regulation of β-adrenoceptors in rat soleus muscle β_3-Adrenoceptors in cDNA expression systems (Thomas *et al.*, 1993; Liggett *et al.*, 1993; Nantel *et al.*, 1993) are resistant to agonist induced desensitisation. This is thought to be related to the lack of phosphorylation sites for protein kinase A (PKA) and β-adrenoceptor kinase (β-ARK) on the 3rd cytoplasmic loop and carboxyterminus of the β_3-adrenoceptor (Emorine *et al.*, 1991; Nantel *et al.*, 1993). No down-regulation was observed in hamsters following noradrenaline infusion (Carpene *et al.*, 1993) but in rat BAT and WAT β_3-mRNA is decreased in response to BRL 28630, isoprenaline or cold exposure (Granneman & Lahners, 1992). In contrast, long term agonist exposure has been reported to upregulate β_3- expression via multiple cAMP response elements in 3T3-F442A cells (Thomas *et al.*, 1993). β_2-Adrenoceptors are readily phosphorylated and desensitised and long term agonist exposure causes loss and decreased transcription of mRNA (Bouvier *et al.*, 1989). Long-term (7–14 day) infusion of (-)-isoprenaline into guinea-pigs and rats causes significant loss of β_1- and almost total loss of β_2-adrenoceptors from the heart (Molenaar *et al.*, 1990 ; Russell *et al.*, 1994). This study examined the effect of 14 day infusion of (-)-isoprenaline to rats on β_2-adrenoceptors and atypical β-adrenoceptor binding in rat soleus muscle. Male Sprague-Dawley rats (280–320g) were anaesthetised with pentobarbitone sodium (25mg/kg i.p.), the back of the neck shaved and a local s.c. injection of lignocaine (2%) given. The skin was cut and a sub-

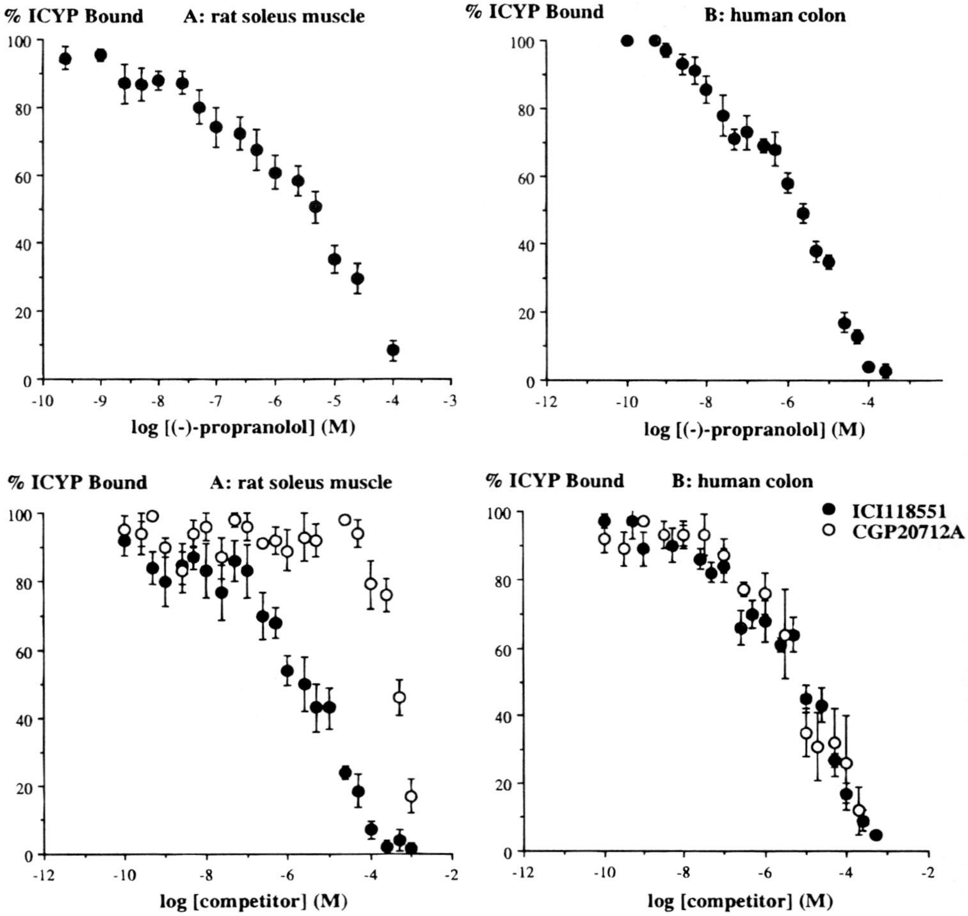

FIGURE 2 Competition for (-)-[^{125}I]-CYP (500pM) binding by (-) propranolol (upper panels) and the β_1-adrenoceptor selective antagonist CGP 20712A and the β_2-adrenoceptor selective antagonist ICI 118551 (lower panels) in homogenates of A, rat soleus and B, human colon. Non-specific binding was defined by alprenolol (500µM).

cutaneous pocket formed by blunt dissection. Osmotic minipumps (ALZET 2002) were implanted to infuse (-)-isoprenaline at 400 µg/kg/hr or vehicle (N/1000 HCl). After 14 days, the animals were sacrificed and the tissues harvested and the saturation binding assays conducted as described above. (-)-Isoprenaline infusion markedly reduced β_2-adrenoceptors in rat soleus homogenates from 23.4 ± 3.8 in control to 7.8 ± 1.1 fmol/mg protein in treated animals (P = 0.002) whereas K_D was not significantly changed (42.8 ± 7.8 and 31.6 ± 3.9 pM respectively; P > 0.05 ; n = 7). When the same homogenates were subjected to the binding protocol for atypical β-adrenoceptors there was no significant change in Bmax (control 96.8 ± 24.7; treated 71.1 ± 10.9 fmol/mg protein; P > 0.05) or K_D (control 311.0 ± 59.1; treated 285.0 ± 59.3pM; P > 0.05; n = 7). (-)-Isoprenaline treatment

TABLE 1
Affinity and pseudo Hill coefficients for compound competing against (-)-[^{125}I]-CYP binding

Competitor	*rat soleus muscle*			*human colon*		
	(n)	*pKD ± S. E*	*nH ± S.E*	*(n)*	*pKD ± S.E*	*nH ± S. E*
(-)-propranolol	7	5.33 ± 0.08	0.38 ± 0.03	4	5.30 ± 0.05	0.42 ± 0.03
ICI118551	4	5.45 ± 0.09	0.40 ± 0.04	4	4.69 ± 0.08	0.39 ± 0.04
CGP20712A	4	3.68 ± 0.06	1.44 ± 0.31	4	5.49 ± 0.09	0.46 ± 0.05
(±)-cyanopindolol	6	7.63 ± 0.09	0.70 ± 0.11	4	7.57 ± 0.05	0.56 ± 0.04
(-)-[^{125}I]-CYP	4	9.28 ± 0.02	0.94 ± 0.03	4	9.16 – 0.34	0.86 ± 0.02
(±)-alprenolol	6	5.66 ± 0.05	0.62 ± 0.04		nd	nd
(±)-tertatolol	6	6.70 ± 0.06	0.66 ± 0.06		nd	nd
nadolol	6	5.09 ± 0.07	0.50 ± 0.04	4	4.96 – 0.09	0.51 – 0.06
(-)-pindolol	7	4.61 ± 0.08	0.64 ± 0.07		nd	nd
(+)-pindolol	7	4.48 ± 0.05	0.83 ± 0.07		nd	nd
(-)-tertatolol	4	6.82 ± 0.04	0.70 ± 0.05	4	7.13 ± 0.08	0.38 ± 0.03
(+)-tertatolol	4	6.40 ± 0.04	0.81 ± 0.06	4	6.16 ± 0.06	0.66 ± 0.06
(-)-alprenolol	4	6.16 ± 0.03	0.88 ± 0.05	4	6.29 ± 0.07	0.45 ± 0.03
(+)-alprenolol	4	5.67 ± 0.02	0.95 ± 0.03	4	5.46 ± 0.06	0.65 ± 0.06

therefore significantly reduced β_2-adrenoceptors without significant alteration in atypical β-adrenoceptor binding.

Characterisation of β-adrenoceptors in human colon Few studies have examined β-adrenoceptors in human colon and the results are equivocal. Spontaneous activity in human colonic circular muscle is inhibited by (-)-isoprenaline and this effect is blocked by (-) propranolol with a pA_2 of 7.7 (McLaughlin *et al.*, 1988), a value higher than that at other gut tissues but lower than the pA_2 expected at human β_1- or β_2-adrenoceptors (8.6–9.1; see Arch & Kaumann, 1993). The same preparation (McLaughlin *et al.*, 1991) did not respond to BRL 37344 and responses to (-)-isoprenaline were blocked by the β_1-adrenoceptor selective antagonist betaxolol. It has been suggested that the human colonic β_3-adrenoceptor may not bind BRL 37344 or that there is poor β_3-adrenoceptor reserve in this tissue (Arch & Kaumann, 1993). To study the binding characteristics, segments of normal human colon were dissected from specimens obtained at surgery for cancer and immediately frozen in liquid N_2. (-)-[^{125}I]-CYP binding was examined in homogenates of human colon using the conditions established in rat soleus. Saturation analysis of (-)-[^{125}I]-CYP binding showed two sites, a high affinity site (K_D 12 ± 5 pM, Bmax 8.2 ± 2.1 fmol/mg protein) and a low affinity site (K_D 691 ± 385 pM; Bmax 53 ± 8 fmol/mg protein) as shown in Figure 1. Competition for (-)-[^{125}I]-CYP binding (500 pM) with (-)-propranolol produced a biphasic competition curve (Fig. 2) with pK_D values of 8.14 ± 0.1 and 5.30 ± 0.05 (n = 4). Examination of a range of competitors, including stereoisomers showed a profile of activity similar to that observed in the soleus muscle with the exception of CGP 20712A which was 65 times more potent in human colon compared to rat soleus and the stereoselectivity indices observed for tertatolol (9.3) and alprenolol (6.8) were greater than in rat soleus (2.6 and 3.1 respectively) as shown in Table 1.

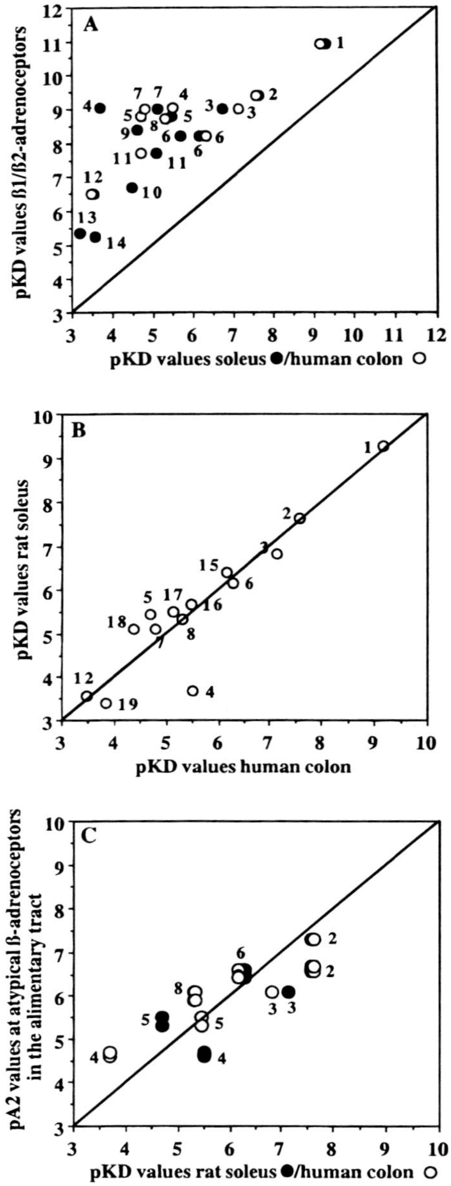

FIGURE 3 Comparison of antagonist affinities at β_1- and β_2-adrenoceptor subtypes and atypical receptors in rat soleus and human colon. In A, atypical sites in rat soleus/human colon c.f. β_1- or β_2-adrenoceptors; B, rat soleus c.f. human colon; C, rat soleus/human colon c.f. atypical β-adrenoceptor antagonist affinities from functional studies in gastrointestinal tract tissues. Key: 1-(-)-[^{125}I]-CYP, 2-(-)CYP, 3-(-) tertatolol, 4-CGP20712A, 5-ICI118551, 6-(-)alprenolol, 7-CGP12177A, 8-(-) propranolol, 9-(-) pindolol, 10-(+) pindolol, 11-nadolol, 12-(-) isoprenaline, 13-(-) noradrenaline, 14-(-) adrenaline, 15-(+) tertatolol, 16-(+) alprenolol, 17-SR58611A, 18-ICID7114, 19-BRL37344

Comparison of antagonist affinities at β_1- and β_2-adrenoceptor subtypes and atypical β-adrenoceptors in rat soleus and human colon The affinities of β-adrenoceptor antagonists at the atypical β-adrenoceptor binding sites in rat soleus and human colon were clearly less than those at typical β-adrenoceptors as shown in Figure 3a. The atypical β-adrenoceptor binding sites in the two tissues are similar as shown in Figure 3b with the exception of CGP 20712A. These binding sites also clearly correlate with pA_2 values obtained in functional studies in oesophagus, gut and adipocytes (Fig. 3c).

Receptor autoradiography

Localisation of β-adrenoceptor subtypes in rat soleus While much is known about the function of atypical β-adrenoceptors little is known about their distribution. The anatomical localisation of (-) propranolol resistant (-)-[^{125}I]-CYP binding has been observed in rat hind limb skeletal muscle and showed a high density of sites associated with the soleus muscle (Molenaar *et al.*, 1991). These studies formed the basis of subsequent receptor binding studies in this tissue (Roberts *et al.*, 1993) but were carried out under conditions that were designed to identify typical β-adrenoceptors. The present study examined the distribution of typical and atypical β-adrenoceptor binding sites in sections of rat hind limb muscle containing soleus using appropriate conditions for each receptor. Sprague-Dawley rats were sacrificed as before, the chest opened and tissues perfused through the left ventricle with 40ml Krebs/PMSF: 0.32M sucrose, 1:1 at 37°C followed by light fixation with 20ml of the same medium containing 0.05% paraformaldehyde and 0.05% gluteraldehyde. Hind limb muscle bundles were removed and rapidly frozen in isopentane cooled in liquid nitrogen. Tissue sections (10μm) were cut on a Reichert-Jung 1800 cryostat and mounted onto subbed slides. Every 10th slide was stained with haematoxylin and eosin to show tissue morphology. Sections for labelling were preincubated at room temperature as previously described (Molenaar *et al.*, 1987). Preincubation solution was replaced with (-)-[^{125}I]- CYP (540pM) in Kreb's phosphate containing 5HT (10μM), ICI 118551 (0.7μM) and CGP 20712A (1μM) and the slides incubated in a sealed humidified chamber at 37°C for 60 min. Non-specific binding was determined with (±)-alprenolol (500μM). After incubation, slides were drained, rinsed and washed (2 by 5 min) at 37°C in buffer. Labelling of sections for localisation of β_1 and β_2-adrenocêptors, was as previously described (Molenaar *et al.*, 1987). Both β_2-adrenoceptors and atypical β-adrenoceptors were found and had a similar distribution with high concentrations of both receptors associated with the soleus muscle and with slow contracting oxidative fibres (Kim *et al.*, 1991) in the plantaris and gastrocnemius (Fig. 4).

Localisation of β-adrenoceptor subtypes in human colon No previous studies have been carried out to localise β-adrenoceptors in human colon. Tissues were prepared in the same way as rat soleus except that human colonic tissue was rapidly frozen but was not perfused or fixed. Labelling was carried out as for rat soleus muscle. Examination of atypical β-adrenoceptor binding revealed high levels of binding to the mucosa, to the medial smooth muscle of submucosal arteries and veins, to localised areas in the circular smooth muscle, the myenteric plexus and to the longitudinal smooth muscle (Fig. 5a). The levels of binding to blood vessels varied, being particularly intense in some large arteries and

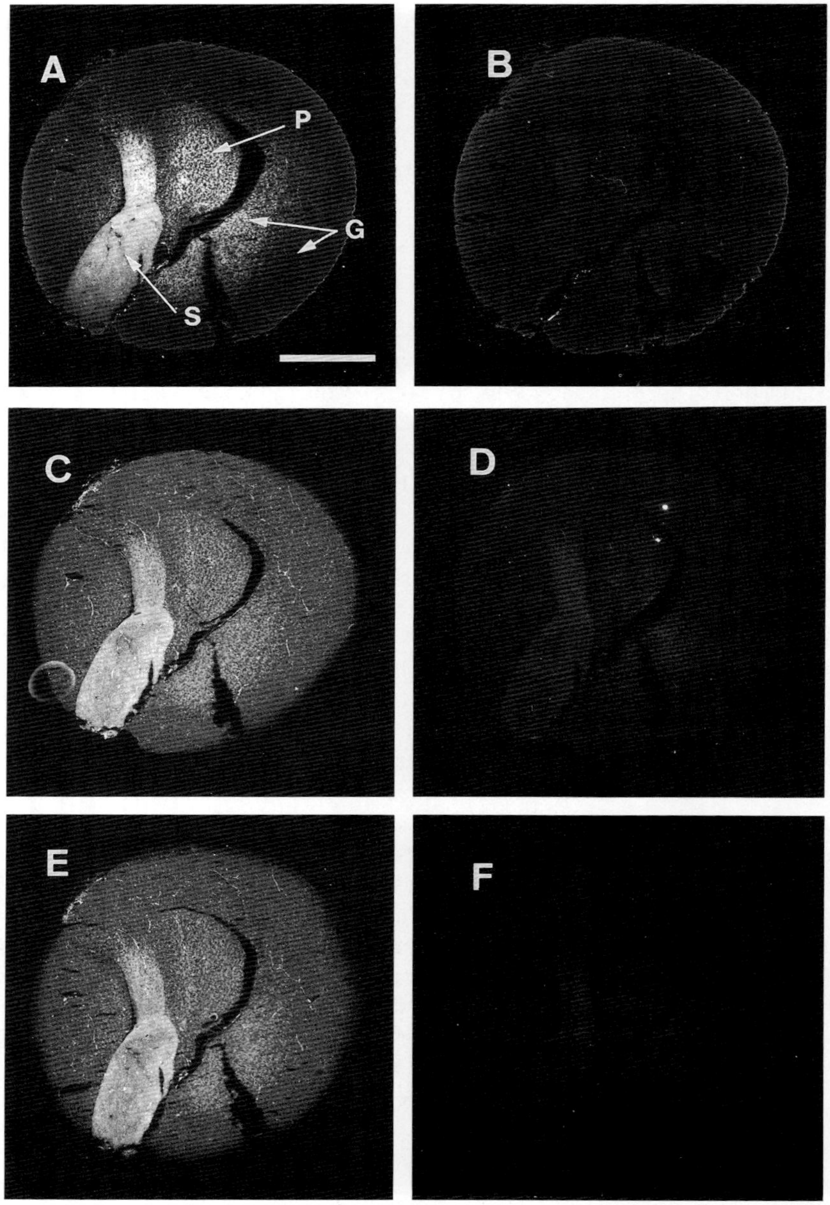

FIGURE 4 Autoradiographic localisation of β_1-, β_2- and atypical binding sites in rat hindlimb skeletal muscle. In A atypical binding ((-)-[^{125}I]-CYP 540pM + 5HT 10μM + ICI 118551 0.7μM + CGP 20712A 1μM) showing binding localised to soleus (S) and also to areas in plantaris (P) and gastrocnemius (G) corresponding to oxidative fibres. In B non-specific binding determined in the presence of alprenolol (500μM). The remaining panels show typical β-adrenoceptor binding; in C, β_1- + β_2-adrenoceptor binding ((-)-[^{125}I]-CYP 50pM + 5HT 10μM); D, β_1-adrenoceptor binding (+ ICI 118551 70nM); E, β_2-adrenoceptor binding (+ CGP 20712A 100nM); and F, non-specific binding ((-) propranolol 1μM). Bar = 4mm.

weak in large veins. Non-specific binding (alprenolol 500μM) was weak and not localised (Fig 5b). The pattern of labelling of β_1- and β_2-adrenoceptors was particularly interesting in human colon as the distribution varied markedly from that observed for the atypical β-adrenoceptor. β_1-Adrenoceptors were located in three main areas—in the muscularis mucosa, in a population of small arteries in the submucosa and in the longitudinal muscle (Fig. 5d). β_2-Adrenoceptors were found at low levels in the mucosa, in blood vessels

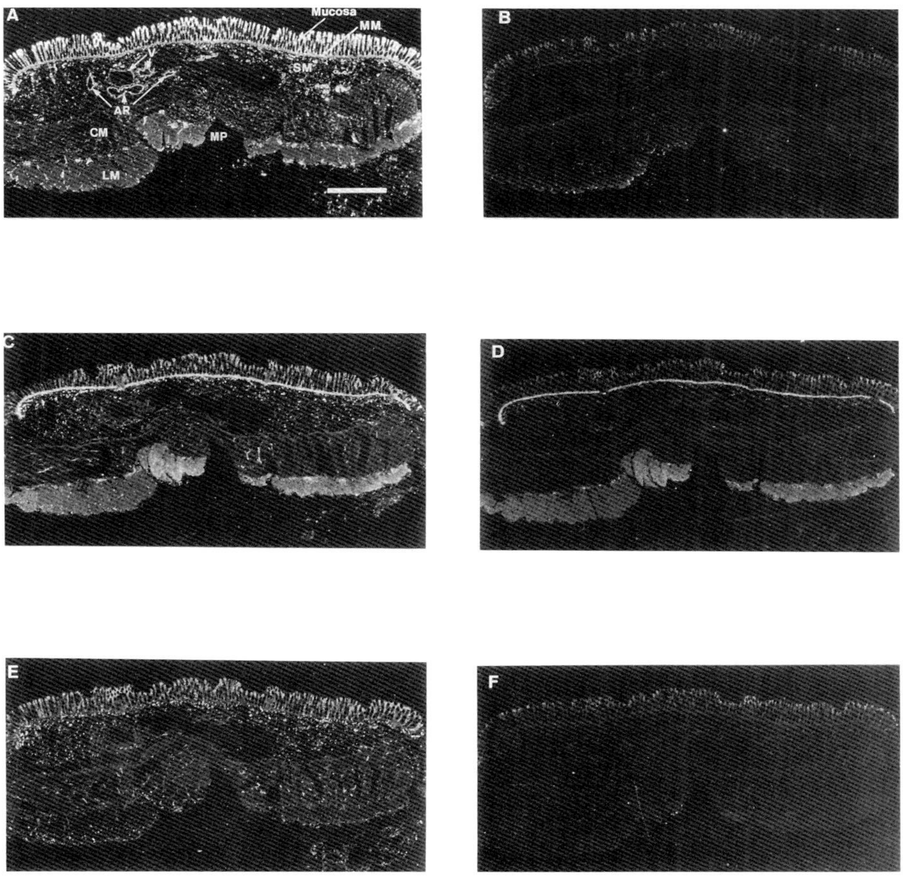

FIGURE 5 Autoradiographic localisation of β_1-, β_2- and atypical binding sites in human colon. In A atypical binding ((-)- [^{125}I]-CYP 540pM + 5HT 10μM + ICI 118551 0.7μM + CGP 20712A 1μM) showing binding localised to the mucosa, medial smooth muscle of submucosal arteries and veins, to localised areas of circular smooth muscle, the myenteric plexus and to longitudinal smooth muscle; B, non-specific binding conducted in the presence of alprenolol (500μM) was weak and not localised. In C, β_1- and β_2-adrenoceptor binding ((-)-[^{125}I]-CYP 50pM + 5HT 10μM); D, β_1-adrenoceptor binding ((-)-[^{125}I]-CYP + 5HT + ICI 118551 70nM); E, β_2-adrenoceptor binding ((-)-[^{125}I]-CYP + 5HT + CGP 20712A 100nM); and F, non-specific binding ((-)-[^{125}I]-CYP + 5HT + propranolol 1μM). β_1-Adrenoceptors were located in three main areas—in the muscularis mucosa, in a population of small blood vessels in the submucosa and in the longitudinal muscle. β_2-Adrenoceptors were found at low levels in the mucosa, in blood vessels in the submucosa and in localised patches in the circular and longitudinal muscle. Non-specific binding ((-) propranolol 1μM) was weak and not localised. Bar = 2mm.

in the submucosa and in localised patches in the circular and longitudinal muscle (Fig. 5e). Non-specific binding ((-) propranolol 1μM) was weak and not localised (Fig. 5f).

Molecular biology

Detection of β_3-adrenoceptor mRNA in rat soleus muscle Molecular evidence for the presence of atypical β-adrenoceptor mRNA in rat soleus is equivocal. Although functional studies show that the β_3-adrenoceptor agonist BRL 28410 stimulates metabolic responses in rat soleus (Challiss *et al.*, 1988) caution has been urged on the grounds that this response may reflect the presence of BAT in the hind limb muscle or an indirect humoral effect (Arch & Kaumann, 1993). Recent studies show that BRL 35135A (pro-drug ester of BRL 37344) increases glucose uptake in skeletal muscle, particularly in oxidative (red) muscle such as soleus and extensor digitorum longus, and in the red portions of quadriceps and gastrocnemius (Abe *et al.*, 1993). Furthermore, BRL 37344 dose-dependently increased glucose uptake in isolated soleus muscle with a maximal response at 100pM (Abe *et al.*, 1993). The functional evidence suggests that β_3-adrenoceptors stimulate glucose uptake in skeletal muscle by an insulin independent mechanism. Although β_3-mRNA has been detected in rat soleus using a cDNA probe based on the human β_3-adrenoceptor sequence (Emorine *et al.*, 1989) this displayed non-specific hybridisation and no signal was seen in Northern blots or with a ribonuclease protection assay using a specific rat cDNA probe (Granneman *et al.*, 1991). In contrast, molecular studies clearly show that β_3-mRNA is present in human colon (Granneman *et al.*, 1993; Krief *et al.*, 1993) and that the signal is not accompanied by UCP (uncoupling protein, a marker for BAT; Krief *et al.*, 1993). In the present studies, RNA was extracted from rat tissues using a modification of the method of Chomczynski and Sacchi (1987), and aliquots (0.5μg) subjected to reverse transcription/polymerase chain reaction (RT/PCR). RT was performed with AMV reverse transcriptase and either oligo dT or a reverse primer based on the 7th transmembrane segment of the rat β_3-adrenoceptor. PCR was performed on 1/10th of the resulting cDNA using the reverse primer and a specific β_3-adrenoceptor primer. PCR was performed for 35 cycles—denature 94°C, anneal 55°C, extend 72°C. Product was separated by gel electrophoresis, blotted and hybridised using a ^{32}P labelled probe complimentary to the 3rd intracellular loop of the β_3-adrenoceptor in a region quite distinct from β_1- and β_2-adrenoceptors. A hybridisation signal, of ≅400 b.p. for β_3-mRNA was detected in BAT and rat soleus muscle as shown in Fig 6.

Localisation of β_3-adrenoceptors in human colon by in situ hybridisation While several studies have demonstrated β_3-mRNA in human colon, the location has not been determined. A riboprobe was prepared to label β_3-adrenoceptor mRNA as follows. A PCR amplified segment of the rat β_3-adrenoceptor (382 b.p.) was cloned into the Sma1 site of pBluescript II KS flanked by T3 and T7 promoters (Stratagene Inc., La Jolla, California, USA). The orientation of the fragment was determined by sequencing. The vector was linearised downstream of the insert to be transcribed with either BamH1 (antisense) or HindIII (sense). Probes were labelled by the incorporation of α^{33}P-UTP during transcription, phenol/chloroform extracted, ethanol precipitated and stored at –70°C. For *in situ* studies probes were subjected to alkaline hydrolysis to give a riboprobe approx 180 b.p. in length. Slide mounted tissue sections were defatted in graded alcohols, fixed in 4% paraformaldehyde and acetylated (Molenaar *et al.*, 1993). The tissues were prehybridised, then hybridised with sense or antisense probes at 53°C for 18h, washed and apposed to hyperfilm β-max for 4 days

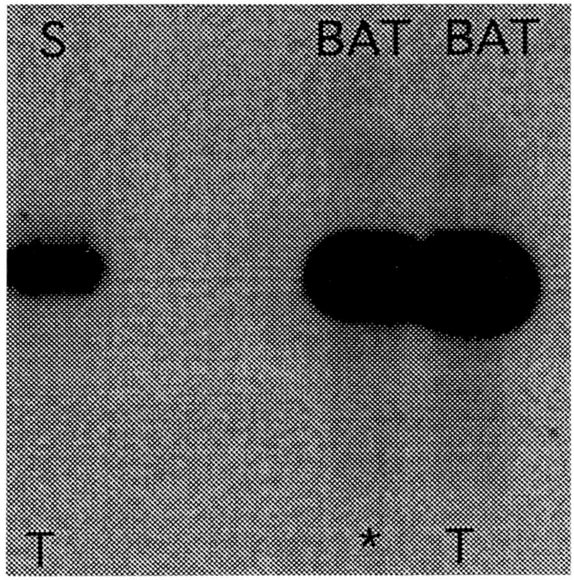

FIGURE 6 Detection of β_3-adrenoceptor mRNA in rat soleus by reverse transcription/PCR. RNA from rat intrascapular brown adipose tissue (BAT) and soleus muscle (S) was reverse transcribed with oligo dT (T) or a specific β_3-primer (*), the product amplified (35 cycles) with specific β_3-adrenoceptor primers, separated by gel electrophoresis and blotted and hybridised using a ^{33}P labelled probe complimentary to a region corresponding to the 3rd intracellular loop of the β_3-adrenoceptor. A hybridisation signal was obtained in both BAT and soleus muscle.

(Molenaar *et al.*, 1993). Sections labelled with the antisense probe showed localisation of β_3-mRNA to the mucosa and to a lesser extent to the longitudinal and circular smooth muscle. Sections incubated with the sense probe showed no localisation of label (Fig 7).

DISCUSSION

There are now several lines of evidence which support the presence of β_3-adrenoceptors in rat soleus muscle. Early functional studies indicated a metabolic response to BRL 28410 in rat hindlimb muscles (Challiss *et al.*, 1988) and this has recently been supported by studies indicating that the β_3-adrenoceptor agonist BRL 35135A(37344 ester) increases glucose uptake by an insulin independent mechanism in rat oxidative (red) muscle including soleus muscle (Abe *et al.*, 1993). Binding studies from several laboratories demonstrate that (-)-[^{125}I]-CYP identifies two sites in soleus muscle homogenates. The high affinity site (pK_D 10.52) has the properties of a β_2-adrenoceptor whereas the low affinity site (pK_D 9.28) resembles the β_3-adrenoceptor in terms of its affinity for (-) propranolol, the relative affinities of antagonists, and the low stereoselectivity indices (Roberts *et al.*, 1993; Sillence *et al.*, 1993). Comparison of the binding sites in soleus muscle with those obtained in functional studies in gastrointestinal tract shows close similarities between the receptors involved. Also there was no down-regulation of atypical β-adrenoceptor binding sites with chronic (14 day) (-)-isoprenaline infusion. Soleus muscle therefore has similar properties

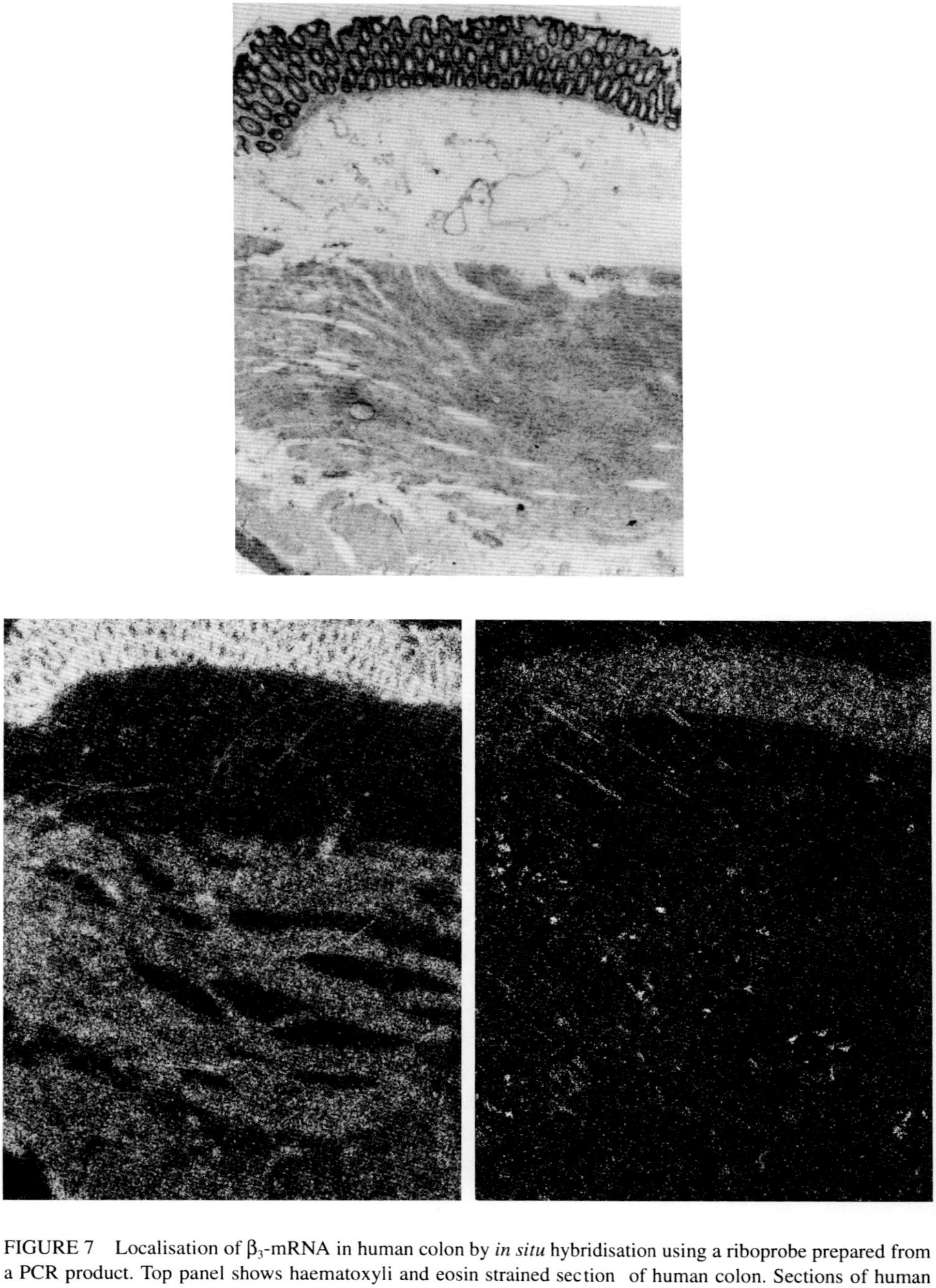

FIGURE 7 Localisation of β_3-mRNA in human colon by *in situ* hybridisation using a riboprobe prepared from a PCR product. Top panel shows haematoxyli and eosin strained section of human colon. Sections of human colon labelled with the antisense probe showed signal in the mucosa and to a lesser extent in longitudinal and circular smooth muscle (bottom left panel). Sections incubated with labelled sense probe showed no localisation of label (bottom right panel).

to cloned β_3-adrenoceptors in expression systems (Nantel *et al.*, 1993). Autoradiographic studies using conditions designed to favour binding of (-)-[^{125}I]-CYP to atypical β-adrenoceptors showed localisation to soleus and regions of plantaris and gastrocnemius muscle. These areas are populated predominantly by oxidative (red) muscle fibres as determined by staining for succinate dehydrogenase (Kim *et al.*, 1991). The presence of β_3-adrenoceptors in soleus muscle was further supported by the detection of β_3-mRNA by RT/PCR. Early studies demonstrating rat β_3-mRNA in Northern blots used a rather non-selective cDNA probe made from a human placental library. This probe also bound to ribosomal RNA and indicated similar amounts of β_3-mRNA in adipose tissue and soleus muscle. Our results would not support this conclusion and used RT/PCR to indicate that while β_3-mRNA was found in soleus the amounts were much less than in adipose tissue. The results are therefore more in accord with later studies which used less sensitive RPA and Northern analysis and were unable to detect β_3-mRNA in skeletal muscle RNA (type unspecified) (Granneman *et al.*, 1991).

In human colon, binding studies reported here show a site with the properties of an atypical β-adrenoceptor resembling that in rat soleus in all but two properties. The human colon binding site showed greater selectivity indices and the β_1-adrenoceptor selective antagonist CGP 20712A competed with much higher affinity than in soleus. The relatively high affinity of CGP 20712A is of interest given that relaxant responses in human colon were blocked by another β_1-adrenoceptor selective antagonist betaxolol (McLaughlin *et al.*, 1989). This may also reflect differences seen in other situations and in particular the relative efficacy of BRL 37344 and CGP 12177 at human and murine receptors (Liggett, 1992; Arch & Kaumann, 1993). There are also differences between the β_3-adrenoceptors in soleus and gastrointestinal tract. In the binding studies reported here the atypical β-adrenoceptor binding was resistant to down regulation, yet it has been repeatedly shown that the gut receptor is desensitised by exposure to β_3-adrenoceptor agonists (see Arch & Kaumann, 1993). These differences remain to be resolved. In human colon it would appear that there is disagreement between molecular and functional approaches in that there has been consistent detection of β_3-mRNA in this tissue (Granneman *et al.*, 1993) yet the relaxation responses do not convincingly follow the pattern seen in other mammalian gastrointestinal tissues. The autoradiographic and *in situ* hybridisation studies reported here go some way to resolving the differences between the functional and molecular approaches. The autoradiographic and *in situ* hybridisation studies showed β_3-adrenoceptors or β_3-mRNA in the mucosa which would explain why β_3-mRNA can be detected in human colon RNA yet relaxation is not obviously mediated by β_3-adrenoceptors. The β_3-adrenoceptors in the mucosa may have a secretory role and atypical β-adrenoceptors have been shown to mediate bicarbonate secretion in rat caecum (Canfield & Abdul-Ghaffar, 1992). In addition it has been recently shown that n-butyrate has a marked down-regulatory effect on β_3-adrenoceptors (Krief *et al.*, 1994) and this short chain fatty acid is a major product of anaerobic fermentation of polysaccharides in caecum and colon. Clearly, further studies of the secretory roles of β_3-adrenoceptors and their regulation by dietary factors is warranted.

In conclusion, there is evidence from binding, functional, autoradiographic and molecular studies for β_3-adrenoceptors in rat soleus and human colon. The binding studies demonstrate sites in both rat soleus and human colon which have many of the characteristics of atypical β-adrenoceptors. The location of the ligand binding in rat soleus and other oxidative muscle fibres is in accord with the metabolic roles now being suggested for β_3-adrenoceptors in this tissue. Localisation of ligand binding and of a β_3-riboprobe to the mucosal surface of the human colon suggests that β_3-adrenoceptors may have secretory or absorptive roles in this tissue.

ACKNOWLEDGMENTS

This work was supported by the National Health and Medical Research Council, National Heart Foundation and SmithKline Beecham Australia Pty Ltd under the factor'f'program. The isomers of tertatolol were kindly supplied by Dr Michel Dubar of Servier Laboratories and BRL 35135A by Dr Mike Cawthorne of SKB. We would like to thank Prof. P. Bhathal and colleagues at the Royal Melbourne Hospital for helpful discussions and supply of human colon. S.J.R. is in receipt of an Australian postgraduate research scholarship and V.B. a National Heart Foundation postgraduate scholarship. P.M. is a NH&MRC Research Fellow.

References

Abe, H., Minokoshi, Y. and Shimazu, T. (1993) Effect of a β_3-adrenergic agonist, BRL 35135A, on glucose uptake in rat skeletal muscle *in vivo* and *in vitro*. *Journal of Endocrinology*, **139,** 479–486.

Arch, J.R. and Kaumann, A.J. (1993) β_3 and atypical β-adrenoceptors. *Medicinal Research Reviews*, **13,** 663–729.

Arch, J.R.S., Cawthorne, M.A., Coney, K.A., Gusterson, B.A., Piercy, V., Sennitt, M.V., Smith, S.A., Wallace, J. and Wilson, S. (1991) β-adrenoceptor-mediated control of thermogenesis, body composition and glucose homeostasis. In N.J. Rothwell and M.J. Stock (Eds.), *Obesity and cachexia* Chichester: John Wiley and sons, pp. 241–268.

Bond, R.A. and Vanhoutte, P.M. (1992) Interaction of tertatolol at the "atypical" or β_3-adrenoceptor in guinea-pig ileum. *General Pharmacology* **23(2),** 171–6.

Bouvier, M., Collins, S., O'Dowd, B.F., Campbell, P.T., deBlasi, A., Kobilka, B.K., MacGregor, C., Irons, G.P., Caron, M.G. and Lefkowitz, R.J. (1989) Two distinct pathways for cAMP mediated down-regulation of the β_2-adrenergic receptor. *Journal of Biological Chemistry*, **264,** 16786–16792.

Canfield, P. and Abdul, G.T. (1992) Stimulation of bicarbonate secretion by atypical β-receptor agonists in rat cecum in vitro. *European Journal of Pharmacology*, **216(2),** 293–7.

Canfield, P. and Paraskeva, P. (1992) β-adrenoceptor agonist stimulation of acid secretion by rat stomach in vitro is mediated by 'atypical' β-adrenoceptors. *British Journal of Pharmacology*, **106(3),** 583–6.

Carpene, C., Galitzky, J., Collon, P., Esclapez, F., Dauzats, M. and Lafontan, M. (1993) Desensitization of β-1 and β-2, but not β-3, adrenoceptor-mediated lipolytic responses of adipocytes after long-term norepinephrine infusion. *Journal of Pharmacology and Experimental Therapeutics*, **265(1),** 237–47.

Challiss, R.A., Leighton, B., Wilson, S., Thurlby, P.L. and Arch, J.R. (1988) An investigation of the β-adrenoceptor that mediates metabolic responses to the novel agonist BRL28410 in rat soleus muscle. *Biochemical Pharmacology*, **37(5),** 947–50.

Chomczynski, P. and Sacchi, N. (1987) A single step method of RNA isolation by acid guanidinium thiocyanate-phenol-chloroform extraction. *Analytical Biochemistry*, **162,** 156–159.

Emorine, L.J., Feve, B., Pairault, J., Briend, S.M., Marullo, S., Delavier, K.C. and Strosberg, D.A. (1991) Structural basis for functional diversity of β_1-, β_2- and β_3-adrenergic receptors. *Biochemical Pharmacology*, **41,** (6–7), 853–9.

Emorine, L.J., Marullo, S., Briend, S.M., Patey, G., Tate, K., Delavier, K.C. and Strosberg, A.D. (1989) Molecular characterization of the human β_3-adrenergic receptor. *Science*, **245**(4922), 1118–21.

Esbenshade, T.A., Han, C., Theroux, T.L., Granneman, J.G. and Minneman, K.P. (1992) Coexisting β_1- and atypical β-adrenergic receptors cause redundant increases in cyclic AMP in human neuroblastoma cells. *Molecular Pharmacology*, **42**(5), 753–9.

Feve, B., Emorine, L.J., Lasnier, F., Blin, N., Baude, B., Nahmias, C., Strosberg, A.D. and Pairault, J. (1991) Atypical β-adrenergic receptor in 3T3-F442A adipocytes. Pharmacological and molecular relationship with the human β 3-adrenergic receptor. *Journal of Biological Chemistry*, **266**(30), 20329–36.

Granneman, J.G. and Lahners, K.N. (1992) Differential adrenergic regulation of β_1- and β_3-adrenoreceptor messenger ribonucleic acids in adipose tissues. *Endocrinology*, **130**(1), 109–14.

Granneman, J.G., Lahners, K.N. and Chaudhry, A. (1991) Molecular cloning and expression of the rat β_3-adrenergic receptor. *Molecular Pharmacology*, **40**(6), 895–9.

Granneman, J.G., Lahners, K.N. and Chaudhry, A. (1993) Characterization of the human β_3-adrenergic receptor gene. *Molecular Pharmacology*, **44**(2), 264–70.

Granneman, J.G., Lahners, K.N. and Rao, D.D. (1992) Rodent and human β_3-adrenergic receptor genes contain an intron within the protein-coding block. *Molecular Pharmacology*, **42**(6), 964–70.

Kim, Y. S., Sainz, R.D., Molenaar, P. and Summers, R.J. (1991) Charcterisation of β_1 and β_2-adrenoceptors in rat skeletal muscles. *Biochemical Pharmacology*, **42**(9), 1783–1789.

Krief, S., Feve, B., Baude, B., Zilberfarb, V., Srosberg, A.D., Pairault, J. and Emorine, L.J. (1994) Transcriptional modulation by n-butyric acid of β_1, β_2 and β_3-adrenergic receptor balance in 3T3-F442A adipocytes. *Journal of Biological Chemistry*, **269**(9), 6664–6670.

Krief, S., Lonnqvist, F., Raimbault, S., Baude, B., Van, S. A., Arner, P., Strosberg, A.D., Ricquier, D. and Emorine, L.J. (1993) Tissue distribution of β_3-adrenergic receptor mRNA in man. *Journal of Clinical Investigation*, **91**(1), 344–9.

Landi, M., Bianchetti, A., Croci, T. and Manara, L. (1992) Phenylethanolaminotetralines compete with [^{3}H]dihydroalprenolol binding to rat colon membranes without evidencing atypical β-adrenergic sites. *Biochemical Pharmacology*, **44**(4), 665–72.

Liggett, S.B., Freedman, N.J., Schwinn, D.A. and Lefkowitz, R.J. (1993) Structural basis for receptor subtype-specific regulation revealed by a chimeric β_3/β_2-adrenergic receptor. *Proceedings of the National Academy of Sciences U S A*, **90**(8), 3665–9.

Lowry, O.H., Roseborough, N.J., Farr, A.L. and Randall, R.J. (1951) Protein measurement with the folin-phenol reagent. *Journal of Biological Chemistry*, **193,** 265–275.

McLaughlin, D.P., Fulton, J.D., MacDonald, A., MacDonald, E. and Scott, P.W. (1988) Effects of catecholamines on human colonic circular smooth muscle. *British Journal of Pharmacology*, **95,** 542P.

McLaughlin, D.P., Fulton, J.D., MacDonald, E., Scott, P.W. and MacDonald, A. (1991) β-adrenoceptors mediating relaxation in human colon *in vitro*. *British Journal of Pharmacology*, **104,** 152P.

McPherson, G.A. (1983) A practical computer-based approach to the analysis of radioligand binding experiments. *Computer Programs in Biomedicine*, **17,** 107–114.

Molenaar, P., Canale, E. and Summers, R.J. (1987) Autoradiographic localisation of β-1 and β-2 adrenoceptors in guinea-pig atrium and regions of the conducting system. *Journal of Pharmacology and Experimental Therapeutics*, **241,** 1048–1064.

Molenaar, P., O'Reilly, G., Sharkey, A., Kuc, R.E., Harding, D.P., Plumpton, C., Gresham, G.A. and Davenport, A.P. (1993) Characterisation and localisation of endothelin receptor subtypes in the human atrioventricular conducting system and myocardium. *Circulation Research*, **72,** 526–538.

Molenaar, P., Roberts, S.J., Kim, Y.S., Pak, H.S., Sainz, R.D. and Summers, R.J. (1991) Localization and characterization of two propranolol resistant (-) [^{125}I]cyanopindolol binding sites in rat skeletal muscle. *European Journal of Pharmacology*, **209**(3), 257–62.

Molenaar, P., Smolich, J.J., Russell, F.D., McMartin, L.R. and Summers, R.J. (1990) Differential regulation of β_1 and β_2 adrenoceptors in guinea-pig atrioventricular conducting system after chronic (-) isoprenaline infusion. *Journal of Pharmacology and Experimental Therapeutics*, **255,** 393–400.

Munson, P.J. and Rodbard, D. (1980) A versatile computerised approach for the characterisation of ligand binding systems. *Analytical Biochemistry*, **107,** 220–239.

Muzzin, P., Revelli, J.P., Fraser, C.M. and Giacobino, J.P. (1992) Radioligand binding studies of the atypical β_3-adrenergic receptor in rat brown adipose tissue using [^{3}H]CGP 12177. *Febs Letters*, **298**(2–3), 162–4.

Muzzin, P., Revelli, J.P., Kuhne, F., Gocayne, J.D., McCombie, W.R., Venter, J.C., Giacobino, J.P. and Fraser, C.M. (1991) An adipose tissue-specific β-adrenergic receptor. Molecular cloning and down-regulation in obesity. *Journal of Biological Chemistry*, **266**(35), 24053–8.

Nahmias, C., Blin, N., Elalouf, J.M., Mattei, M.G., Strosberg, A.D. and Emorine, L.J. (1991) Molecular characterization of the mouse β_3-adrenergic receptor: relationship with the atypical receptor of adipocytes. *Embo. Journal*, **10**(12), 3721–7.

Nantel, F., Bonin, H., Emorine, L.J., Zilberfarb, V., Strosberg, A.D., Bouvier, M. and Marullo, S. (1993) The human β_3-adrenergic receptor is resistant to short term agonist-promoted desensitization. *Molecular Pharmacology*, **43**(4), 548–55.

Peres-Polon, V.L. and Correa, F.M.A. (1984) Central mechanisms of the isoprenaline-induced hypotension in anaesthetised and concious rats. *General Pharmacology*, **15,** 505–509.

Roberts, S.J., Molenaar, P. and Summers, R.J. (1993) Characterization of propranolol-resistant (-)-[^{125}I]- cyanopindolol binding sites in rat soleus muscle. *British Journal of Pharmacology*, **109**(2), 344–52.

Russell, F.D., Kompa, A.R., Molenaar, P. and Summers, R.J. (1994) Regulation of β-adrenoceptors in the guinea-pig sino-atrial node. *Naunyn-Schmiedeberg's Archives of Pharmacology*, **349,** 463–472.

Sillence, M.N., Moore, N.G., Pegg, G.G. and Lindsay, D.B. (1993) Ligand binding properties of putative β_3-adrenoceptors compared in brown adipose tissue and in skeletal muscle membranes. *British Journal of Pharmacology*, **109**(4), 1157–63.

Tate, K.M., Briend, S.M., Emorine, L.J., Delavier, K.C., Marullo, S. and Strosberg, A.D. (1991) Expression of three human β-adrenergic-receptor subtypes in transfected Chinese hamster ovary cells. *European Journal of Biochemistry*, **196**(2), 357–61.

Thomas, R.F. and Liggett, S.B. (1993) Lack of β_3-adrenergic receptor mRNA expression in adipose and other metabolic tissues in the adult human. *Molecular Pharmacology*, **43**(3), 343–8.

van der Vliet. A., Rademaker, B. and Bast, A. (1990) A β adrenoceptor with atypical characteristics is involved in the relaxation of the rat small intestine. *Journal of Pharmacology and Experimental Therapeutics*, **255**(1), 218–26.

Wilson, C., Pettinger, S.J., Hayward, J., Birtles, S. and Holloway, B.R. (1992) Cardiovascular effects of the thermogenic β-adrenoceptor agonist ICI D7114 in concious telemetered dogs. *British Journal of Pharmacology*, **105,** 234P.

ARYLOXYPROPANOLAMINOTETRALINS ARE THE FIRST SELECTIVE ANTAGONISTS FOR ATYPICAL (β_3) β-ADRENOCEPTORS

LUCIANO MANARA[1]*, DOMENICO BADONE[1], MARCO BARONI[1], GIOVANNI BOCCARDI[1], ROBERTO CECCHI[1], TIZIANO CROCI[1], ANTONINA GIUDICE[1], UMBERTO GUZZI[1] and GÉRARD LE FUR[2]

[1]*Research Centre SANOFI MIDY S.p.A., Via G.B. Piranesi 38, 20137 Milan, Italy*
[2]*SANOFI Recherche, 32/34 Rue Marbeuf, 75008 Paris, France*

SR 58894A, 3-(2-allylphenoxy)-1-[(1S)-1,2,3,4-tetrahydronaphth-1-ylamino]-(2S)-2-propanol hydrochloride, and SR 59230A, 3-(2-ethylphenoxy)-1-[(1S)-1,2,3,4-tetrahydronaphth-1-ylamino]-(2S)-2-propanol oxalate are representatives of the novel β_3-subtype selective adrenoceptor antagonists aryloxypropanolaminotetralins (APAT). Comparative assays in isolated tissues each accounting for a given β-adrenergic response, attributed to either recognized (β_1 and β_2, guinea pig atrium and trachea, respectively) or atypical (β_3, rat proximal colon) receptor subtypes, disclosed the high antagonist potency (pA_2 8.1 to 8.8) and substantial selectivity (at least 40 times) of the two SR compounds for the putative β_3-subtype. Only a few of over forty conventional β-blockers, tested similarly *in vitro*, had measurable antagonist potency at the atypical β-sites at lower than micromolar concentrations and all of them were far more potent antagonists of β_1 and/or β_2 responses. SR 59230A had scant potency *in vitro* in competition binding assays with the appropriate radiolabelled ligands, each indicative of non β-adrenergic sites for given neurotransmitters or autacoids; it was orally effective, potent and β_3-selective in an *in vivo* rat model designed to rate the antagonism of responses putatively elicited at the different β-adrenoceptor subtypes. The APAT thus appear to be the first selective antagonists for β_3-adrenoceptors and provide final unambiguous evidence of the distinctive functional features of those abundant in the rat proximal colon.

KEY WORDS: β_3-adrenoceptors, β_3-selective blocking agents, aryloxypropanolaminotetralins, atypical β-adrenoceptors

INTRODUCTION

Beta-adrenoceptors not conforming to the established β_1 or β_2 subtypes have been referred to as atypical or β_3, following their recognition on the grounds of: a) agonist responses poorly susceptible to conventional β-blockers; b) newly synthesized selective agonists; c) structure analysis by molecular biology techniques (for a review, see Arch and Kaumann, 1993). However no potent and selective antagonist has yet provided final, unambiguous evidence of the distinctive functional features of the newer subtype accounting for further β-adrenoceptor multiplicity.

We have previously prepared phenylethanolaminotetralin (PEAT) agonists that are highly potent and selective for the atypical β-adrenoceptors abundant in rat proximal colon and

*Correspondence to: L. Manara, M.D., Research Centre SANOFI MIDY S.p.A., Via G.B. Piranesi 38, 20137 Milan, Italy.

white adipocytes (Cecchi *et al.*, 1994); we characterized these receptors in functional assays with the conventional antagonists alprenolol and propranolol (Bianchetti and Manara, 1990). The two antagonists had higher, and equal, affinity at β_1- and β_2-adrenoceptors, yet their affinity at the atypical sites was not only lower, as expected, but also different from each other. This strongly suggested that the rat atypical sites are indeed distinct functional entities (Bianchetti and Manara, 1990). In addition, we pinpointed the remarkable pharmacological homology (highly correlated rank order of potencies of agonists and identical pA_2 of alprenolol) of rat colon and adipocyte atypical β-adrenoceptors (Landi *et al.*, 1993), in agreement with molecular biology evidence that the β_3-adrenoceptor cloned from rat colonic tissue has the same sequence as rat adipocyte cDNA (Bensaid *et al.*, 1993).

Clearly β-adrenoceptor antagonists with inverse subtype selectivity, i.e. with greater affinity at the putative β_3-site than for either the β_1 or the β_2, would serve best for their unquestionable functional identification. We therefore set out to obtain such antagonists, and give the first account of our accomplishment in the present report.

MATERIALS AND METHODS

Functional assays in isolated tissues

The first 3cm segment of the rat (male Crl:CD BR) proximal colon, starting from the ileocecal junction, was mounted longitudinally in a 20ml organ bath containing Krebs-Ringer solution maintained at 37°C and aerated with 5% CO_2 in O_2. Spontaneous motility was recorded isotonically under a constant load of 1 g in the presence of 10 μM phentolamine, 0.5 μM desmethylimipramine and 30 μM hydrocortisone. Quantitative computer analysis was based on a motility index (expressed as the percentage difference) established by comparing the area under the spontaneous contraction waves (computed as a time integral of the region between the waves and the baseline tonus, sampled every 15 sec) over 10-min periods, before and after addition of cumulative concentrations of the agonist.

Spontaneously beating right atria from guinea-pigs (male DH IFFA-CREDO) were set up in Krebs solution maintained at 30°C, aerated with 5% CO_2 in O_2 and preincubated (30 min) with 12 μM phenoxybenzamine before the test. Antagonists were added 30 min before cumulative concentrations of isoprenaline (contact time for each concentration 2.5 min). Chronotropic responses were expressed as a percentage of the maximal agonist effect.

Tracheal chains from guinea-pigs (male DH IFFA-CREDO) mounted in pairs from the same animal were set up in Krebs solution and preincubated (30 min) with 12 μM phenoxybenzamine before the test; cumulative concentration-response curves for salbutamol with or without one concentration of antagonist (contact time 30 min) were compared directly in the two preparations. Salbutamol-induced relaxation (contact time between doses was 5 min or longer until the effect reached a steady state) was recorded isotonically under a constant load of 0.5 g.

Radioligand binding assays

Membranes were prepared from cerebral tissue stored at –80°C and binding assays were done according to Mennini *et al.*, (1993).

Data analysis

The pA_2, as defined by Arunlakshana and Schild (1959), was obtained from the concentrations of the agonist producing 50% maximal response calculated from concentration-response curves in the absence or presence of the antagonist by plotting the log (agonist dr-1) against the negative log of the antagonist concentration.

Computer analysis was done as described by Tallarida and Murray (1987). Radioligand binding inhibition curves of SR 59230A were analyzed by the "All-fit" program (De Lean *et al.*, 1978).

Chemicals

SR 58894A and SR 59230A (EP-B-375560; US 5,254,595) whose structures appear in Fig. 1, and (+)alprenolol, were synthesized in the Chemistry Section of the SANOFI MIDY S.p.A. Research Centre. Radiolabelled ligands for competition binding assays were from commercial sources. All the conventional β-adrenoceptor blocking agents tested were kindly donated as indicated in "Acknowledgements".

RESULTS

The representative APAT SR 58894A and SR 59230A were compared *in vitro* with alprenolol in isolated tissue preparations each accounting for a given β-adrenergic response attributed to either established (β_1 and β_2, guinea-pig atrium and trachea respectively), or atypical (β_3, rat proximal colon) β-adrenoceptors (Bianchetti and Manara, 1990). The lowest pA_2 for alprenolol (Table 1) was scored in the rat proximal colon, in line with the generally accepted notion that atypical β-adrenoceptors are moderately susceptible to conventional β-blocking agents (Arch and Kaumann, 1993). The opposite was the case for both SR 58894A and SR 59230A, whose rat colon pA_2 values were substantially higher than that of alprenolol in the same tissue, or their own in guinea-pig atrium and trachea (Table 1).

Beside alprenolol, we likewise tested over 40 conventional β-adrenoceptor blocking agents in the rat colon, guinea-pig atrium and trachea. Only a few of them had measurable antagonist potency at the putative β_3-subtype receptors at lower than micromolar concentrations and they were all far more potent antagonists of β_1- and/or β_2-responses (Table 2).

Finally, SR 59230A had generally scant potency *in vitro* in competition binding assays with several radiolabelled ligands, each indicative of non β-adrenergic sites for given neurotransmitters or autacoids (Table 3).

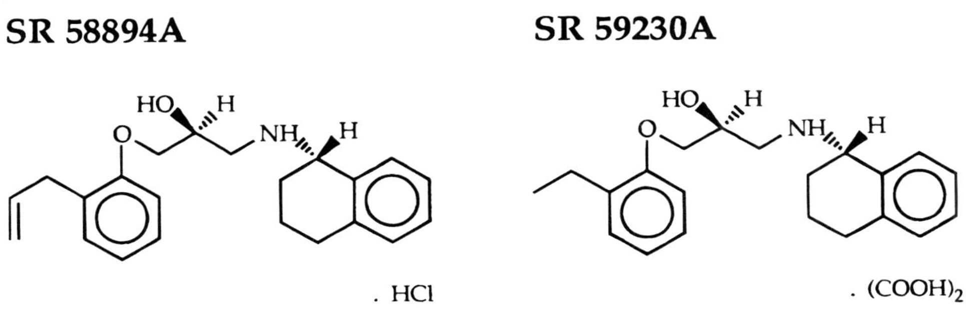

FIGURE 1

TABLE 1

In vitro quantitative antagonism of β-adrenergic responses accounting for atypical (β_3) or previously recognized β-adrenoceptor subtypes.

	β_3 Rat proximal colon		β_1 Guinea-pig atrium		β_2 Guinea-pig trachea	
	pA_2	slope	pA_2	slope	pA_2	slope
Alprenolol	6.82 ± 0.15	1.14 ± 0.14	8.50 ± 0.11	1.10 ± 0.12	8.90 ± 0.10	1.1 ± 0.10
SR 58894A	8.06 ± 0.40	1.06 ± 0.40	6.88 ± 0.09	n.c.[a]	~ 6.4	n.c.[a]
SR 59230A	8.76 ± 0.17	0.93 ± 0.09	7.09 ± 0.06	n.c.[a]	6.61 ± 0.14	n.c.[a]

The pA_2 (± S.E.) was obtained as indicated in "Materials and Methods" from concentration-response curves of the appropriate agonist for each tissue: colon, selective β_3-agonist SR 58611A (Bianchetti and Manara, 1990); guinea-pig atrium, isoprenaline; trachea, salbutamol. The following concentrations of antagonists (μM, added 30 min before the agonist) were used, each giving one cumulative agonist concentration response curve: alprenolol, 0.3, 1, 3, 10; SR 58894A, 0.03, 0.1, 0.3; SR 59230A, 0.003, 0.01, 0.03, 0.1.
n.c.[a]: non competitive antagonism, slope ≠ 1 ($p < 0.05$).

TABLE 2

Antagonism by SR 59230A and conventional β-blockers of *in vitro* responses accounting for atypical (β_3) or previously recognized β-adrenoceptor subtypes

	β_3 Rat proximal colon		β_1 Guinea-pig atrium IC_{50}, nM		β_2 Guinea-pig trachea	
SR 59230A	40	(33–48)	408	(287–528)	648	(535–786)
Penbutolol	55	(40–76)	15	(11–22)	1	
Carazolol	150	(122–184)	1.5	(0.9–2.1)	0.34	(0.2–0.6)
Carvedilol	224	(146–343)	~20		~20	
Bupranolol	385	(262–568)	18	(9–27)	4	(3–5)
Adimolol	574	(348–945)	447	(324–580)	29	(23–36)
Alprenolol	805	(628–1,030)	19	(10–28)	3.9	(2.5–5.3)
Pindolol	1,140	(790–1,650)	8.2	(6.2–10.1)	2.3	(1.7–2.9)
Propranolol	3,080	(2,470–3,840)	15	(10–21)	5.4	(3.6–7.2)

The concentrations inhibiting the agonist response by 50% (IC_{50}, in parentheses 95% confidence limits) in guinea-pig atria and trachea were determined as described in "Materials and Methods", from concentration-response curves using cumulative concentrations of agonist. In rat colon, cumulative concentrations of the antagonist (one every 10 min) were added to the bath 10 min after the selective atypical β-adrenoceptor agonist SR 58375A (Bianchetti and Manara, 1990) at a concentration (0.3 μM) causing 80 to 90% reduction of spontaneous motility. The IC_{50} of the following compounds in rat proximal colon was ≥ 1 μM (maximal concentration tested): (+)alprenolol, amosulalol, atenolol, BAAM, betaxolol, bevantolol, bisoprolol, bornaprolol, bucindolol, bufuralol, bunolol, celiprolol, cetamolol, CGP 20712A, CGP 12177A, cicloprolol, cloranolol, dilevalol, epanolol, HI 42BS, ICI 118,551, indenolol, KOE 3290, medroxalol, mepindolol, metipranolol, mindolilol, moprolol, nebivolol, nipradilol, oxprenolol, pacrinolol, pargolol, spirendolol, sulfinalol, talinolol; MK761 IC_{50} ~ 1 μM.

DISCUSSION

Despite the structural diversity of atypical or putative β_3-adrenoceptors, indicated by molecular biology sequence analysis, and the imposing pharmacological evidence in their favour (Arch and Kaumann, 1993), their functional identity has remained elusive because of the lack of selective antagonists. We have obtained confirmation of the current belief that conventional β-blocking agents antagonize atypical β-adrenoceptor responses with substantially less potency than those elicited at either the β_1 or β_2 sites. Of the over 40 compounds we tested, most were hardly effective as β_3-adrenoceptor antagonists (rat colon) at lower

TABLE 3
Competition binding of SR 59230A at several non-β-adrenergic sites in different assay conditions

Site	*IC_{50}, µM*	*[^{3}H]-ligand*	*Cerebral tissue*[a]
Seroronin			
5-HT_{1A}	0.22	8-OH-DPAT	Hippocampus
5-HT_{1B}	0.53	5-HT	Striatum
5-HT_{1C}	> 10	Mesulergine	Porcine frontal cortex
5-HT_{1D}	> 10	5-HT	Porcine striatum
5-HT_2	5.5	Ketanserin	Cortex
5-HT_3	> 10	BRL 43694	Cortex
5-HT uptake	0.26	Citalopram	Cortex
Dopamine			
D_1	> 10	SCH 23390	Striatum
D_2	> 10	Spiperone	Striatum
Adrenergic			
α_1	0.7	Prazosin	Cortex
α_2	> 10	p-Aminoclonidine	Cortex
Histamine			
H_1	> 10	Mepyramine	Cortex

[a]from rats, unless otherwise indicated

than micromolar concentrations and the few exceptions were far more potent at β_1- or β_2-adrenoceptors.

Our approach to designing new antagonists set out from our PEAT agonists which had higher selectivity for putative β_3-adrenoceptors in the rat colon than the "lipolytic" β_3-agonists introduced formerly; also unlike these, PEAT have slight, if any, stimulating activity *in vitro* in rodent (Bianchetti and Manara, 1990; Manara *et al.*, 1990) and human (Martin *et al.*, 1994) β_2-responsive preparations and do not produce tremor or tachycardia in man, even at the highest dose tested (Landi *et al.*, 1993). Thus the aminotetralin moiety of PEAT conferred similar subtype recognition properties to the aryloxypropanolamines whose β-adrenergic blocking action is well known (Ruffolo, 1991); the two representative APAT compounds SR 58894A and SR 59230A were at least 40 times more potent antagonists of β-adrenergic responses in the rat colon (β_3), than in either guinea-pig atrium or trachea (β_1 and β_2 respectively). Additional tests with SR 59230A in competition binding assays showing its virtually negligible affinity for non-β-adrenergic sites further supported the selectivity of the APAT in a more general context.

These findings indicate that our novel potent β-blocking compounds are definitely selective for β_3-adrenoceptors and provide final unambiguous evidence of the distinctive functional features of those abundant in the rat proximal colon. Our first *in vivo* results, to be presented in detail elsewhere, are encouraging and show that SR 59230A is orally effective and β_3-selective in a rat model suitable for assessing the antagonism of responses putatively elicited at the different β-adrenoceptors subtypes (ED_{50}, mg/kg: 3.6, 48 and 56, respectively for antagonism of SR 58611A colonic motility inhibition, salbutamol bronchodilation and isoprenaline tachycardia).

We are continuing our efforts to ascertain the therapeutic potential of the β_3-selective antagonists, but are confident they already provide a new, essential tool for conclusive functional identification of β-adrenoceptors not conforming to the generally recognized β_1- and β_2-subtypes.

ACKNOWLEDGMENTS

We wish to thank: Dr. Tiziana Mennini, Istituto di Ricerche Farmacologiche "Mario Negri", Milan, Italy, for the binding assays (Table 3); the companies, Boehringer Ingelheim, Germany; Boehringer Ingelheim S.p.A., Italy, Boehringer Mannheim, Germany; Bristol-Myers Squibb, U.S.A., Ciba-Geigy, Switzerland; Gedeon Richter, Hungary; Hoechst, Germany; Hoffmann-La Roche AG, Switzerland; ICI, England; Janssen, Belgium; Kowa, Japan; Marion Merrell Dow, USA; Merck, Germany; Merck Sharp & Dohme, USA; Parke-Davis S.p.A., Italy; Poli, S.p.A., Italy; Rhône-Poulenc Santé, France; Rorer, USA; Sandoz Pharma, Switzerland; Schering, Germany; Schering-Plough, USA; Schwarz Pharma, Germany; Sterling Drug, USA; Synthélabo Recherche, France; Veb Arzneimittel, Germany; Wyeth-Ayerst, USA; Yamanouchi Pharmaceutical Co., Japan; Zambon Group, Italy, for kindly giving their β-adrenoceptor blocking agents appearing in Table 2; A. Aureggi and M. Meleri for skillful technical help.

References

Arch, J.R.S. and Kaumann, A.J. (1993) β_3 and atypical β-adrenoceptors. *Medicinal Research Reviews*, **13**, 663–729.

Arunlakshana, O. and Schild, H.O. (1959) Some quantitative uses of drug antagonists, *British Journal of Pharmacology*, **14**, 48–57.

Bensaid, M., Kaghad M., Rodriguez, M., Le Fur G. and Caput, D. (1993) The rat β_3-adrenergic receptor gene contains an intron. *FEBS Letters*, **318**, 223–226.

Bianchetti, A. and Manara, L. (1990) *In vitro* inhibition of intestinal motility by phenylethanolaminotetralines: evidence of atypical β-adrenoceptors in rat colon. *British Journal of Pharmacology*, **100**, 831–839.

Cecchi, R., Croci, T., Boigegrain R., Boveri, S., Baroni, M., Boccardi, G., Guimbard, J.P. and Guzzi, U. (1994) Synthesis and β-adrenergic activity of atypical β-adrenergic phenylethanolaminotetralin stereoisomers, *European Journal of Medicinal Chemistry*, **29**, 259–267.

De Lean, A., Munson, P.J. and Rodbard, D. (1978) Simultaneous analysis of families of sigmoidal curves: application of bioassay, radioligand assay, and physiological dose-response curves, *American Journal of Physiology*, **235**, E97–E102.

Landi, M., Croci, T. and Manara, L. (1993) Similar atypical β-adrenergic receptors mediate *in vitro* rat adipocyte lipolysis and colonic motility inhibition. *Life Sciences*, **53**, PL 297–302.

Manara L., Guzzi, U., Aureggi G., Croci T. and Bianchetti A. (1990) On the selectivity of in vitro inhibition of rat colon motility by β-adrenergic substituted phenylethanolamines. *European Journal of Pharmacology*, **187**, 563.

Martin C., Naline, E., Bakdach H. and Advenier, C. (1994) β_3-adrenoceptor agonists, BRL 37344 and SR 58611A, do not induce relaxation of human, sheep and guinea-pig airway smooth muscle *in vitro*. *European Respiratory Journal* (in press).

Mennini T., Bernasconi, P. and Fiori, M.G. (1993) Neurotoxicity of nonionic low-osmolar contrast Media—A receptor binding study. *Investigative Radiology*, **28**, 821–827.

Ruffolo, R.R., Jr. (1991) Chirality in α-and β-adrenoceptor agonists and antagonists. *Tetrahedron*, **47**, 9953–9980.

Tallarida, R.J. and Murray, R.B. (1987) *Manual of Pharmacologic Calculations with Computer Programs*, Springer-Verlag, New York, pp. 58–61.

CLINICAL SIGNIFICANCE OF β-ADRENOCEPTORS

OTTO-ERICH BRODDE

Institute of Pharmacology Martin-Luther-University of Halle-Wittenberg, Magdeburger Str. 4, D-06097 Halle, Germany

KEY WORDS: β-adrenoceptors, congestive heart failure, G-proteins, desensitization, receptor down-regulation, receptor internalization

INTRODUCTION

It is now generally accepted that adrenergic receptors rather than being static entities are dynamically regulated by a variety of drugs, hormones, pathological and physiological conditions (Motulsky and Insel, 1982; Stiles *et al.*, 1984; Lefkowitz and Caron 1985; Brodde, 1989). This has initially been shown in animal studies and cultured cell systems; however, with the combined application of biochemical and physiological *in vitro* and clinical pharmacological *in vivo* techniques it is possible to study adrenoceptor regulation also in humans. This article focusses on regulation of β-adrenoceptor (AR) function in man.

1) Desensitization

One important clinically relevant consequence of β-AR regulation is the phenomenon of desensitization, i.e. the fact that after long-term exposure of a cell to a β-AR agonist the subsequent response of the cell to that agonist is blunted; this blunting is often accompanied by a decrease in the number of cell-surface receptors (Harden, 1983; Hertel and Perkins, 1984; Hausdorff *et al.*, 1990). Initially circulating lymphocytes containing a homogeneous population of β_2-ARs (Brodde *et al.*, 1987) have been used as a model to study such an agonist-induced β-AR desensitization in humans, and numerous studies in patients and healthy volunteers have shown, that chronic treatment with β_2-AR agonists down-regulates lymphocyte β_2-AR number (Motulsky and Insel, 1982; Brodde, 1989; Brodde *et al.*, 1988). However, β-AR down-regulation following chronic agonist treatment can be also demonstrated in a solid human tissue: in women undergoing chronic β_2-adrenergic therapy to prevent pre-term labor myometrial β-AR density was markedly lower than in non-treated women (Berg *et al.*, 1985; Michel *et al.*, 1989; Brodde *et al.*, 1994b); on the other hand, the amount of immunodetectable G_i and G_s-protein was not changed (Fig. 1). Chronic agonist treatment leads, however, not only to a down-regulation of β-AR number, but also to desensitization of β-AR *in vivo* function: in healthy volunteers chronically (9–14 days) treated with the β_2-AR agonists procaterol (Brodde *et al.*, 1990) or terbutaline (Brodde *et al.*, 1992) β_2-AR mediated physiological effects (isoprenaline-infusion induced decrease in

Correspondence to: Otto-Erich Brodde, Institute of Pharmacology Martin Luther-University of Halle-Wittenberg, Magdeburger Str. 4, D-06097 Halle, Germany.

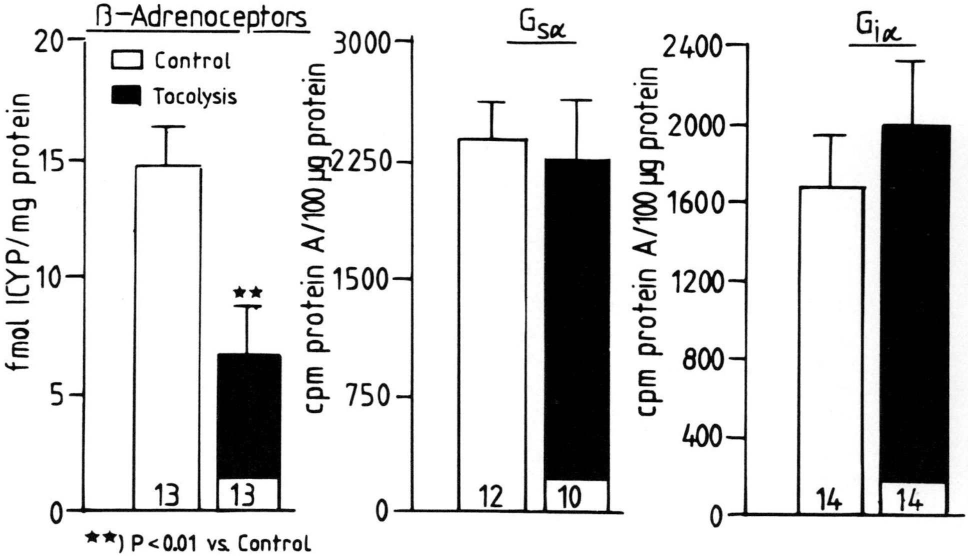

FIGURE 1 Effect of chronic (at least 5 days) β_2-adrenergic tocolytic therapy with i.v.-fenoterol on β-adrenoceptor density and the amount of immunodetectable $G_{s\alpha}$ and $G_{i\alpha}$ in myometrial membranes of pregnant women. Ordinate: β-adrenoceptor density in fmol (–)-[^{125}I]iodocyanopindolol (ICYP) bound/mg protein (left panel); amount of immunodetectable $G_{s\alpha}$ (middle panel) and $G_{i\alpha}$ (right panel) in cpm [^{125}I]protein A/100 µg protein. Means ± S.E.M.; number of experiments at the bottom of the columns. From Brodde *et al.*, 1994b.

diastolic blood pressure [P_{diast}] and increase in plasma noradrenaline) were markedly attenuated while β_1-AR mediated effects (isoprenaline-induced increase in systolic blood pressure [P_{syst}] and exercise-induced increase in heart rate [HR]) were unchanged. Isoprenaline-induced HR-increase that in man is mediated by (cardiac) β_1-and β_2-AR stimulation to about the same degree (McDevitt, 1989; Brodde, 1991), was also attenuated but to a lesser extent than the pure β_2-AR mediated effects. In contrast, after 14 days treatment with the β_1-AR partial agonist xamoterol all β_1-AR mediated effects were depressed while β_2-AR mediated effects were not affected (Brodde *et al.*, 1990). Again, isoprenaline-induced HR-increase was attenuated to a lesser extent than the pure β_1-AR mediated effects. Thus, it appears that in general long-term treatment of humans with β-agonists causes β-AR desensitization, but in β-AR subtype-selective manner.

However, not only exogenously applied but also endogenously chronically enhanced catecholamines can cause desensitization of β-ARs. An important clinical example for a disease state that is characterized by chronic activation of the sympatho-adrenal-system is chronic heart failure (Packer, 1992). Various authors have shown that in patients suffering from chronic heart failure plasma noradrenaline levels are elevated, and this increase seems to be related to the severity of the disease (Cohn, 1990). In fact it has been suggested that plasma noradrenaline levels may serve as a predictor of the prognosis of the patients (Cohn *et al.*, 1984; Rector *et al.*, 1987). Thus, in patients with chronic heart failure cardiac β-ARs are chronically exposed to high concentrations of endogenous noradrenaline; hence it is not surprising that under these circumstances cardiac β-ARs are down-regulated. This has first been demonstrated by Bristow *et al.*, (Bristow *et al.*, 1982) in 1982 and subsequently confirmed by many authors. At present there seems to be general agreement that a common feature of chronic heart failure is a decrease of cardiac β-AR function and the extent of this decrease is directly related to the severity of the disease (Brodde, 1991; Bristow *et al.*, 1990; Brodde, 1994). Moreover, since the human heart contains only a very few, if any, spare receptors for β-AR mediated positive inotropic effects (Kaumann *et al.*, 1989; Schwinger *et al.*, 1990; Brown *et al.*, 1992) the decrease in β-AR number is accompanied by a simultaneous decrease in β-AR-mediated positive inotropic effects. In addition to the reduction in β-AR number and function various authors have shown that in patients with chronic heart failure the amount of the inhibitory G-protein G_i is increased (Feldman, 1991). The increase in G_i leading to inhibition of cAMP formation might explain why in chronic heart not only β-AR mediated positive inotropic effects but also those of all other drugs acting via cAMP (including PDE-inhibitors) are blunted (Fig. 2).

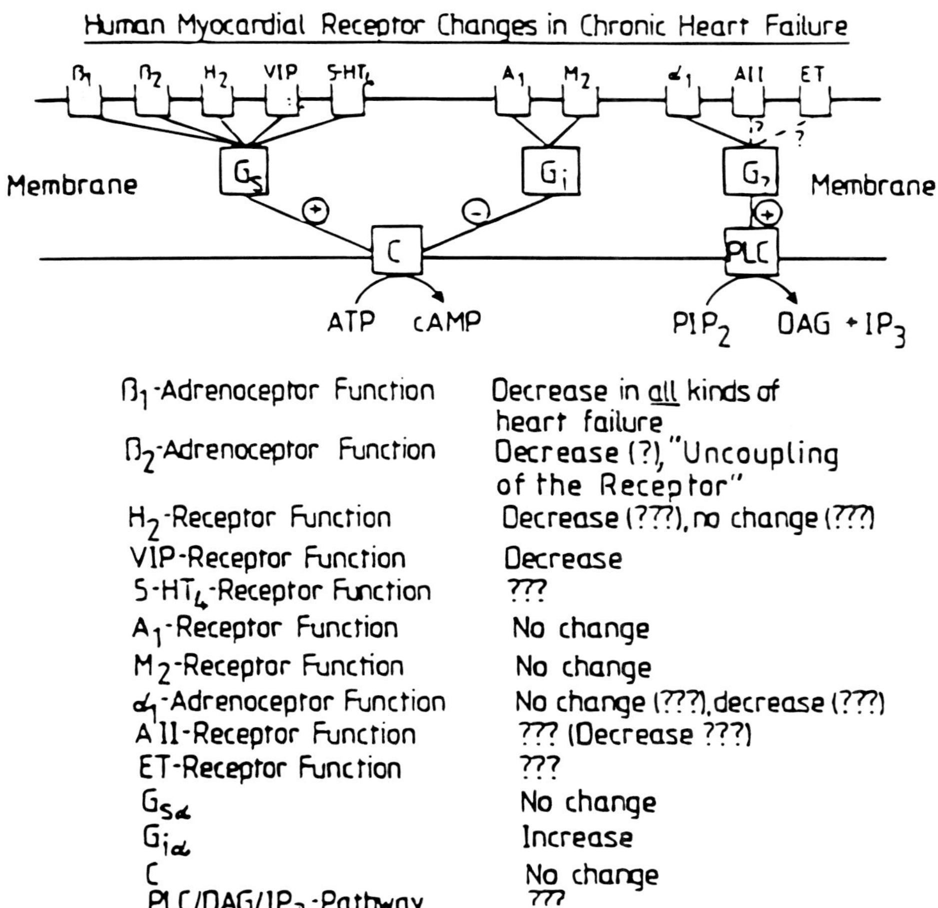

FIGURE 2 Human myocardial receptor changes in chronic heart failure.

Another more physiological setting that is associated with a reduced responsiveness of many hormone receptors is ageing (Roth, 1979). We recently studied whether in ageing also in human heart β-AR function might be impaired (Brodde *et al.*, 1994a). In right atria from patients undergoing coronary artery bypass grafting without apparent heart failure β-AR number did not change with age but activation of adenylate cyclase by various receptor-mediated (isoprenaline, histamine, serotonin) and non-receptor mediated activators (GTP, NaF, forskolin and Mn^{2+}) was markedly decreased indicating that with advanced age the activity of the catalytic unit of the adenylate cyclase is reduced. In addition there was a small but significant increase in cardiac G_i- protein. Preliminary experiments revealed that these biochemical changes are accompanied by changes in the functional responsiveness of β-ARs in ageing: the β_1-AR-mediated isoprenaline-induced P_{syst}-increase was markedly reduced; in addition, isoprenaline-induced HR-increase was in ageing blunted but to a lesser extent than the β_1-AR mediated P_{syst}-increase. On the other hand, the β_2-AR mediated P_{diast}-decrease was only marginally reduced (Table 1). Thus, the ageing human heart shows some similarities with the failing human heart: in both settings β-AR-mediated effects and all other cAMP-dependent effects are depressed and G_i-protein is increased. The mechanism of ageing induced changes in β-AR function, however, must be different from that in chronic heart failure: in heart failure, but not in ageing, β-AR number is reduced; on the other hand, an impairement of the activity of the catalytic unit of the adenylate cyclase seems to be the prominent change in the ageing human heart while in chronic heart failure adenylate cyclase activity is unchanged (cf. Fig. 2).

2) Sensitization

The opposite of desensitization is sensitization, i.e. the fact that after reduced exposure of β-ARs to catecholamines (either by denervation or by treatment with β-AR antagonists) the subsequent response to agonists is enhanced (Trendelenburg, 1966), and this is often accompanied by an increase in β-AR number. Autonomic dysfunction is a disease state that is characterized by very low endogenous catecholamine levels. We have studied in two patients suffering from sympathetic dysfunction (diabetic neuropathy) β-AR function. In these two patients lymphocyte β_2-AR density was increased and isoprenaline-induced HR-increase was exaggerated when compared to age-matched healthy controls (Fig. 3) indicating β-AR up-regulation and denervation supersensitivity.

A similar up-regulation of β-ARs can be observed in patients after chronic treatment with β-AR antagonists. This process occurs—similar to the agonist-induced desensitization (see above)—in a subtype-selective manner: chronic treatment of patients without apparent heart failure undergoing coronary artery bypass grafting with non-selective β-AR antagonists such as propranolol or sotalol caused increases in right atrial β_1- and β_2-ARs (Michel *et al.*, 1988) and in saphenous vein β_2-ARs (Brodde *et al.*, 1989), while chronic treatment with β_1-AR antagonists such as bisoprolol, atenolol or metoprolol increased cardiac β_1-ARs but did not affect cardiac or saphenous vein β_2-ARs (Michel *et al.*, 1988; Brodde *et al.*, 1989). Interestingly, on isolated electrically driven right atria from β_1-AR antagonist treated patients the positive inotropic effect of noradrenaline via β_1-AR stimulation was—even with the increased β-AR number—not changed while that of the β_2-AR agonists salbutamol and procaterol were significantly enhanced (Hall *et al.*, 1990; Motomura *et al.*, 1990); moreover, while in non-β-blocker-treated patients the positive inotropic effect of salbutamol

TABLE 1
Age-dependency of human β_1- and β_2-adrenoceptor function

	EC_{20}-Values (ng/kg/min for 8min each) Young Subjects (19–26 Yrs, n = 9)	Old Subjects (47–58 Yrs, n = 5)
Isoprenaline-Induced Increase in Systolic Blood Pressure (β_1-AR mediated)	17.5 ± 2.2	> 35**)
Isoprenaline-Induced Decrease (β_2-AR-mediated) in Diastolic Blood Pressure	17.0 ± 1.8	21.0 ± 1.9
Isoprenaline-Induced Increase in Heart Rate (β_1- and β_2-AR-mediated)	18.9 ± 1.1	25.9 ± 2.0*)

Means ± S.E.M. n = number of experiments.
**) $p < 0.01$, *) $p < 0.05$ vs. young subjects.
Brodde, unpublished observations.

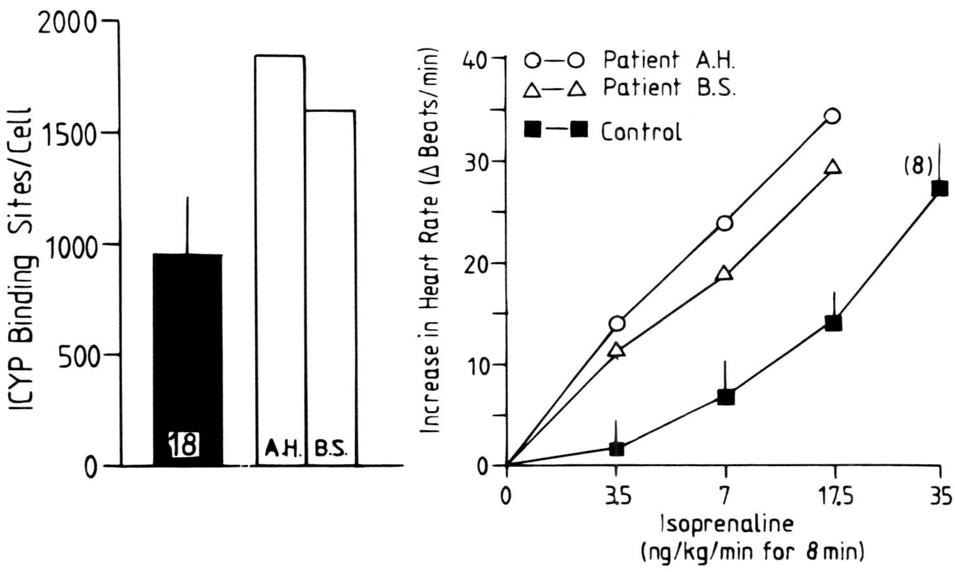

FIGURE 3 Lymphocyte β-adrenoceptors (left) and chronotropic response to isoprenaline (right) in two patients suffering from autonomic dysfunction. Ordinate (left): β_2-Adrenoceptor density—assessed by (–)-[^{125}I]iodocyanopindolol (ICYP) binding in intact lymphocytes—in ICYP binding sites/cell; right: increase in heart rate in **Δ** beats/min. Abscissa (right): dose of isoprenoline in ng/kg/min for 8 min each.

amounted to about 39% and that of procaterol to about 75% of that of isoprenaline, in β-blocker treated patients their positive inotropic effects increased up to 80 and 100%, respectively (Hall *et al.*, 1990; Motomura *et al.*, 1990). Thus, it seems that in human right atrium chronic β-AR blockade causes a selective up-regulation of β_1-ARs and simultaneously a sensitization of β_2-ARs. Whether such a phenomenon occurs also in vivo is at present a matter of controversy: in patients with chronic stable angina pectoris who have been

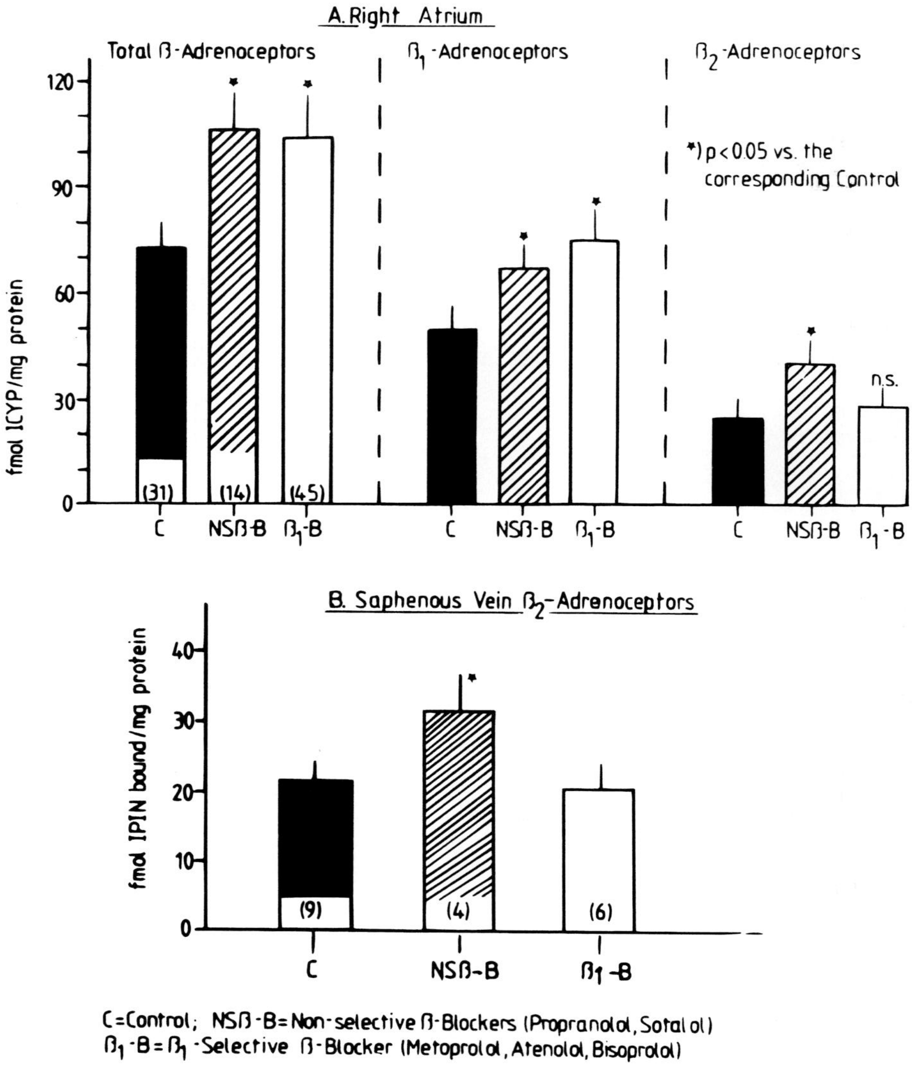

FIGURE 4 Effects of chronic β–AR antagonist treatment on β–andrenoceptor number in right atria (upper panel) and saphenous veins (lower panel) from patients without apparent heart failure undergoing coronary artery bypass grafting. Ordinate, upper panel: right atrial total β–, β_1– and β_2 adrenoceptor number – assessed by (–)–[^{125}I]iodocyanopindolol (ICYP) binding in membranes – in fmol ICYP/mg protein; lower panel: saphenous vein β_2–adrenoceptor number – assessed by (–)–[^{125}I]iodopindolol (IPIN) binding in membranes – in fmol IPIN/mg protein. Means ± S.E.M.; number of experiments in parenthesis.
Modified from Refs. Michel *et al.*, 1988; Brodde *et al.*, 1989; Motomura *et al.*, 1990.

chronically treated with atenolol the positive chronotropic effect induced by intracoronary injection of salbutamol was markedly enhanced (Hall *et al.*, 1991). In healthy volunteers chronically treated with the β_1-selective antagonist bisoprolol, however, published data are controvers: while Hall *et al.* (Hall *et al.*, 1993) found a small sensitizing effect of bisoprolol on the positive chronotropic effect of salbutamol we recently failed to demonstrate any sensitizing effect of chronic bisoprolol treatment on the positive chronotropic or inotropic effect evoked by intravenous terbutaline via β_2-AR stimulation (Schäfers *et al.*, 1994).

Finally, we were interested to study whether the transplanted human heart as a denervated organ (Regitz *et al.*, 1990; Bristow, 1990) might also develop up-regulation of β-ARs. We, therefore, performed serial measurements of β-ARs in right ventricular endomyocardial biopsies of heart transplant recipients in weekly, and later monthly intervals for up to 18 months post-transplantation (Brodde *et al.*, 1991). In these patients β-AR number was significantly higher than in their previously explanted (diseased) hearts, but remained during the whole study period unchanged indicating no development of β-AR up-regulation. Interestingly, however, with increasing time after transplantation β-AR subtypes showed a redistribution towards β_2-ARs: while immediately after transplantation β_1-:β_2-AR ratio was 80:20%, after 18 months it amounted to 65:35%. These results—no β-AR up-regulation, but a shift in the β_1-:β_2-AR ratio towards β_2–ARs-have been subsequently confirmed by Steinfath *et al.* (Steinfath *et al.*, 1992) and Farrukh *et al.* (Farrukh *et al.*, 1993). The mechanism underlying this redistribution is not known; however, it is interesting to note that we recently observed in rats, chronically treated with cyclosporin A, a similar shift in the cardiac β_1-:β_2-AR ratio towards β_2-ARs (Brodde *et al.*). Since all heart transplant recipients are treated with cyclosporin A, it is tempting to speculate that the immunosuppressive agent might be the cause of this cardiac β-AR subtype redistribution.

References

Berg, G., Andersson, R.G.G., Ryden, G. (1985) β-Adrenergic receptors in human myometrium during pregnancy: changes in the number of receptors after β-mimetic treatment. *Am. J. Obstet. Gynecol.*, **151,** 392–396.

Bristow, M.R. (1990) The surgically denervated, transplanted human heart. *Circulation*, **82,** 658–660.

Bristow, M.R., Ginsburg, R., Minobe, W., Cubicciotti, R.S., Sageman, W.S., Lurie, K., Billingham, M.E., Harrison, D.C. and Stinson, E.B. (1982) Decreased catecholamine sensitivity and β-adrenergic-receptor density in failing human hearts. *N. Engl. J. Med.*, **307,** 205–211.

Bristow, M.R., Hershberger, R.E., Port, J.D., Gilbert, E.M., Sandoval, A., Rasmussen, R., Cates, A.E. and Feldman, A.M. (1990) β-Adrenergic pathways in nonfailing and failing human ventricular myocardium. *Circulation*, **82,** I-12–I-25.

Brodde, O.-E. (1989) β-Adrenoceptors. In: Williams M, Glennon RA, Timmermanns P.B.M.W.M., eds. Receptor pharmacology and function. Marcel Dekker, New York, 207–255.

Brodde, O.-E. (1991) β_1- and β_2-Adrenoceptors in the human heart: properties, function, and alterations in chronic heart failure. *Pharmacol. Rev.*, **43,** 203–242.

Brodde, O.-E. (1994) β-Adrenoceptors in cardiac disease. *Pharmacol. Ther.*, **60,** 405–430.

Brodde, O.-E., Adamczyk, M., Busch, F., Bossaller, C. Duske, E., Fleck, E., Götze, S. and Auch-Schwelk, W. Selective down-regulation of rat cardiac β_1-adrenoceptors by cyclosporine A: prevention by diltiazem or ACE-inhibitors. *J. Am. Coll. Cardiol.*, in press.

Brodde, O.-E., Beckeringh, J.J. and Michel, M.C. (1987) Human heart β-adrenoceptors: a fair comparison with lymphocyte β-adrenoceptors? *Trend Pharmacol. Sci.*, **8,** 403–407.

Brodde, O.-E., Daul, A., Michel-Reher, M., Boomsma, F., Man in't Veld, A.J., Schlieper, P. and Michel, M.C. (1990) Agonist—induced desensitization of β-adrenoceptor function in humans. Subtype-selective reduction in β_1- or β_2-adrenoceptor-mediated physiological effects by xamoterol or procaterol. *Circulation*, **81,** 914–921.

Brodde, O.-E., Khamssi, M. and Zerkowski, H.-R. (1991) β-Adrenoceptors in the transplanted human heart: unaltered β-adrenoceptor density, but increased proportion of β_2-adrenoceptors with increasing posttransplant time. *Naunyn-Schmiedeberg's Arch. Pharmacol.*, **344,** 430–436.

Brodde, O.-E., Petrasch, S., Bauch, H.-J., Daul, A., Gnadt, M., Oefler, D. and Michel, M.C. (1992) Terbutaline-induced desensitization of β_2-adrenoceptor in vivo function in humans: attenuation by ketotifen. *J. Cardiovasc. Pharmacol.*, **20,** 434–439.

Brodde, O.-E. and Wang, X.L. (1988) β-Adrenoceptor changes in blood lymphocytes and altered drug responsiveness. *Ann. Clin. Res.*, **20,** 311–323.

Brodde, O.-E., Zerkowski, H.-R., Doetsch, N. and Khamssi, M. (1989) Subtype-selective up-regulation of human saphenous vein β_2-adrenoceptors by chronic β-adrenoceptor antagonist treatment. *Naunyn-Schmiedeberg's Arch. Pharmacol.*, **339,** 479–482.

Brodde, O.-E., Zerkowski, H.-R., Schranz, D., Broede-Sitz, A., Michel-Reher, M. and Schäfer-Beisenbusch E. (1994a) Age-dependent decrease of β-adrenoceptor function in human right atrium. *Br. J. Pharmacol.*, **112(Suppl),** 38 P. (abstract).

Brodde, O.-E., Zieger, W., Broede-Sitz, A., Michel-Reher, M. and Schäfer-Beisenbusch E. (1994b) β_2-Adrenergic tocolytic therapy decreases human myometrial β_2-adrenoceptor number but does not affect G_S- or G_i-proteins. *Canad. J. Physiol. Pharmacol.*, **722(Suppl 1),** 534 (abstract).

Brown, L., Deighton, N.M., Bals, S., Söhlmann, W., Zerkowski, H.-R., Michel, M.C. and Brodde, O.-E. (1992) Spare receptors for β-adrenoceptor-mediated positive inotropic effects of catecholamines in the human heart. *J. Cardiovasc. Pharmacol.*, **19,** 222–232.

Cohn, J.N., Levine, T.B., Olivari, M.T., Garberg, V., Lura, D., Francis, G.S., Simon, A.B. and Rector, T. (1984) Plasma norepinephrine as a guide to prognosis in patients with chronic congestive heart failure. *N. Engl. J. Med.*, **311,** 819–823.

Cohn, J.N. (1990) Abnormalities of peripheral sympathetic nervous system control in congestive heart failure. *Circulation*, **82,** I-59–I-67.

Farrukh, H.M., White, M., Port, J.D., Handwerger, D., Larrabee, P, Klein, J., Roden, R.A., Skerl, L., Renlund, D.G., Feldman, A.M. and Bristow, M.R. (1993) Up-regulation of β_2-adrenergic receptors in previously transplanted, denervated nonfailing human hearts. *J. Am. Coll. Cardiol.*, **22,** 1902–1908.

Feldman, A.M. (1991) Experimental issues in assessment of G protein function in cardiac disease. *Circulation*, **84,** 1852–1861.

Hall, J.A., Kaumann, A.J. and Brown, M.J. (1990) Selective β_1-adrenoceptor blockade enhances positive inotropic responses to endogenous catecholamines mediated through β_2-adrenoceptors in human atrial myocardium. *Circulation Res.*, **66,** 1610–1623.

Hall, J.A., Petch, M.C. and Brown, M.J. (1991) In vivo evidence of cardiac β_2-adrenoceptor sensitization due to atenolol treatment. *Circulation Res.*, **69,** 959–964.

Hall, J.A., Ferro, A., Dickerson, J.E.C. and Brown, M.J. (1993) β-Adrenoceptor subtype cross regulation in the human heart. *Br. Heart J.*, **69,** 332–337.

Harden, T.K. (1983) Agonist-induced desensitization of the β-adrenergic receptor-linked adenylate cyclase. *Pharmacol. Rev.*, **35,** 5–32.

Hausdorff, W.P., Caron, M.G. and Lefkowitz, R.J. (1990) Turning off the signal: desensitization of β-adrenergic receptor function. *FASEB J.*, **4,** 2881–2889.

Hertel, C. and Perkins, J.P. (1984) Receptor-specific mechanisms of desensitization of β-adrenergic receptor function. *Mol. Cell. Endocrinol.*, **37,** 245–256.

Kaumann, A.J., Lemoine, H., Schwederski-Menke, U. and Ehle, B. (1989) Relations between β-adrenoceptor occupancy and increases of contractile force and adenylate cyclase activity induced by catecholamines in human ventricular myocardium. Acute desensitization and comparison with feline ventricle. *Naunyn-Schmiedeberg's Arch. Pharmacol.*, **339,** 99–112.

Lefkowitz, R.J. and Caron, M.G. (1985) Adrenergic receptors: molecular mechanisms of clinically relevant regulation. *Clin. Res.*, **33,** 395–406.

McDevitt, D.G. (1989) In vivo studies on the function of cardiac β-adrenoceptors in man. *Eur. Heart J.*, **10(suppl. B),** 22–28.

Michel, M.C., Pingsmann, A., Beckeringh, J.J., Zerkowski, H.-R., Doetsch, N. and Brodde, O.-E. (1988) Selective regulation of β_1- and β_2-adrenoceptors in the human heart by chronic β-adrenoceptor antagonist treatment. *Br. J. Pharmacol.*, **94,** 685–692.

Michel, M.C. Pingsmann, A. Nohlen, M. Siekmann U, Brodde, O.-E. (1989) Decreased myometrial β-adrenoceptors in women receiving β_2-adrenergic tocolytic therapy: Correlation with lymphocyte β-adrenoceptors. *Clin. Pharmacol. Ther.*, **45,** 1–8.

Motomura, S., Deighton, N.M., Zerkowski, H.-R., Doetsch, N., Michel, M.C. and Brodde, O.-E. (1990) Chronic β_1-adrenoceptor antagonist treatment sensitizes β_2-adrenoceptors but desensitizes M-2 muscarinic receptors in the human right atrium. *Br. J. Pharmacol.*, **101,** 362–369.

Motulsky, H.J. and Insel, P.A. (1982) Adrenergic receptors in man: direct identification, physiologic regulation and clinical alterations. *N. Engl. J. Med.*, **307,** 18–29.

Packer, M. (1992) Pathophysiology of chronic heart failure. *Lancet*, **340,** 88–92.

Rector, T.S., Olivari, M.T., Levine, T.B., Francis, G.S. and Cohn, J.N. (1987) Predicting survival for an individual with congestive heart failure using the plasma norepinephrine concentration. *Am. Heart J.*, **114,** 148–152.

Regitz, V., Bossaller, C., Strasser, R., Schüler, S., Hetzer, R. and Fleck, E. (1990) Myocardial catecholamine content after heart transplantation. *Circulation*, **82,** 620–623.

Roth, G.S. (1979) Hormone receptor changes during adulthood and senescence: significance for aging research. *Fed. Proc.*, **38,** 1910–1914.

Schäfers, R.F., Adler, S., Daul, P., Zeitler, G., Vogelsang, M., Zerkowski, H-R., Brodde, O.-E. (1994) Positive inotropic effects of the β_2-adrenoceptor agonist terbutaline in the human heart: effects of long-term β_1-adrenoceptor antagonist treatment. *J. Am. Coll. Cardiol.*, **23,** 1224–1233.

Schwinger, R.H.G., Böhm, M. and Erdmann, E. (1990) Evidence against spare or uncoupled β-adrenoceptors in the human heart. *Am. Heart J.*, **119,** 899–904.

Steinfath, M., von der Leyen, H., Hecht, A., Neumann, K.H., Schmitz, W., Scholz, H., Haverich, A. and Heublein, B. (1992) Decrease in β_1- and increase in β_2-adrenoceptors in long-term follow-up after orthotopic cardiac transplantation. *J. Mol. Cell Cardiol.*, **24,** 1189–1198.

Stiles, G.L., Caron, M.G., Lefkowitz, R.J. (1984) β-Adrenergic receptors. Biochemical mechanisms of physiological regulation. *Physiol. Rev.*, **64,** 661–743.

Trendelenburg, U. (1966) Mechanisms of supersensitivity and subsensitivity to sympathomimetic amines. *Pharmacol. Rev.*, **18,** 629–640.

V. CONCLUSIONS

ken cell preparations be employed and that these be purified to some degree to contain predominantly the parts of the cell membranes which contain the receptive sites. This produces two fundamental constraints.

First it limits the tissues used to those in which a high enough concentration of the receptive sites can be achieved. Secondly, it dissociates the receptive site from the response so that it is no longer known with certainty how many, if any, are the sites involved in initiating the response; this can only be inferred from the similarity of the binding affinities to those obtained from studying agonist / response / antagonist interactions, and the hope that there is not a high proportion of these binding sites which is not associated with the response. Where the tissue is heterogeneous in its cell population it is also not possible to know with certainty from which cells the receptive sites have originated. A further limitation, which emerged in practice, is that agonists are more demanding as ligands, their binding being highly dependent on precise local conditions, presumably since it involves an active influence on the site rather than the passive binding expected of an antagonist. For these reasons, work which followed this line became dissociated from tissue responses, from accurate knowledge of the cells of origin and from agonists. Definition of receptors became almost entirely dependent on antagonists and indeed the primary pharmacological consideration became the interaction between different antagonists, a phenomenon unknown in the pharmacology of tissue responses. In turn, all cloned adrenoceptors owe their discovery to this source.

This is the background to the current debate, which is attempting to reconcile different means of receptor classification. Perhaps at the front of our minds should be the likeliehood that following the ligand binding site/ molecular biology line of thought has led to an understanding of phenomena which do not necessarily coincide with the sum of the receptors defined by the "past" methods (and which have, of course, continued to be used; there was no moratorium on functional work following the implementation of ligand binding). It is possible both, i. that some receptors have not yet been encompassed at this level of understanding and, ii. that some of the phenomena uncovered may be additional to those defined by functional studies. This makes further discussion much easier. Figure 1 shows a schematic view of the relationship of the different approaches: viewing the different approaches through Venn diagrams allows recognition of the extent to which a given "receptor" fits the various subsets of valid criteria.

The difficulty of identifying discrete populations of receptors in ex-vivo tissue or of knowing which are their cells of origin has been circumvented by employing cell lines, seeking cells whose membranes contain receptive sites with ligand binding properties similar to adrenoceptor subtypes. This is intended to allow homogeneous populations of receptive sites to be studied in detail, characterised by comparison with other sites and harvested. Similarly cells can be transfected with cloned receptors and their role in function can be inferred if they can find their way to the appropriate place in the cell and find other parts of the transduction system, endogenous to the cell or added by the investigator. Native and cloned receptive sites are correlated by ligand binding characteristics. These methods can facilitate the study of the signalling process in these cells for their receptor populations and can lead to the development of tools for the study of the signalling processes in functional cells in real tissue. It would of course be unrealistic to believe that the signalling processes in such systems were ubiquitous for similar receptors in real tissue. Since these approaches

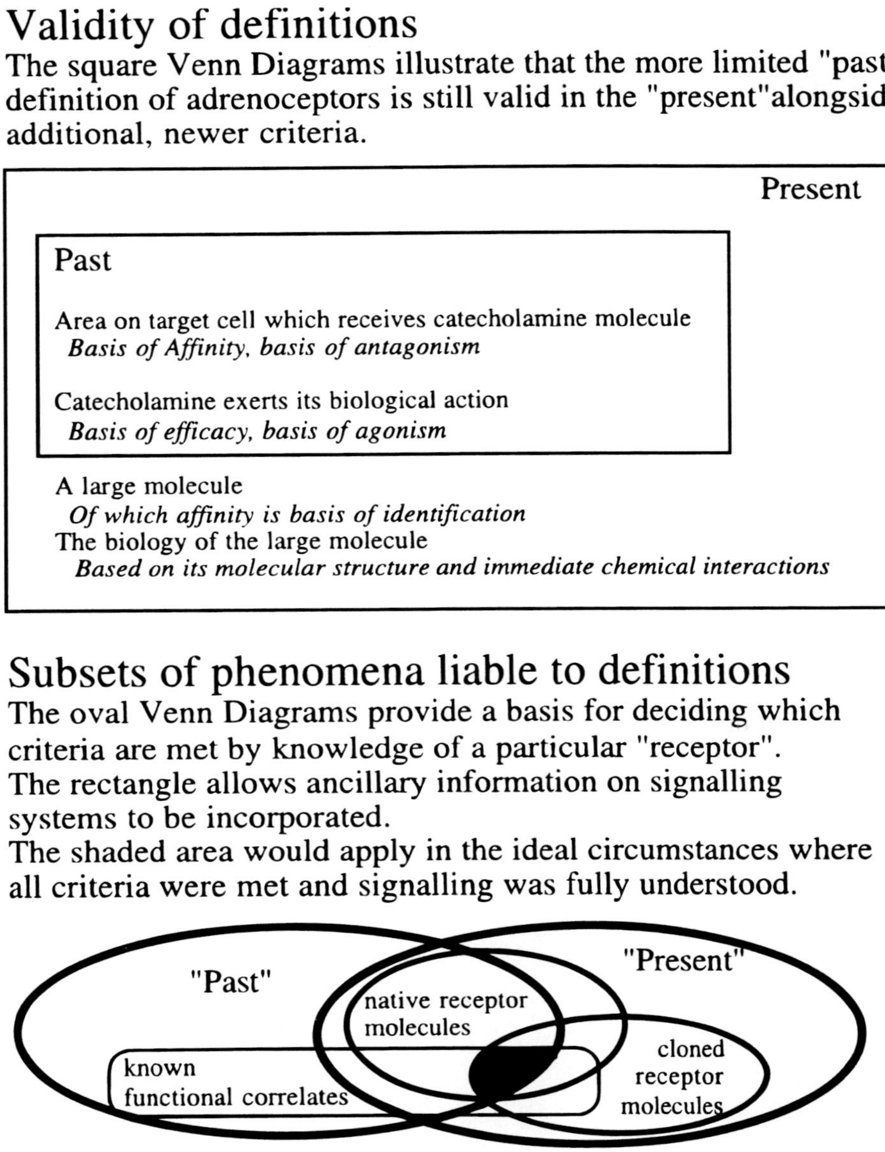

FIGURE 1 has no further legend.

continue the narrowing of the field of study, the phenomena need not necessarily coincide precisely with known functional receptors.

Thus, historically, classical and radioligand binding methods have provided the strongest evidence for subtype heterogeneity. However, as employed so far, none of these methods has provided reliable information on location or distribution of receptors within tissues except for the circumstantial evidence from the type of response initiated by agonists or blocked by antagonists. It is particularly disappointing that although high affinity radiolabelled ligands for receptor subtypes are available and well-characterised in binding assays, for technical reasons they have not proved useful in localisation. Cytochemical methods would be of value. There was evidence from this meeting of a growing interest in localisation of receptors, particularly as a guide to the possible function of "receptor subtype" which have been identified by their structure and receptive sites but whose function remains to be found.

This interest in localisation was part of a trend, clear from the papers delivered and the discussion thus informed at this meeting, that the reconciliation of the long-standing functional pharmacological approach and the more recent ligand binding identification / genetic approach is now conceivable but not yet realised, particularly in relation to alpha-adrenoceptors. However there emerged a strong will to achieve this and some pointers towards it, viz.

1. A will to bring together the classification methods was clear. Perhaps the most significant advance is the admission that neither molecular methods nor classical ones alone can presently cover everything. This came out clearly from the presentations by D.E. Clarke and I. Muramatsu (this volume) and applies to both α_1 and α_2 (see Ford, Williams, Blue & Clarke, 1994 for α_1; MacKinnon, Spedding & Brown, 1994 for α_2). The expansion and contraction of the number of subtypes continued. There is a welcome reduction in the number of α_1 subtypes brought about by the merger of "classical" α_{1A} and the cloned α_{1C} (D.E. Clarke, this volume; R.M. Graham, this volume; Ford *et al.*, 1994). However, the incorporation of the "classically" established α_{1L} subtype (Drew, 1985; Flavahan & Vanhoutte, 1986) and developed by I. Muramatsu (this volume; Muramatsu *et al.*, 1990) continues to evade ligand binding and molecular biology, presumably because its affinity for existing labelled α_1 ligands is low. The sum of these changes is that the number of α_1 subtypes remains at approximately four.
2. Another interesting trend which forces the different technologies together is the re-emergence of agonist ligands. There has recently been a reluctance to use agonists to classify alpha-adrenoceptor subtypes which was not the case before the dominance of ligand binding (see Ruffolo, 1985). There is now a revived confidence in the use of "agonists" as functional probes for differentiating subtypes (I. Muramatsu; P.W. Abel; Posters 27 and 59: all this volume) and as ligands, despite the technical binding problems (R.M. Graham; Posters 28 and 48). The main standard agonists of interest were the natural ligand, noradrenaline, the commonly used α_1 -agonists phenylephrine and methoxamine. It is interesting that oxymetazoline has featured largely in α_1 and α_2 ligand binding, despite being an agonist, presumably because it is one of the few ligands to distinguish between subtypes of α_1 and α_2 adrenoceptors; it is also notable that it is now rarely used as an agonist; in the first illustration of post-junctional α_2-adrenoceptors (Docherty *et al.*, 1979; 1980) its potency at both subtypes was a key observation; its potency at pre- and post-junctional α_2 in that early work would now correspond to its high affinity for α_{2D}, which mediates transmitter release in the rat (K. Starke: Poster 33: both this volume).
3. A further obstacle to comprehensive receptor classification has been species differences between receptors which otherwise broadly fit a subtype. Starke and coworkers have overcome this, combining classical pharmacological methods with the drugs in-

dicated by binding data, on the autoreceptors on sympathetic nerve terminals from several species (Starke, this volume; Posters 37, 55). This sets the scene for re-analysis of other pharmacological data which predates the current knowledge of receptor subtypes. For example, longstanding pharmacological evidence for differences between the rat and rabbit in both pre-junctional and post-junctional. α_2-adrenoceptors (Alabaster *et al.*, 1986; Bulloch *et al.*, 1987), may now be explained by α_{2D} in rat and α_{2A} in rabbit as found by Starke *et al.* for α_2 autoceptors. On the other hand species difference raises other problems for research in the immediate future; for example genetic manipulation tends to rely on a limited range of species, particularly the mouse (see Kobilka, this volume). For various historical reasons, mainly the size of tissues, there is very little functional pharmacological information on mouse α-adrenoceptors, with the limited exception of the vas deferens, and even in this latter case there is evidence of considerable inter-strain difference. It would be valuable to take a coordinated approach to the functional pharmacology of this species.

4. The localisation of receptors is an issue in which interest has risen and techniques are developing. The interest comes from several directions. First, the search for functional correlates for binding sites which, so far, have none would be sharpened by a knowledge of their location. Secondly, the availability of molecular probes for the receptor/transducer molecules and their RNA gives rise to hopes for immunological probes and in situ hybridisation (this volume: R.M. Graham; K.R. Lynch; Poster 43). Thirdly the increased use of transfected cells has accelerated the ability to study the lifecycle of receptor/transducers and associated systems (this volume: M.L. Toews; B. Kobilka; L.E. Limbird; Posters 15 & 21) making it an imperative to localise the receptor/transducer molecule and the receptive sites within the cell and to know whether binding to the receptive site is possible in different circumstances and locations. Fourthly, the technology for localisation of fluorescent probes within tissues has advanced and a fluorescent ligand probe for α-adrenoceptors has been found, which can permit the localisation and study of the receptive site (Poster 16). The two separate approaches to localisation of the α-adrenoceptors, namely, staining for antibodies to part of the receptor/transducer molecule (K.R. Lynch) and visualisation of fluorescent ligand probes (see below) concentrated at the receptive site, so far demonstrate a common property of "punctate staining", suggesting clustering of receptors at points on the cell surface.

The ability to cross-correlate these two methods, colocalising the living receptive site (by fluorescent ligand) with post-fixation antibody staining for components of the receptor/transducer molecule holds great promise for linking the molecular structure to the functional event since the fluorescent ligand can be displaced by competitor agonist or antagonist ligands and related to activation or blockade of cellular function. There follows a summary of this new method for sub-cellular localisation and real-time binding studies of α-adrenoceptors, involving confocal microscopy and the probe BODIPY FL-prazosin, a fluorescent, competitive α-adrenoceptor antagonist.

The method can be applied to cell preparations or intact tissues. Transfected fibroblasts make an interesting model. For example, fluorescent ligands showing receptors in living fibroblasts enable study of ACh receptor clustering (Phillips *et. al.*, 1991). We have, therefore, tested the system on fibroblasts transfected with α-adrenoceptor subtypes then on living smooth muscle tissues known to possess well-characterised functional α-adrenoceptors.

MATERIALS AND METHODS

The probe used was BODIPY FL-prazosin, "green" (abbreviation = BFLP); Molecular Probes. Its structure is shown in figure 2.

BFLP antagonism/binding was assessed using: standard analysis of concentration-response curves to phenylephrine, on functioning smooth muscle tissues; $[H]^3$-prazosin ligand binding; and confocal microscopy (Odyssey LSCM). BFLP binding was visualised using the 488nm line of an Argon-Ion laser with a 515nm LP emission filter.

Incubation conditions for confocal analysis. Isolated blood vessels or rat anococcygeus were incubated in Krebs' at 37°C, gassed with 95% O_2 / 5% CO_2 and labelled with BFLP (1nM–100nM) for up to 30 minutes. Cell cultures of rat fibroblasts transfected with α_{1a}-adrenoceptors were labelled in culture medium at room temperature. Some tissues and cells were pre-incubated in phenoxybenzamine, phentolamine, prazosin or YM 12617 (1nM–10μM). Fluorescence visualised in the continued presence of the ligand allowed real-time competition studies.

RESULTS

Functional antagonism: In rings of rat aorta BFLP (1nM–100nM) caused a reversible, parallel, concentration-dependent rightward shift of the CRC to phenylephrine: Schild plot slope 0.94, pA_2 8.25.

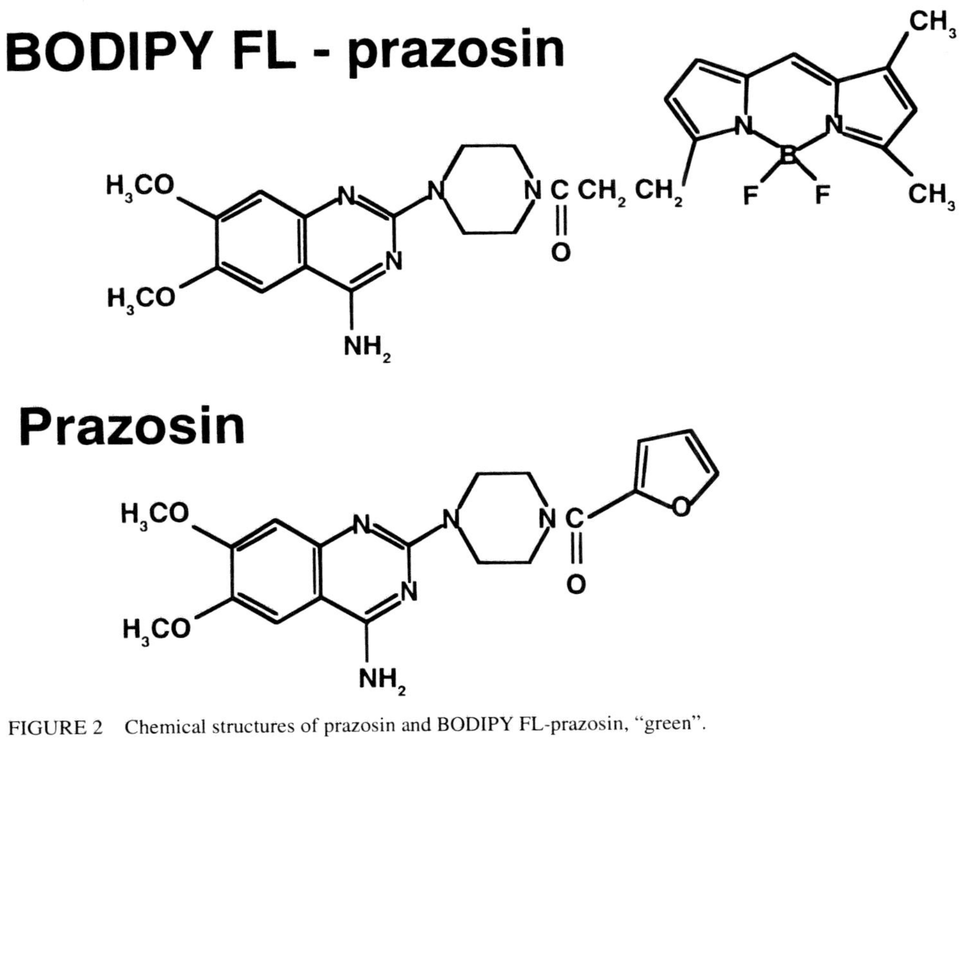

FIGURE 2 Chemical structures of prazosin and BODIPY FL-prazosin, "green".

Ligand binding: On membranes prepared from α_{1a}- transfected fibroblasts the estimated pK_i for BFLP was 7.97 against $[H]^3$-prazosin. BFLP was not as selective as prazosin, having less affinity at α_1 subtypes but more at α_2. This promiscuity can be used as an asset since within a preparation the same label can be used to label all α_1 subtypes and the individual subtypes can then be characterised first by their affinity for BFLP and secondly by the affinity of displacing competitors.

Confocal microscopy: On transfected fibroblasts, rat aorta, mesenteric artery and anococcygeus smooth muscle, BFLP produced discrete "hot spots" of fluorescence distributed along the length of the s.m. cells or around the surface of the fibroblasts. This fluorescent staining could be prevented by phenoxybenzamine or "antagonised" by phentolamine, prazosin or YM 12617 at concentrations consistent with α_1-adrenoceptor binding sites (figure 3). Time course of fluorescence attenuation by competitors and restoration of fluorescence after photobleaching indicated dynamic, reversible binding. 3D reconstructions of transfected fibroblasts showed the "hot spots" to be on the surface but not inside the cells.

Differential location of receptors in different tissues: In rat aorta most of the fluorescence was located on the cells of the inner layers of smooth muscle close to the intima, with little or no staining in the outer-most layers of muscle. Staining of rat basilar artery was poor. The adventitia and smooth muscle of the rat mesenteric artery exhibited a high density of "hot spots". No staining on endothelium was found in any of these vessels. These results are consistent with the known sensitivities of these vessels to α_1 mediated agonism. Rat anococcygeus showed clusters throughout the smooth muscle layers, displaceable by YM 12617 and staining of "surface cells" believed to be ganglion cell bodies, displaceable by rauwolscine, consistent with α_1 and α_2 adrenoceptors, respectively.

DISCUSSION

The most important aspect of this method is that binding is occurring in real time, i.e. binding is live and dynamic. Competition between ligands can be visualised simultaneously at many sites within a functional preparation. This allows quantification of binding characteristics including dynamic aspects as well as steady state.

The method lends itself to ready quantification. Confocal microscopy provides a 2 or 3 dimensional digital image which has intensity as well as position for each point. The amount of light emitted by each defined area or object is related to the amount of ligand bound. This allows a large number of approaches to quantification. Examples are: a. Intensity profiles, where the 2 dimensional image is displayed in 3 dimensions as an intensity histogram. b. Binary imaging, where thresholding is used to identify areas or objects above a given intensity or size. These have been employed to show differential effects of the irreversible ligand phenoxybenzamine and the competitive and reversible YM 12617 (figure 3).

The basic conclusions from this initial work are: BFLP (green) is a reversible, competitive, functional antagonist at α_1-adrenoceptors and has exploitable affinity also at α_2. It produces fluorescent "hot spots" on living cells which we suggest indicate binding to clusters of α_1- and α_2-adrenoceptors. Binding can be quantified by microscopy, allowing localisation within a tissue and correlation with cell characteristics. Variations in vascular receptor distribution is consistent with known responsiveness and may be related to degree of

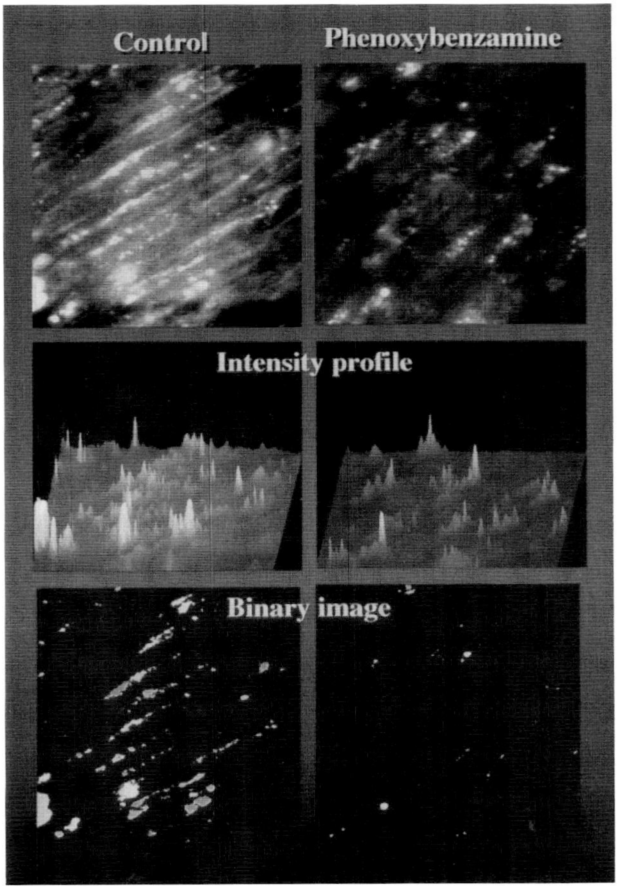

FIGURE 3 Localisation of α_1-adrenoceptors on vascular smooth muscle cells, within the media of a rat mesenteric resistance artery, by fluorescent ligand binding (left panels); the effects upon this of phenoxbenzamine (right panels) and illustration of the types of quantification which are possible. Both tissues were exposed to BODIPY FL-prazosin (0.1μM). Phenoxybenzamine (PBZ) pre-treatment was 10μM for 45 minutes.
Top panels: Original images from the confocal microscope. Images from control and phenoxybenzamine treated tissues were collected under identical conditions (i.e. frame average, brightness, contrast and laser intensity). Both images were collected using a x60, oil immersion objective (NA 1.3), image size 73μm square.
Middle panels: Visual display of the intensity profile. Metamorph image processing software was used to create a graphical profile of the intensity level of each image, which is displayed as a histogram. The PBZ profile exhibits a lower level of total fluorescence and has fewer peaks than the control. The maximum intensity of peaks or the total fluorescence from a given area can be used to quantify the amount of binding.
Bottom panels: Binary imaging allows estimation of numbers and dimensions of discrete objects such as the intense "hot spots" of α-adrenoceptor ligand found here. For comparison it is necessary, therefore, to "threshold" the image at an equivalent intensity level to produce binary images. In this case thresholding was performed at an intensity level of 113 and above out of a maximum intensity of 255 arbitrary units. The resulting binary images are shown on the bottom panels. Object analysis reported 73 clusters in the control segment and 11 clusters in the PBZ treated segments. This provides an additional approach to quantification which can be useful when there are "hot spots" with different characteristics which might be useful to analyse separately, e.g. with different receptor subtypes. (See Color Plate I).

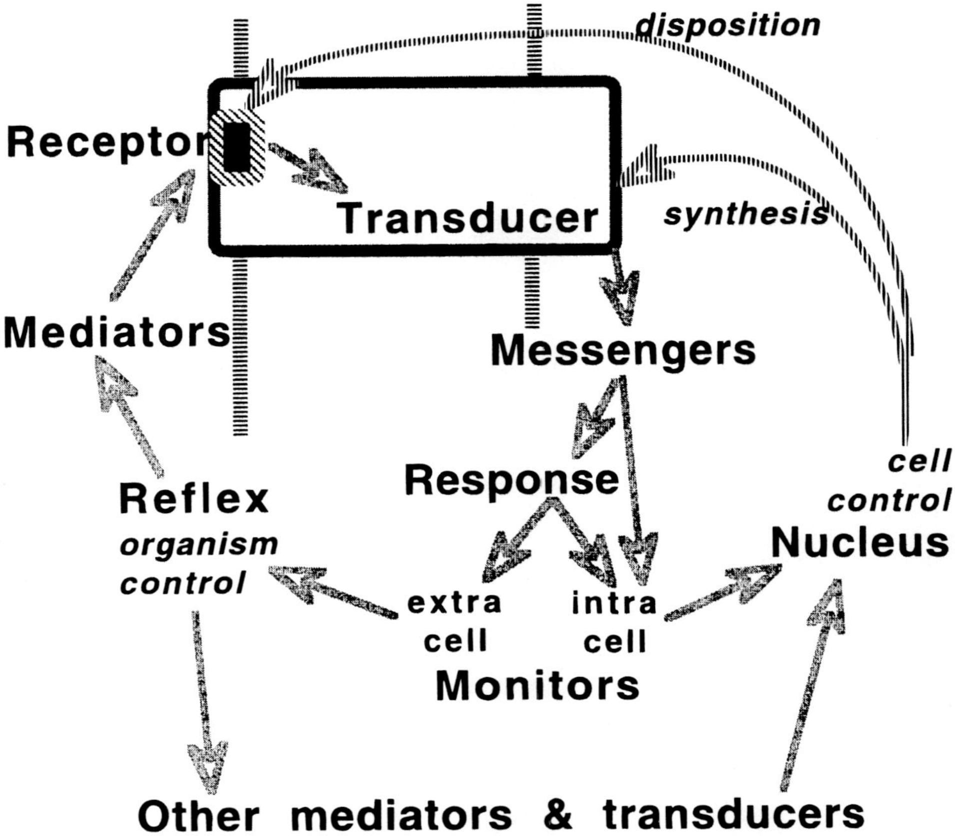

FIGURE 4 Position of adrenoceptor molecule (receptor/transducer) at the intersection of extracellular and intracellular reflexes.

sympathetic innervation. The compound is a novel tool for functional study of α_1-and α_2-adrenoceptors and opens a new path to the study of adrenoceptors and a precedent for the study of other receptor types.

In conclusion, adrenoceptor research has reached an intersesting stage in which the reductionism inherent in cellular and molecular studies has produced probes which can aid understanding of the role of receptors in tissue and organ function. Receptors sit in key positions between the organism's brain and the cell's ability to conduct its life cycle and its acute functions, and are the interlocking point between extracellular and intracellular reflexes, the latter informing the nucleus how to play its role. Perhaps the most exciting prospect for adrenoceptor research lies in using the tools which are, or are nearly, to hand to understand the dual process by which the same receptors control acute and cell life-cycle processes and as part of this same process have their own acute function and life cycles determined (figure 4).

SUBJECT INDEX—PHARMACOLOGY OF ADRENOCEPTORS

Page numbers followed by 'f' indicate figures, those followed by 't' indicate tables.

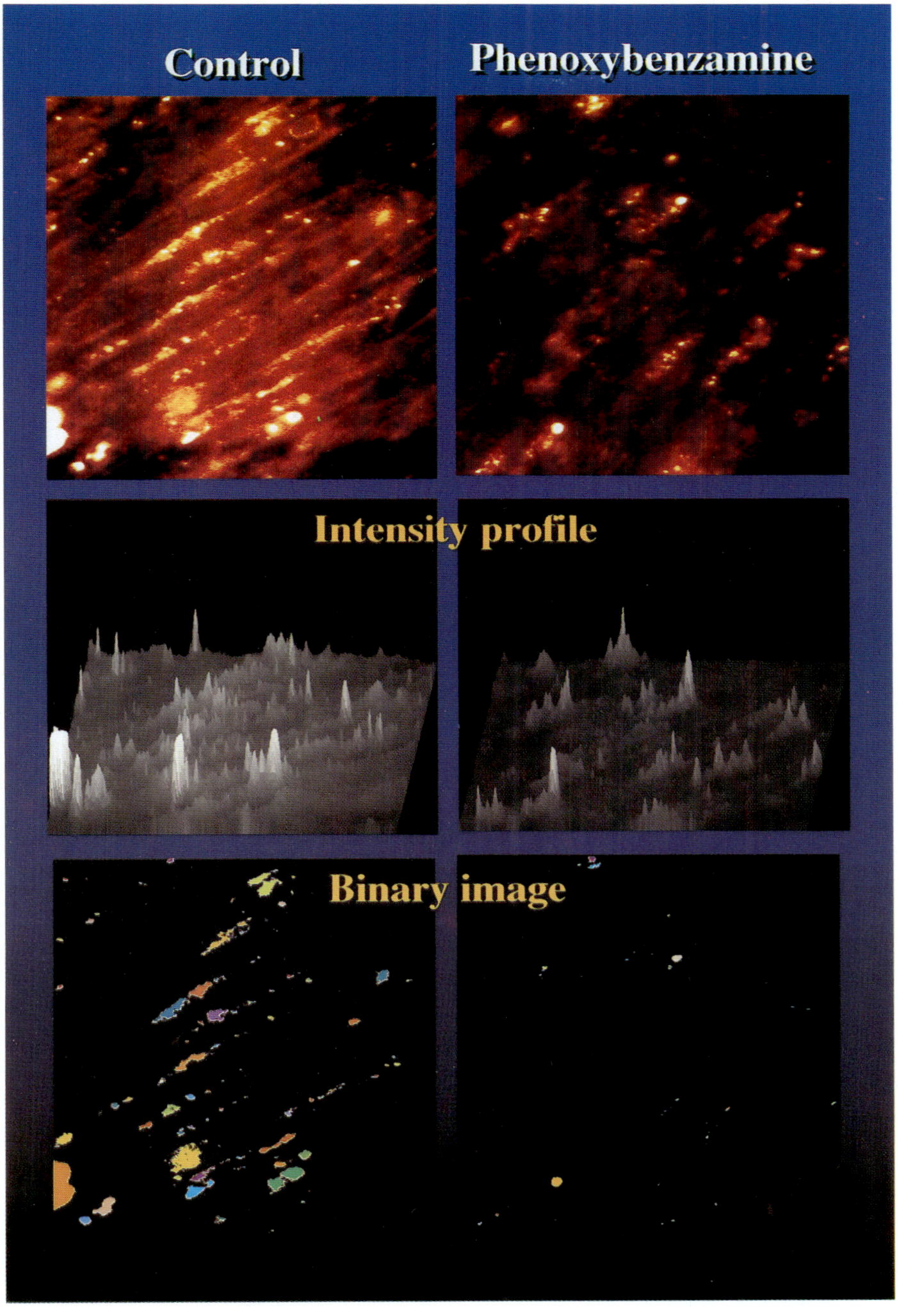

PHARMACOLOGY COMMUNICATIONS, VOLUME 6, NUMBERS 1-3, COLOR PLATE 1. See J. C. McGrath and C. J. Daly, Figure 3.

AUTHOR'S INDEX—PHARMACOLOGY OF ADRENOCEPTORS